Annals of Mathematics Studies

Number 84

KNOTS, GROUPS, AND 3-MANIFOLDS

Papers Dedicated to the Memory of R. H. Fox

EDITED BY

L. P. NEUWIRTH

PRINCETON UNIVERSITY PRESS

AND

UNIVERSITY OF TOKYO PRESS

PRINCETON, NEW JERSEY

1975

Published in Japan exclusively by
University of Tokyo Press;
In other parts of the world by
Princeton University Press

Printed in the United States of America
by Princeton University Press, Princeton, New Jersey

Library of Congress Cataloging in Publication data will
be found on the last printed page of this book

CONTENTS

INTRODUCTION vii

BIBLIOGRAPHY, RALPH HARTZLER FOX viii

Knots and Links

SYMMETRIC FIBERED LINKS 3
by Deborah L. Goldsmith

KNOT MODULES 25
by Jerome Levine

THE THIRD HOMOTOPY GROUP OF SOME HIGHER DIMENSIONAL KNOTS 35
by S. J. Lomonaco, Jr.

OCTAHEDRAL KNOT COVERS 47
by Kenneth A. Perko, Jr.

SOME KNOTS SPANNED BY MORE THAN ONE UNKNOTTED SURFACE OF MINIMAL GENUS 51
by H. F. Trotter

GROUPS AND MANIFOLDS CHARACTERIZING LINKS 63
by Wilbur Whitten

Group Theory

HNN GROUPS AND GROUPS WITH CENTRE 87
by John Cossey and N. Smythe

QUOTIENTS OF THE POWERS OF THE AUGMENTATION IDEAL IN A GROUP RING 101
by John R. Stallings

KNOT-LIKE GROUPS 119
by Elvira Rapaport Strasser

3-Dimensional Manifolds

ON THE EQUIVALENCE OF HEEGAARD SPLITTINGS OF CLOSED, ORIENTABLE 3-MANIFOLDS 137
by Joan S. Birman

BRANCHED CYCLIC COVERINGS 165
by Sylvain E. Cappell and Julius L. Shaneson

ON THE 3-DIMENSIONAL BRIESKORN MANIFOLDS M(p, q, r) 175
by John Milnor

SURGERY ON LINKS AND DOUBLE BRANCHED COVERS OF OF S^3 227
by José M. Montesinos

PLANAR REGULAR COVERINGS OF ORIENTABLE CLOSED SURFACES 261
by C. D. Papakyriakopoulos

INFINITELY DIVISIBLE ELEMENTS IN 3-MANIFOLD GROUPS 293
by Peter B. Shalen

INTRODUCTION

The influence of a great teacher and a superb mathematician is measured by his published work, the published works of his students, and, perhaps foremost, the mathematical environment he fostered and helped to maintain. In this last regard Ralph Fox's life was particularly striking: the tradition of topology at Princeton owes much to his lively and highly imaginative presence. Ralph Fox had well defined tastes in mathematics. Although he was not generally sympathetic toward topological abstractions, when questions requiring geometric intuition or algebraic manipulations arose, it was his insights and guidance that stimulated deepened understanding and provoked the development of countless theorems.

This volume is a most appropriate memorial for Ralph Fox. The contributors are his friends, colleagues, and students, and the papers lie in a comfortable neighborhood of his strongest interests. Indeed, all the papers rely on his work either directly, by citing his own results and his clarifications of the work of others, or indirectly, by acknowledging his gentle guidance into and through the corpus of mathematics.

The reader may gain an appreciation of the range of Fox's own work from the following bibliography of papers published during the thirty-six years of his mathematical life.

L. Neuwirth

PRINCETON, NEW JERSEY

OCTOBER 1974

BIBLIOGRAPHY, RALPH HARTZLER FOX (1913 - 1973)

1936

(with R. B. Kershner) Transitive properties of geodesics on a rational polyhedron, *Duke Math. Jour.* 2, 147-150.

1940

On homotopy and extension of mappings, *Proc. Nat. Acad. Sci. U.S.A.* 26, 26-28.

1941

Topological invariants of the Lusternik-Schnirelmann type, *Lecture in Topology*, Univ. of Mich. Press, 293-295.

On the Lusternik-Schnirelmann category, *Ann. of Math.* (2), 42, 333-370.

Extension of homeomorphisms into Euclidean and Hilbert parallelotopes, *Duke Math. Jour.* 8, 452-456.

1942

A characterization of absolute neighborhood retracts, *Bull. Amer. Math. Soc.* 48, 271-275.

1943

On homotopy type and deformation retracts, *Ann. of Math.* (2) 44, 40-50.

On the deformation retraction of some function spaces associated with the relative homotopy groups, *Ann. of Math.* (2) 44, 51-56.

On fibre spaces, I, *Bull. Amer. Math. Soc.*, 49, 555-557.

On fibre spaces, II, *Bull. Amer. Math. Soc.*, 49, 733-735.

1945

On topologies for function spaces, *Bull. Amer. Math. Soc.*, 51, 429-432.

Torus homotopy groups, *Proc. Nat. Acad. Sci. U.S.A.*, 31, 71-74.

1947

On a problem of S. Ulam concerning Cartesian products, *Fund. Math.*, 34, 278-287.

1948

On the imbedding of polyhedra in 3-spaces, *Ann. of Math.* (2) 49, 462-470.

Homotopy groups and torus homotopy groups, *Ann. of Math.* (2) 49, 471-510.

(with Emil Artin) Some wild cells and spheres in three-dimensional space, *Ann. of Math.*, (2) 49, 979-990.

1949

A remarkable simple closed curve, *Ann. of Math.*, (2) 50, 264-265.

1950

On the total curvature of some tame knots, *Ann. of Math.*, (2) 52, 258-260.

(with William A. Blankinship) Remarks on certain pathological open subsets of 3-space and their fundamental groups, *Proc. Amer. Math. Soc.*, 1, 618-624.

1951

(with Richard C. Blanchfield) Invariants of self-linking, *Ann. of Math.* (2) 53, 556-564.

1952

Recent development of knot theory at Princeton, *Proceedings of the International Congress of Mathematicians*, Cambridge, Mass., Vol. 2, 453-457, *Amer. Math. Soc.*, Providence, R. I.

On Fenchel's conjecture about F-groups, *Mat. Tidsskr.* B. 1952, 61-65.

On the complementary domains of a certain pair of inequivalent knots, *Nederl. Akad. Wetensch. Proc. Ser. A. 55-Indagationes Math.*, 14, 37-40.

1953

Free differential calculus, I, Derivation in the free group ring, *Ann. of Math.*, (2) 57, 547-560.

1954

Free differential calculus, II, The isomorphism problem of groups, *Ann. of Math.*, (2) 59, 196-210.

(with Guillermo Torres) Dual presentations of the group of a knot, *Ann. of Math.*, (2) 59, 211-218.

1956

Free differential calculus, III. Subgroups *Ann. of Math.*, (2) 64, 407-419.

1957

Covering spaces with singularities. *Algebraic Geometry and Topology.* A symposium. Princeton University Press. 243-257.

1958

Congruence classes of knots, *Osaka Math. Jour.*, **10**, 37-41.

On knots whose points are fixed under a periodic transformation of the 3-sphere, *Osaka Math. Jour.*, **10**, 31-35.

(with K. T. Chen and R. G. Lyndon) Free differential calculus, IV. The quotient groups of the lower central series, *Ann. of Math.*, (2) **68**, 81-95.

1960

Free differential calculus, V. The Alexander matrices re-examined, *Ann. of Math.*, (2) 71, 408-422.

The homology characters of the cyclic coverings of the knots of genus one, *Ann. of Math.*, (2) 71, 187-196.

(with Hans Debrunner) A mildly wild imbedding of an n-frame, *Duke Math. Jour.*, 27, 425-429.

1962

"Construction of Simply Connected 3-Manifolds." *Topology of 3-Manifolds and Related Topics*, Englewood Cliffs, Prentice-Hall, 213-216.

"A Quick Trip Through Knot Theory." *Topology of 3-Manifolds and Related Topics*, Englewood Cliffs, Prentice-Hall, 120-167.

"Knots and Periodic Transformations." *Topology of 3-Manifolds and Related Topics*, Englewood Cliffs, Prentice-Hall, 177-182.

"Some Problems in Knot Theory." *Topology of 3-Manifolds and Related Topics*, Englewood Cliffs, Prentice-Hall, 168-176.

(with O. G. Harrold) "The Wilder Arcs." *Topology of 3-Manifolds and Related Topics*, Englewood Cliffs, Prentice-Hall, 184-187.

(with L. Neuwirth) The Braid Groups, *Math. Scand.*, **10**, 119-126.

1963

(with R. Crowell) *Introduction to Knot Theory*, New York, Ginn and Company.

1964

(with N. Smythe) An ideal class invariant of knots, *Proc. Am. Math. Soc.*, 15, 707-709.

Solution of problem P79, *Canadian Mathematical Bulletin*, 7, 623-626.

1966

Rolling, *Bull. Am. Math. Soc.* (1) 72, 162-164.

(with John Milnor) Singularities of 2-spheres in 4-space and cobordism of knots, *Osaka Jour. of Math.* 3, 257-267.

1967

Two theorems about periodic transformations of the 3-sphere, *Mich. Math. Jour.*, 14, 331-334.

1968

Some n-dimensional manifolds that have the same fundamental group, *Mich. Math. Jour.*, 15, 187-189.

1969

A refutation of the article "Institutional Influences in the Graduate Training of Productive Mathematicians." *Ann. Math. Monthly*, 76, 1968-70.

1970

Metacyclic invariants of knots and links, *Can. Jour. Math.*, 22, 193-201.

1972

On shape, *Fund. Math.*, 74, 47-71.

Knots and Links

SYMMETRIC FIBERED LINKS

Deborah L. Goldsmith

0. *Introduction*

The main points of this paper are a construction for fibered links, and a description of some interplay between major problems in the topology of 3-manifolds; these latter are, notably, the Smith problem (can a knot be the fixed point set of a periodic homeomorphism of S^3), the problem of which knots are determined by their complement in the 3-sphere, and whether a simply connected manifold is obtainable from S^3 by surgery on a knot.

There are three sections. In the first, symmetry of links is defined, and a method for constructing fibered links is presented. It is shown how this method can sometimes be used to recognize that a symmetric link is fibered; then it reveals all information pertaining to the fibration, such as the genus of the fiber and the monodromy. By way of illustration, an analysis is made of the figure-8 knot and the Boromean rings, which, it turns out, are symmetric and fibered, and related to each other in an interesting way.

In Section II it is explained how to pass back and forth between different ways of presenting 3-manifolds.

Finally, the material developed in the first two sections is used to establish the interconnections referred to earlier. It is proved that completely symmetric fibered links which have repeated symmetries of order 2 (e.g., the figure-8 knot) are characterized by their complement in the 3-sphere.

I would like to thank Louis Kauffman and John W. Milnor for conversations.

I. Symmetric fibered links

§1. *Links with rotational symmetry*

By a *rotation of* S^3 we mean an orientation preserving homeomorphism of S^3 onto itself which has an unknotted simple closed curve A for fixed point set, called the *axis* of the rotation. If the rotation has finite period n, then the orbit space of its action on S^3 is again the 3-sphere, and the projection map $p: S^3 \to S^3$ to the orbit space is the n-fold cyclic branched cover of S^3 along $p(A)$.

An oriented link $L \subset S^3$ has a *symmetry of order* n if there is a rotation of S^3 with period n and axis A, where $A \cap L = \phi$, which leaves L invariant. We will sometimes refer to the rotation as the symmetry, and to its axis as the axis of symmetry of L.

The oriented link $L \subset S^3$ *is said to be completely symmetric relative to an oriented link* L_0, if there exists a sequence of oriented links $L_0, L_1, \cdots, L_n = L$ beginning with L_0 and ending with $L_n = L$, such that for each $i \neq 0$, the link L_i has a symmetry of order $n_i > 1$ with axis of symmetry A_i and projection $p_i: S^3 \to S^3$ to the orbit space of the symmetry, and $L_{i-1} \simeq p_i(L_i)$. If L_0 is the trivial knot, then L is called a *completely symmetric link.* The number n is the *complexity* of the sequence. Abusing this terminology, we will sometimes refer to a completely symmetric link L of complexity n (relative to L_0) to indicate the existence of such a sequence of complexity n.

Figure 1 depicts a completely symmetric link L of complexity 3, having a symmetry of order 3.

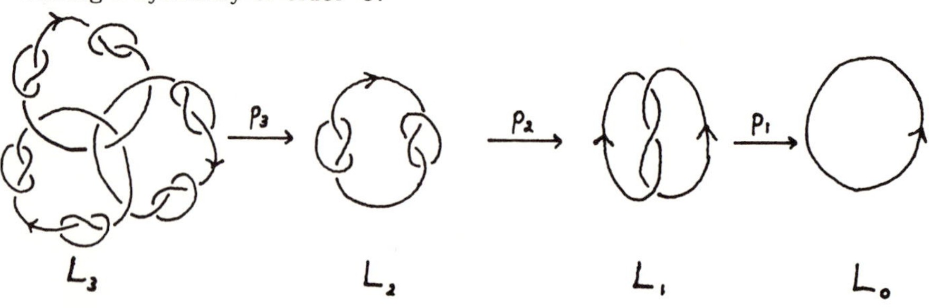

Fig. 1.

§2. *Symmetric fibered links*

An oriented link $L \subset S^3$ is *fibered* if the complement $S^3 - L$ is a surface bundle over the circle whose fiber F over $1 \in S^1$ is the interior of a compact, oriented surface F with $\partial F = L$.

Such a link L is a *generalized axis* for a link $L' \subset S^3 - L$ if L' intersects each fiber of the bundle $S^3 - L$ transversely in n points. In the classical case (which this generalizes) a link $L' \subset R^3$ is said to have the z-axis for an axis if each component L'_i has a parametrization $L'_i(\theta)$ by which, for each angle θ_0, the point $L'_i(\theta_0)$ lies inside the half-plane $\theta = \theta_0$ given by its equation in polar coordinates for R^3. We will define L to be an *axis* for $L' \subset S^3$ if L is a generalized axis for L' and L is an unknotted simple closed curve.

We wish to investigate sufficient conditions under which symmetric links are fibered.

LEMMA 1 (A construction). *Let* L' *be a fibered link in the* 3-*sphere and suppose* $p : S^3 \to S^3$ *is a branched covering of* S^3 *by* S^3, *whose branch set is a link* $B \subset S^3 - L'$. *If* L' *is a generalized axis for* B, *then* $L = p^{-1}(L')$ *is a fibered link.*

Proof. The complement $S^3 - L'$ fibers over the circle with fibers $\mathring{F}_s$, $s \in S^1$, the interior of compact, oriented surfaces F_s such that $\partial F_s = L'$. Let $\hat{F}_s = p^{-1}(F_s)$ be the inverse image of the surface F_s under the branched covering projection. Then $\partial \hat{F}_s = L$ and $\hat{F}_s - L$, $s \in S^1$, is a locally trivial bundle over S^1 by virtue of the homotopy lifting property of the covering space $p : S^3 - (L \cup p^{-1}(B)) \to S^3 - (L' \cup B)$. Thus $S^3 - L$ fibers over S^1 with fiber, the interior of the surface $\hat{F}_1$.

REMARK. An exact calculation of genus $(\hat{F}_1)$ follows easily from the equation $\chi(\hat{F}_1 - p^{-1}(B)) = n\chi(F_1 - B)$ for the Euler characteristic of the covering space $\hat{F}_1 - p^{-1}(B) \to F_1 - B$. For example, if $p : S^3 \to S^3$ is a regular branched covering, L has only one component and k is the number of points in the intersection $B \cap F_1$ of B with the surface F_1, we can derive the inequality: genus $(\hat{F}_1) \geq n$ genus $(F_1) + \frac{1}{2} + \frac{n(k-2)}{4}$. From this it follows that if $k > 1$, or genus $(F_1) > 0$, then genus $(\hat{F}_1) > 0$ and L is knotted.

Recall that a completely symmetric link $L \subset S^3$ (relative to L_0) is given by a sequence of links $L_0, L_1, \cdots, L_n = L$ such that for each $i \neq 0$, the link L_i has a symmetry of order n_i with axis of symmetry A_i, and such that $p_i : S^3 \to S^3$ is the projection to the orbit space of the symmetry.

THEOREM 1. *Let* $L \subset S^3$ *be a completely symmetric link relative to the fibered link* L_0, *defined by the sequence of links* $L_0, L_1, \cdots, L_n = L$. *If for each* $i \neq 0$, *the projection* $p_i(L_i)$ *of the link* L_i *is a generalized axis for the projection* $p_i(A_i)$ *of its axis of symmetry, then* L *is a non-trivial fibered link.*

Proof. Apply Lemma 1 repeatedly to the branched coverings $p_i : S^3 \to S^3$ branched alorg the trivial knot $p_i(A_i)$ having $p_i(L_i) \simeq L_{i-1}$ for generalized axis.

The completely symmetric links L which are obtained from a sequence $L_0, L_1, \cdots, L_n = L$ satisfying the conditions of the theorem, where L_0 is the trivial knot, are called *completely symmetric fibered links*.

EXAMPLES. In Figure 2 we see a proof that the figure-8 knot L is a completely symmetric fibered knot of complexity 1, with a symmetry of order 2. It is fibered because $p(A)$ is the braid $\sigma_2^{-1}\sigma_1$ closed about the axis $p(L)$. The shaded disk F with $\partial F = L_0$ intersects $p(A)$ in three points; hence the shaded surface $\hat{F} = p^{-1}(F)$, which is the closed fiber of the fibration of $S^3 - L$ over S^1, is the 2-fold cyclic branched cover of the disk F branching along the points $F \cap p(A)$, and has genus 1.

In Figure 3, it is shown that the Boromean rings L is a completely symmetric fibered link of complexity 1, with a symmetry of order 3.

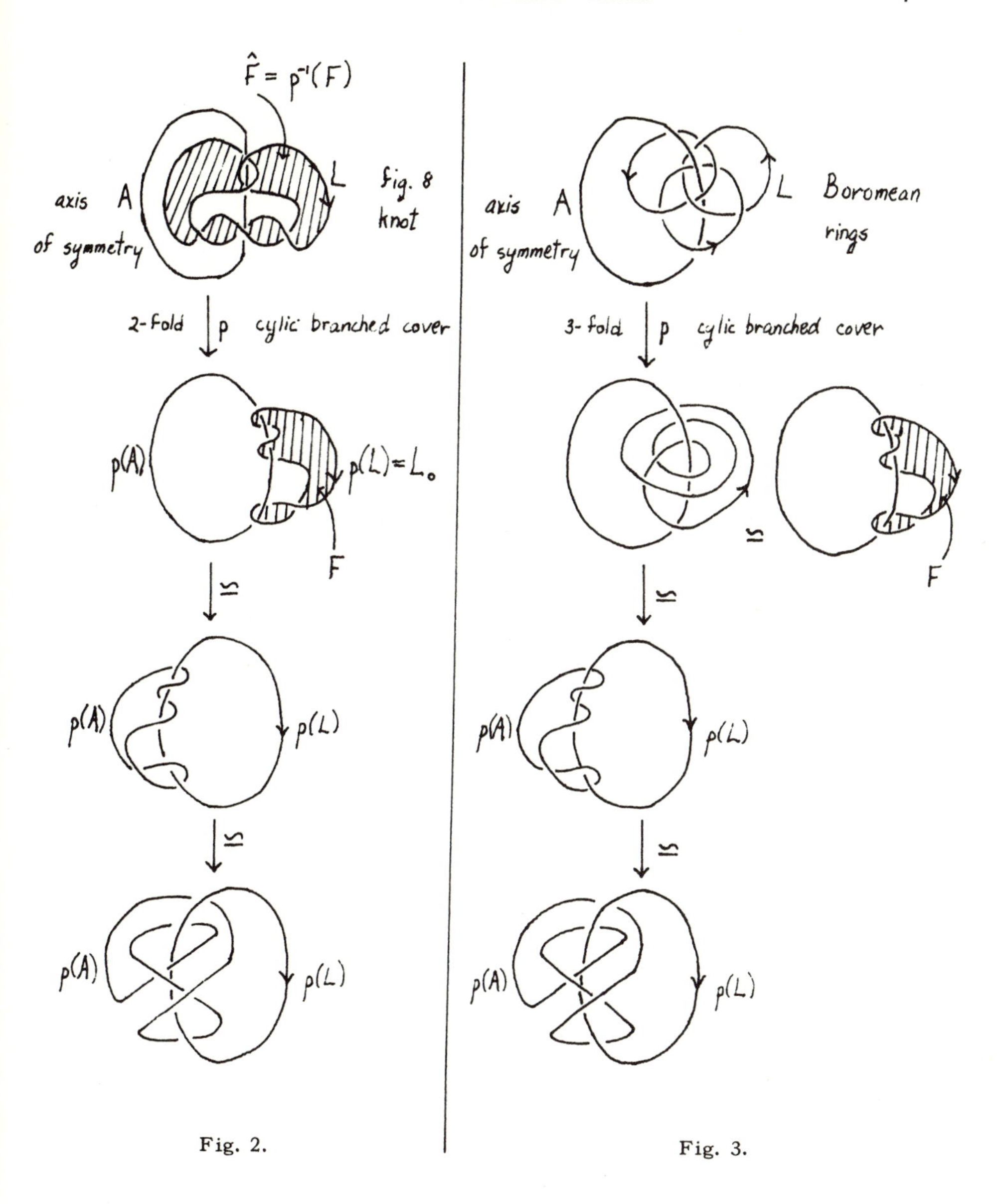

Fig. 2. Fig. 3.

This link is fibered because p(A) is the braid $\sigma_2^{-1}\sigma_1$ closed about the axis p(L). The surface $\hat{F} = p^{-1}(F)$ which is the closed fiber of the fibration of $S^3 - L$ over S^1 is not shaded, but is precisely the surface obtained by Seifert's algorithm (see [12]). It is a particular 3-fold cyclic branched cover of the disk F (shaded) branching along the three points $F \cap p(A)$, and has genus 1.

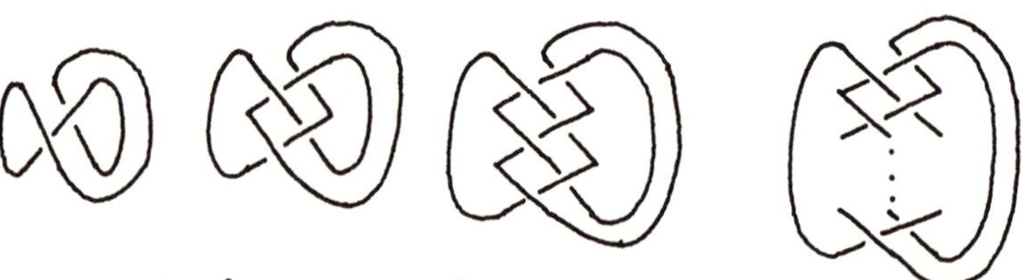

Fig. 4.

Finally, we see from Figure 4 that these two examples are special cases of a class of completely symmetric fibered links of complexity 1 with a symmetry of order n, obtained by closing the braid b^n, where $b = \sigma_2^{-1}\sigma_1$.

II. Presentations of 3-manifolds

There are three well-known constructions for a 3-manifold M: M may be obtained from a Heegaard diagram, or as the result of branched covering or performing "surgery" on another 3-manifold. A specific construction may be called a *presentation*; and just as group presentations determine the group, but not vice-versa, so M has many Heegaard, branched covering and surgery presentations which determine it up to homeomorphism.

Insight is gained by changing from one to another of the three types of presentations for M, and methods for doing this have been evolved by various people; in particular, given a Heegaard diagram for M, it is known how to derive a surgery presentation ([9]) and in some cases, how to present M as a double branched cover of S^3 along a link ([2]). This section deals with the remaining case, that of relating surgery and branched covering constructions.

§1. *The operation of surgery*

Let C be a closed, oriented 1-dimensional submanifold of the oriented 3-manifold M, consisting of the oriented simple closed curves $c_1, \cdots, c_k$. An oriented 3-manifold N is said to be obtained from M by surgery on C if N is the result of removing the interior of disjoint, closed tubular neighborhoods T_i of the c_i's and regluing the closed neighborhoods by orientation preserving self-homeomorphisms $\phi_i : \partial T_i \to \partial T_i$ of their boundary. It is not hard to see that N is determined up to homeomorphism by the homology classes of the image curves $\phi_i(m_i)$ in $H_1(\partial T_i; \mathbf{Z})$, where m_i is a meridian on ∂T_i (i.e., m_i is an oriented simple closed curve on ∂T_i which spans a disk in T_i and links c_i with linking number +1 in T_i). If γ_i is the homology class in $H_1(\partial T_i; \mathbf{Z})$ represented by $\phi_i(m_i)$, then let $M(C; \gamma_1, \cdots, \gamma_k)$ denote the manifold N obtained according to the above surgery procedure.

When it is possible to find a longitude ℓ_i on ∂T_i (i.e., an oriented simple closed curve on ∂T_i which is homologous to c_i in T_i and links c_i with linking number zero in M), then γ_i will usually be expressed as a linear combination $rm_i + s\ell_i$, $r, s \in \mathbf{Z}$, of these two generators for $H_1(\partial T_i; \mathbf{Z})$, where the symbols m_i and ℓ_i serve dually to denote both the simple closed curve and its homology class. An easy fact is that for a knot C in the homology 3-sphere M, $M(C; rm + s\ell)$ is again a homology sphere exactly when $r = \pm 1$.

§2. *Surgery on the trivial knot in* S^3

An important feature of the trivial knot $C \subset S^3$ is that any 3-manifold $S^3(C; m + k\ell)$, $k \in \mathbf{Z}$, obtained from S^3 by surgery on C is again S^3. To see this, decompose S^3 into two solid tori sharing a common boundary, the tubular neighborhood T_1 of C, and the complementary solid torus T_2. Let $\hat{\phi} : T_2 \to T_2$ be a homeomorphism which carries m to the curve $m + k\ell$; then $\hat{\phi}$ extends to a homeomorphism $\phi : S^3 \to S^3(C; , + k\ell)$.

Now suppose $B \subset S^3$ is some link disjoint from C. The link $B \subset S^3(C; m + k\ell)$ is generally different from the link $B \subset S^3$. Specifically, B is transformed by the surgery to its inverse image $\phi^{-1}(B)$ under the identification $\phi : S^3 \to S^3(C; m + k\ell)$. The alteration may be described in the following way:

Let B be transverse to some cross-sectional disk of T_2 having ℓ for boundary. Cut S^3 and B open along this disk, and label the two copies the negative side and the positive side of the disk, according as the meridian m enters that side or leaves it. Now twist the negative side k full rotations in the direction of $-\ell$, and reglue it to the positive side. The resulting link is $\phi^{-1}(B)$.

For example, if B is the n-stringed braid $b \in B_n$ closed about the axis C, where B_n is the braid group on n-strings, and if c is an appropriate generator of center (B_n), then $B \subset S^3(C; m+k\ell)$ is the closed braid $\overline{b \cdot c^k}$. Figure 5 illustrates this phenomenon. In Figure 6 it is shown how to change a crossing of a link B by doing surgery on an unknotted simple closed curve C in the complement of B.

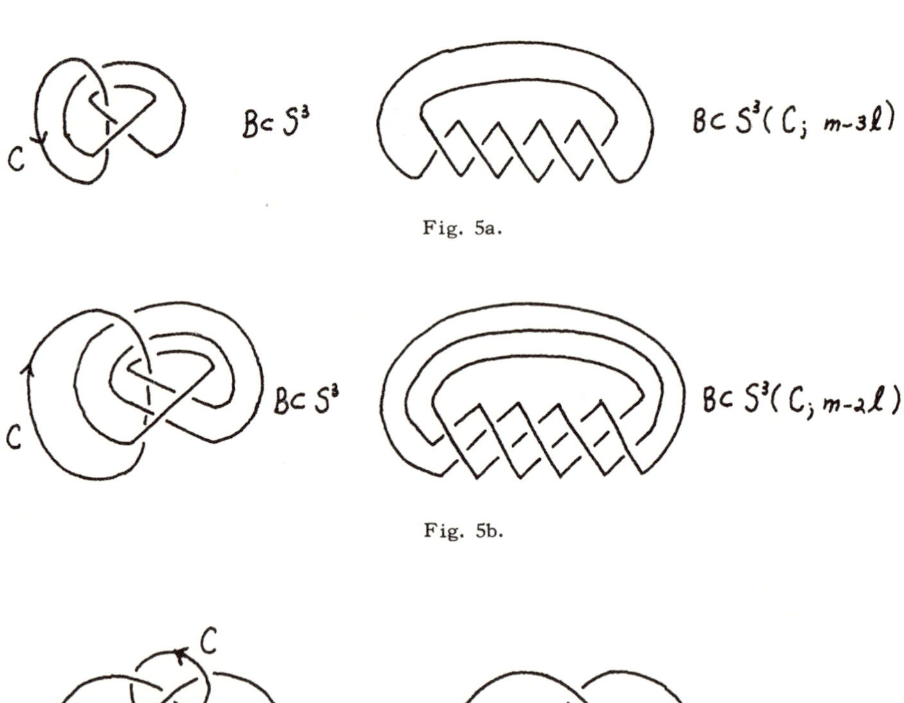

Fig. 5a.

Fig. 5b.

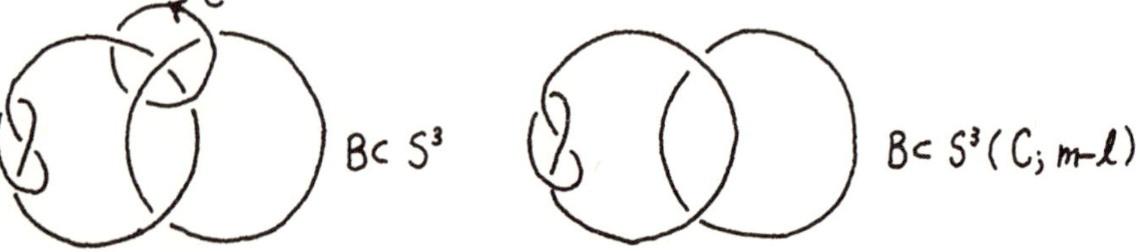

Fig. 6.

§3. *The branched covering operation*

For our purposes, a map $f: N \to M$ between the 3-manifolds N and M is a branched covering map with branch set $B \subset M$, if there are triangulations of N and M for which f is a simplicial map where no simplex is mapped degenerately by f, and if B is a pure 1-dimensional subcomplex of M such that the restriction

$$f \mid N - f^{-1}(B) : N - f^{-1}(B) \to M - B$$

is a covering (see [5]). The foldedness of the branched covering f is defined to be the index of the covering $f \mid N - f^{-1}(B)$.

We will only consider the case where the branch set $B \subset M$ is a 1-dimensional submanifold, and the foldedness of f is a finite number, n. Then f and N are determined by a representation $\pi_1(M-B) \to S(n)$ of the fundamental group of the complement of B in M to the symmetric group on n numbers (see [4]). Given this representation, the manifold N is constructed by forming the covering space $f': N' \to M - B$ corresponding to the subgroup of $\pi_1(M-B)$ represented onto permutations which fix 1, and then completing to $f: N \to M$ by filling in the tubular neighborhood of B and extending f' to f.

A regular branched covering is one for which $f': N' \to M - B$ is a regular covering, or in other words, one for which the subgroup of $\pi_1(M-B)$ in question is normal. Among these are the cyclic branched coverings, given by representations $\pi_1(M-B) \to Z_n$ onto the cyclic group of order n, such that the projection $f: N \to M$ is one-to-one over the branch set. Since Z_n is abelian, these all factor through the first homology group

$$\pi_1(M-B) \to H_1(M-B; Z) \to Z_n \ .$$

Does there always exist an n-fold cyclic branched covering $N \to M$ with a given branch set $B \subset M$? The simplest case to consider is the one in which M is a homology 3-sphere. Here $H_1(M-B; Z) \simeq Z \oplus Z \oplus \cdots \oplus Z$ is generated by meridians lying on tubes about each of the components of the branch set. Clearly all representations of $H_1(M-B; \mathbf{Z})$ onto $\mathbf{Z}_n$ which come from cyclic branched coverings are obtained by linearly extending arbitrary assignments of these meridians to ± 1. This guarantees the existence of many n-fold cyclic branched coverings of M branched along B, except in the case $n = 2$, or in case B has one component, when there is only one.

Should M not be a homology sphere, an n-fold cyclic covering with branch set B will exist if each component of B belongs to the n-torsion of $H_1(M; \mathbf{Z})$, but this condition is not always necessary.

§4. *Commuting the two operations*

If one has in hand a branched covering space, and a surgery to be performed on the base manifold, one may ask whether the surgery can be lifted to the covering manifold in such a way that the surgered manifold upstairs naturally branched covers the surgered manifold downstairs. The answer to this is very interesting, because it shows one how to change the order in which the two operations are performed, without changing the resulting 3-manifold.

Let $f: N \to M$ be an n-fold branched covering of the oriented 3-manifold M along $B \subset M$ given by a representation $\phi: \pi_1(M-B) \to S(n)$, and let $M(C; \gamma_1, \cdots, \gamma_k)$ be obtained from M by surgery on $C \subset M$, where $C \cap B = \phi$. Note that the manifold $N - f^{-1}(C)$ is a branched covering space of $M - C$ branched along $B \subset M - C$, and is given by the representation

$$\tilde{\phi} = \phi \circ i : \pi_1(M - [C \cup B]) \to S(n) ,$$

where $i: \pi_1(M - [C \cup B]) \to \pi_1(M-B)$ is induced by inclusion. Now let the components of $f^{-1}(T_i)$ be the solid tori $\hat{T}_{ij}$, $j = 1, \cdots, n_i$, $i = 1, \cdots, k$; on the boundary of each tube choose a single oriented, simple closed curve in the inverse image of a representative of γ_i, and denote its homology class in $H_1(\partial \hat{T}_{ij}; \mathbf{Z})$ by $\hat{\gamma}_{ij}$.

THEOREM 2. *Suppose* $\gamma_{i_1},\cdots,\gamma_{i_r}$ *are precisely the classes among* $\gamma_1,\cdots,\gamma_k$ *which have a representative all of whose lifts are closed curves; let* $B' = C - \bigcup_{j=1}^{r} c_{i_j}$. *Then* $f: N \to M$ *induces a branched covering*

$$f': N(f^{-1}(C); \gamma_{ij}\ j=1,\cdots,n_i,\ i=1,\cdots,k) \to M(C;\gamma_1,\cdots,\gamma_k)$$

of the surgered manifolds, branched along $B \cup B' \subset M(C;\gamma_1,\cdots,\gamma_k)$. *The associated representation is* $\phi': \pi_1(M(C;\gamma_1,\cdots,\gamma_k) - [B \cup B']) \to S(n)$, *defined by the commutative diagram*

$$\pi_1(M-[C\cup B]) \xrightarrow{i} \pi_1(M(C;\gamma_1,\cdots,\gamma_k)-[B\cup B']) \xrightarrow{\phi'} S(n), \qquad \tilde{\phi}: \pi_1(M-[C\cup B]) \to S(n)$$

and off of a tubular neighborhood $f^{-1}(\cup T_i)$ *of the surgered set, the maps* f *and* f' *agree.*

Proof. One need only observe that the representation $\tilde{\phi}$ does indeed factor through $\pi_1(M(C;\gamma_1,\cdots,\gamma_k)-[B\cup B'])$ because of the hypothesis that there exist representatives of $\gamma_{i_1},\cdots,\gamma_{i_r}$ all of whose lifts are closed curves.

The meaning of this theorem should be made apparent by what follows.

EXAMPLE. It is known that the dodecahedral space is obtained from S^3 by surgery on the trefoil knot K; in fact, it is the manifold $S^3(K; m-\ell)$. We will use this to conclude that it is also the 3-fold cyclic branched cover of S^3 along the $(2,5)$ torus knot, as well as the 2-fold cyclic branched cover of S^3 along the $(3,5)$ torus knot (see [6]). These presentations are probably familiar to those who like to think of this homology sphere as the intersection of the algebraic variety $\{x \in \mathbb{C}^3 : x_1^2 + x_2^3 + x_3^5 = 0\}$ with the 3-sphere $\{x \in \mathbb{C}^3 : |x| = 1\}$.

According to Figure 7, the trefoil knot K is the inverse image of the circle C under the 3-fold cyclic branched cover of S^3 along the trivial knot B. By Theorem 2, $S^3(K; m-\ell)$ is the 3-fold cyclic branched cover of $S^3(C; m-3\ell)$ branched along $B \subset S^3(C; m-3\ell)$. Since C is the trivial knot, $S^3(C; m-3\ell)$ is the 3-sphere, and $B \subset S^3(C; m-3\ell)$ is the (2,5) torus knot, as in Figure 5a. We deduce that the dodecahedral space is the 3-fold cyclic branched cover of S^3 along the (2,5) torus knot.

A similar argument is applied to Figure 8, in which the trefoil knot is depicted as the inverse image of a circle C under the double branched cover of S^3 along the trivial knot B. By Theorem 2, the space $S^3(K; m-\ell)$ is then the 2-fold cyclic branched cover of $S^3(C; m-2\ell)$ along $B \subset S^3(C; m-2\ell)$, which according to Figure 5b is the (3,5) torus

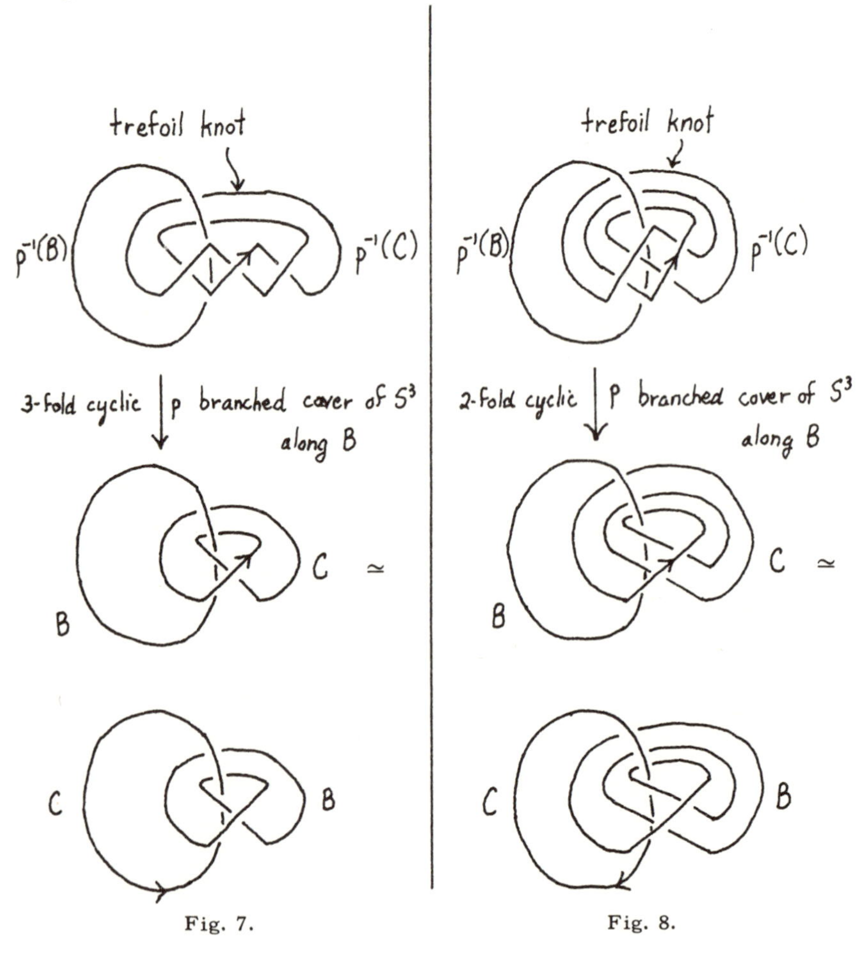

Fig. 7. Fig. 8.

knot. Hence the dodecahedral space is the 2-fold cyclic branched cover of S^3 along the (3,5) torus knot.

The following definition seems natural at this point:

DEFINITION. Let L be a link in a 3-manifold M which is left invariant by the action of a group G on M. Then any surgery $M(L;\gamma_1,\cdots,\gamma_k)$ in which the collection $\{\gamma_1,\cdots,\gamma_k\}$ of homology classes is left invariant by G, is said to be equivariant with respect to G.

The manifold obtained by equivariant surgery naturally inherits the action of the group G.

THEOREM 3 (An algorithm). *Every n-fold cyclic branched cover of S^3 branched along a knot K may be obtained from S^3 by equivariant surgery on a link L with a symmetry of order n.*

Proof. The algorithm proceeds as follows.

Step 1. Choose a knot projection for K. In the projection encircle the crossings which, if simultaneously reversed, cause K to become the trivial knot K′.

Step 2. Lift these disjoint circles into the complement $S^3 - K$ of the knot, so that each one has linking number zero with K.

Step 3. Reverse the encircled crossings. Then orient each curve c_i so that the result of the surgery $S^3(c_i; m \mp \ell)$ is to reverse that crossing back to its original position (see Figure 6).

Step 4. Let $C = \bigcup_{i=1}^{k} c_i$ be the union of the oriented circles in $S^3 - K'$, and let $p: S^3 \to S^3$ be the n-fold cyclic branched cover of S^3 along the trivial knot K′. Then if $L = p^{-1}(C)$, it follows from Theorem 2 that the n-fold cyclic branched cover of S^3 along K is the manifold $S^3(L; r_1 m_1 \mp \ell_1, \cdots, r_k m_k \mp \ell_k)$ obtained from S^3 by equivariant surgery on the link L, which has a symmetry of order n.

EXAMPLE (Another presentation of the dodecahedral space). In Figure 9, let $p: S^3 \to S^3$ be the 5-fold cyclic branched cover of S^3 along the

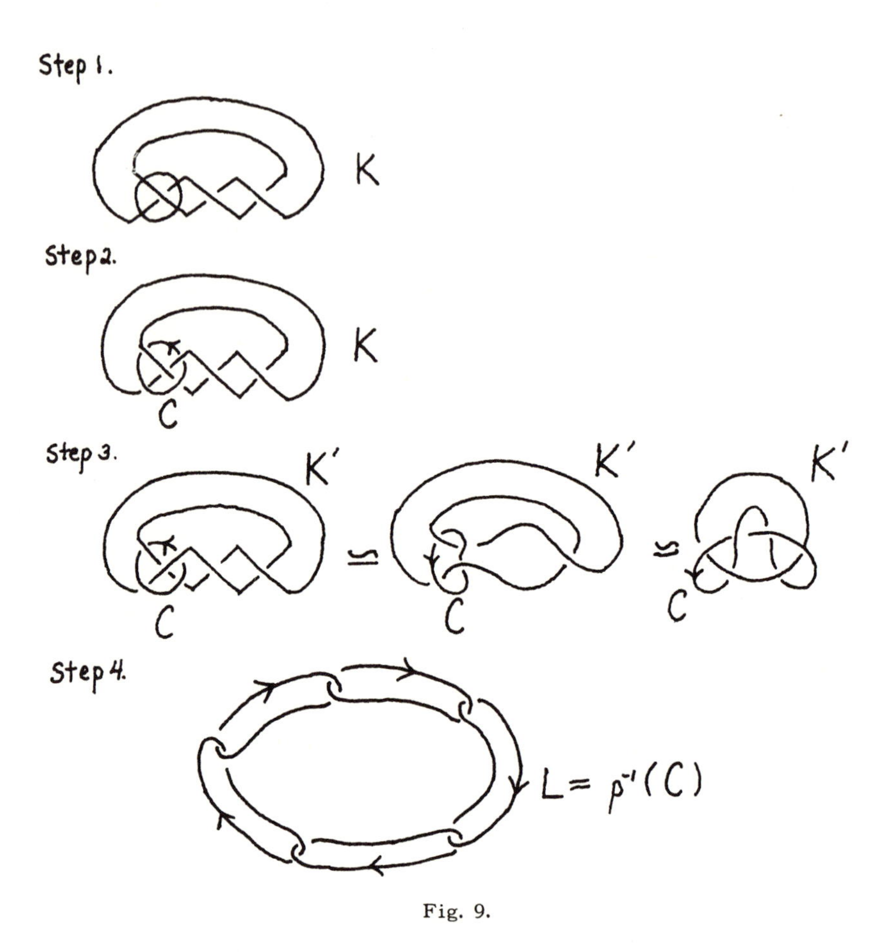

Fig. 9.

trivial knot K′. Then if $L = p^{-1}(C)$ as in step 4 of Figure 9, the 5-fold cyclic branched cover of S^3 along the (2, 3) torus knot K is the manifold $S^3(L; m_1 - \ell_1, \cdots, m_5 - \ell_5)$ obtained from S^3 by equivariant surgery on the link L.

III. Applications

We will now derive properties of the special knots constructed in Section I. Recall that a knot K is characterized by its complement if no surgery $S^3(K; m + k\ell)$, $k \in Z$ and $k \neq 0$, is again S^3. A knot K is said to have property P if and only if no surgery $S^3(K; m + k\ell)$, $k \in Z$ and

$k \neq 0$, is a simply connected manifold. A fake 3-sphere is a homotopy 3-sphere which is not homeomorphic to S^3.

THEOREM 4. *Let* K *be a completely symmetric fibered knot defined by the sequence of knots* $K_0, K_1, \cdots, K_n = K$, *such that each* K_i, $i \neq 0$, *is symmetric of order* $n_i = 2$. *Then* K *is characterized by its complement.*

THEOREM 5. *Let* K *be a completely symmetric fibered knot of complexity* 1, *defined by the sequence* $K_0, K_1 = K$, *where* K *is symmetric of order* $n_1 = n$. *If* K *is not characterized by its complement, then there is a transformation of* S^3 *which is periodic of period* n, *having knotted fixed point set. If a fake 3-sphere is obtained from* S^3 *by surgery on* K, *then there is a periodic transformation of this homotopy sphere of period* n, *having knotted fixed point set.*

THEOREM 6. *Let* K *be a completely symmetric fibered knot. Then if* K *does not have property* P, *there exists a non-trivial knot* $K' \subset S^3$ *such that for some* $n > 1$, *the* n*-fold cyclic branched cover of* S^3 *branched along* K′ *is simply connected.*

It should be pointed out that the property of a knot being characterized by its complement is considerably weaker than property P. For example, it is immediate from Theorem 4 that the figure-8 knot is characterized by its complement, while the proof that it has property P is known to be difficult (see [7]).

The following lemmas will be used to prove Theorems 4-6.

LEMMA 2. *The special genus of the torus link of type* (n, nk), $k \neq 0$, *is bounded below by*

$$\frac{n^2|k|-4}{4} \qquad \textit{if } n \textit{ even}$$

$$\frac{|k|(n^2-1)}{4} \qquad \textit{if } n \textit{ odd, } k \textit{ even}$$

$$\frac{(n-1)(|k|(n+1)-2)}{4} \qquad \textit{if } n \textit{ odd, } k \textit{ odd}\ .$$

Proof. The special genus of an oriented link L is defined here to be the infimum of all geni of connected, oriented surfaces F locally flatly embedded in D^4, whose oriented boundary ∂F is the link $L \subset \partial D^4$. This special genus, which will be denoted $g^*(L)$, satisfies an inequality

$$|\sigma(L)| \leq 2g^*(L) + \mu(L) - \eta(L)$$

where $\sigma(L)$ is the signature, $\mu(L)$ is the number of components and $\eta(L)$ is the nullity of the link L (see [8] or [10]). The lemma will be proved by calculating $\sigma(L)$, $\mu(L)$ and $\eta(L)$, where L is the torus link of type (n, nk), $k > 0$ (see [6]); then the result will automatically follow for torus links of type (n, nk), $k < 0$, since these are mirror images of the above.

In what follows, assume $k > 0$.

(i)
$$\sigma(L) = \frac{2-n^2k}{2} \qquad \text{if } n \text{ even}$$

$$\frac{k(1-n^2)}{2} \qquad \text{if } n \text{ odd}\ .$$

The signature $\sigma(L)$ is the signature of any 4-manifold which is the double branched cover of D^4 along a spanning surface F of L having the properties described above (see [8]). The intersection of the algebraic variety $\{x \in C^3 : x_1^n + x_2^{nk} + x_3^2 = \delta\}$, for small δ, with the 4-ball $\{x \in C^3 : |x| \leq 1\}$ is such a 4-manifold. Its signature is calculated by Hirzebruch ([3]) to be $\sigma^+ - \sigma^-$, where

$\sigma^+ = \#\{(i_1,i_2): 0 < i_1 < n,\ 0 < i_2 < nk\}$ such that $\quad 0 < \frac{i_1}{n} + \frac{i_2}{nk} + \frac{1}{2} < 1 \quad (\text{mod } 2)$

$\sigma^- = \#\{(i_1,i_2): 0 < i_1 < n,\ 0 < i_2 < nk\}$ such that $\ -1 < \frac{i_1}{n} + \frac{i_2}{nk} + \frac{1}{2} < 0 \quad (\text{mod } 2).$

In other words, if we consider the lattice points $\left\{\left(\frac{i_1}{n}, \frac{i_2}{nk}\right): 0 < i_1 < n, 0 < i_2 < nk\right\}$ in the interior of the unit square of the xy-plane, and divide the unit square into positive and negative regions as in Figure 10, then σ^+ is the total number of points interior to the positive regions, σ^- is the number of points interior to the negative region, and their difference $\sigma^+ - \sigma^-$ is given by the formulae in (i).

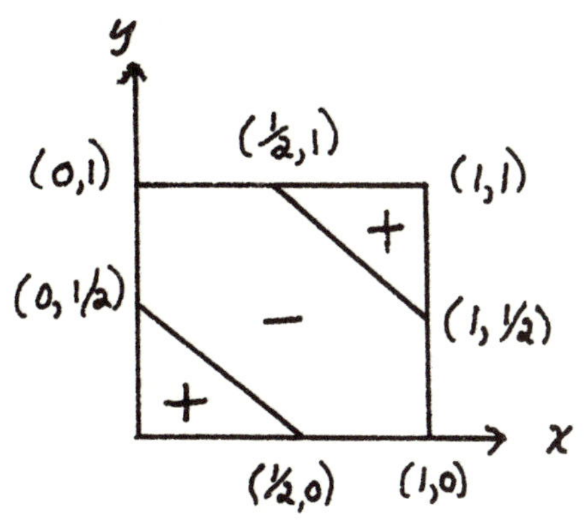

Fig. 10.

(ii) $\qquad \eta(L) = n-1 \quad$ if n even

$\qquad\qquad\quad n \quad$ if n odd, k even

$\qquad\qquad\quad 1 \quad$ if n odd, k odd .

The nullity of a link L is defined to be one more than the rank of the first homology group $H_1(M; R)$ of the double branched cover M of S^3 branched along L; it follows that $\eta(L)$ is independent of the orientation of L. The result in (ii) can be easily obtained from any of the known methods for calculating nullities (see [11]).

(iii) $$\mu(L) = n \ .$$

Substituting these quantities into the inequality gives the desired lower bounds for $g^*(L)$. Note that except for the $(2, \pm 2)$ torus links, none of the non-trivial torus links of type (n,nk) has special genus 0.

In the next few paragraphs, B_n denotes the braid group on n strings; a single letter will be used to signify both an equivalence class of braids, and a representative of that equivalence class; and the notation $\overline{b}$ will stand for the closure of the braid b (i.e., the link obtained by identifying the endpoints of b).

LEMMA 3. *If* $b \in B_n (n \geq 3)$ *is a braid with* n *strings which closes to the trivial knot, and* $c \in B_n$ *is a generator of the center of the braid group* B_n, *then the braid* $b \cdot c^k$, $k \in Z$ *and* $k \neq 0$, *closes to a non-trivial knot.*

Proof. First observe that if b_1 and b_2 are n-stranded braids which have identical permutations and which close to a simple closed curve such that $g^*(\overline{b}_1) = g_1$ and $g^*(\overline{b}_2) = g_2$, then the closed braid $\overline{b_1^{-1} \cdot b_2}$ is a link of n components whose special genus $g^*(\overline{b_1^{-1} \cdot b_2}) \leq g_1 + g_2$. This is illustrated schematically by Figure 11. Imagine that the two abutting cubes are 4-dimensional cubes I_1^4 and I_2^4, that their boundaries are S^3, and that the closed braid $\overline{b}_i (i = 1, 2)$ is positioned in I_i^4 as shown, with the intersection $\overline{b}_1 \cap \overline{b}_2$ consisting of n arcs. Span each closed braid $\overline{b}_i$ by a connected, oriented, locally flatly embedded surface of genus g_i in the cube I_i^4. The union of the two surfaces is then a surface in $I^4 = I_1^4 \cup I_2^4$, whose boundary in the 3-sphere ∂I^4 is the closed braid $\overline{b_1^{-1} \cdot b_2}$.

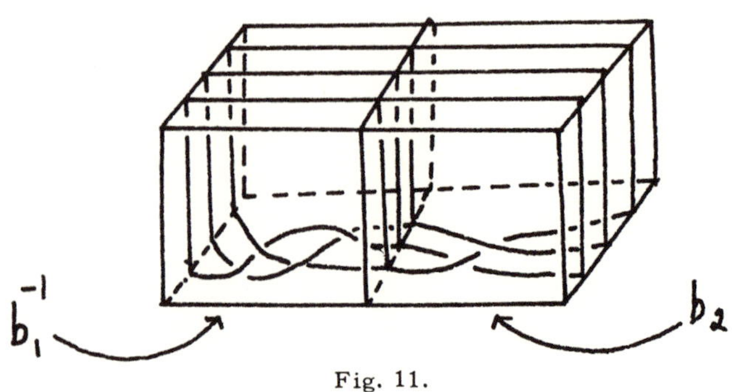

Fig. 11.

The boundary $\overline{b_1^{-1} \cdot b_2}$ has n components because the braid $b_1^{-1} \cdot b_2$ with n strings has the trivial permutation; hence attaching the two surfaces at n places along their boundaries does not increase the genus beyond the sum $g_1 + g_2$. The conclusion that $g^*(\overline{b_1^{-1} \cdot b_2}) \leq g_1 + g_2$ is immediate.

Now suppose the conclusion of the lemma is false; i.e., for some braid $b \in B_n$ and $k \in \mathbf{Z}$, $k \neq 0$, both b and $b \cdot c^k$ close to a trivial knot. Applying the result with $b_1 = b$ and $b_2 = b \cdot c^k$, we reach a contradiction of Lemma 2, which is that $g^*(\overline{c^k}) \leq 0 + 0$, where $\overline{c^k}$ is the torus link of type (n, nk) $n \geq 3$. Therefore Lemma 3 must be true.

Now for the proofs of the theorems:

Proof of Theorem 4. Let $K' = p_n(K)$ and $B = p_n(A_n)$. Then K is the inverse image $p_n^{-1}(K')$ of the completely symmetric fibered knot K' under a 2-fold cyclic branched cover $p_n : S^3 \to S^3$ branched along the unknotted simple closed curve B having K' for generalized axis. The knot $K' \simeq K_{n-1}$ also has repeated symmetries of order 2, and its complexity is one less than that of K. Suppose K is not characterized by its complement. Then a 3-sphere $S^3(K; m + k\ell)$, $k \in \mathbf{Z}$ and $k \neq 0$, may be obtained from S^3 by surgery on K. According to Theorem 2, this 3-sphere is the 2-fold cyclic branched cover of $S^3(K'; m + 2k\ell)$ branched along $B \subset S^3(K'; m + 2k\ell)$. By Waldhausen ([13]), $S^3(K'; m + 2k\ell)$ must be S^3 and $B \subset S^3(K; m + 2k\ell)$ must be unknotted.

We will proceed by induction on the complexity of K. If K has complexity 1, then B is some braid $b \in B_n$ closed about the axis K'. Since K' is unknotted, $S^3(K'; m + 2k\ell)$ is again S^3, and $B \subset S^3(K'; m + 2k\ell)$ is the closed braid $\overline{b \cdot c^{2k}}$ in S^3, for some generator c of center (B_n) (recall Section II, §2). This simple closed curve is knotted, by Lemma 3, which is a contradiction.

Next suppose that every knot of complexity $n < N$ meeting the requirements of the lemma is characterized by its complement, and let K have complexity N. From the induction hypothesis it follows that K' is characterized by its complement, and that $S^3(K'; m+2k\ell)$ cannot be S^3, which is a contradiction.

Hence K must have been characterized by its complement.

Proof of Theorem 5. Let $B = p_1(A_1)$ and $K' = p_1(K)$. Then $p_1 : S^3 \to S^3$ is an n-fold cyclic branched cover of S^3 along the trivial knot B, such that B is a braid $b \in B_n$ closed about the axis K', and $K = p^{-1}(K')$. If K is not characterized by its complement in S^3, then $S^3(K; m+k\ell)$ is the 3-sphere for some $k \in \mathbf{Z}$, $k \neq 0$. It follows from Theorem 2 that S^3 is the n-fold cyclic branched cover of $S^3(K; m+nk\ell)$ branched along $B \subset S^3(K'; m+nk\ell)$. Now since $K' \simeq K_0$ is unknotted, the manifold $S^3(K'; m+nk\ell)$ is S^3 and the simple closed curve $B \subset S^3(K'; m+nk\ell)$ is the closed braid $\overline{b \cdot c^{nk}}$, for some generator c of the center of the braid group B_n. This closed braid is knotted by Lemma 3!

Similarly, if a fake 3-sphere $S^3(K; m+k\ell)$ may be obtained from S^3 as the result of surgery on K, then this homotopy 3-sphere is the n-fold cyclic branched cover of the 3-sphere along the knot $\overline{b \cdot c^{nk}}$.

Proof of Theorem 6. The knot K is defined by a sequence $K_0, K_1, \cdots, K_j = K$. Let $K'_{i-1} = p_i(K_i)$ and $B_{i-1} = p_i(A_i)$. Then there are n_i-fold cyclic branched coverings $p_i : S^3 \to S^3$ branched along the unknotted simple closed curves B_i having K'_i for generalized axis, $0 < i \leq j$, such that $K_i = p_i^{-1}(K'_{i-1})$. If K does not have property P, then a homotopy sphere $S^3(K; m+k\ell)$, $k \in \mathbf{Z}$ and $k \neq 0$, may be obtained from S^3 by surgery on K. This homotopy sphere is the n_j-fold cyclic branched cover of $S^3(K'_{j-1}; m+n_jk\ell)$ branched along $B_{j-1} \subset S^3(K'_{j-1}; m+n_jk\ell)$. It is easy to show that the manifold $S^3(K'_{j-1}; m+n_jk\ell) \simeq S^3(K_{j-1}; m+n_jk\ell)$ is simply connected, and so on, down to $S^3(K_1; m+n_j \cdots n_3n_2k\ell)$. Now $S^3(K_1; m+n_j \cdots n_2k\ell)$ is the n_1-fold cyclic branched cover of the manifold $S^3(K'_0; m+n_j \cdots n_2n_1k\ell)$ branched along $B_0 \subset S^3(K'_0; m+n_j \cdots n_2n_1k\ell)$. Let B_0 be the braid $b \in B_n$ closed about the axis K'_0. Then the homotopy sphere $S^3(K_1; m+n_j \cdots n_3 n_2 k\ell)$ is the n_1-fold cyclic branched cover of S^3 branched along the knot $\overline{b \cdot c^{n_j \cdots n_2 n_1 k}}$, where, as usual, c is some generator of center (B_n).

UNIVERSITY OF CHICAGO

BIBLIOGRAPHY

[1]. Artin, E., "Theory of Braids." Ann. of Math. 48, 101-126 (1947).

[2]. Birman, J., and Hilden, H., "Heegaard Splittings of Branched Covers of S^3." To appear.

[3]. Brieskorn, E., "Beispiele zur Differentialtopologie von Singularitäten." Inventiones math. 2, 1-14 (1966).

[4]. Fox, R. H., "Quick Trip Through Knot Theory." Topology of 3-Manifolds and Related Topics, Ed. M. K. Fish, Jr., Prentice Hall (1962).

[5]. ———, "Covering Spaces with Singularities." Algebraic Geometry and Topology – A Symposium in Honor of S. Lefschetz, Princeton University Press (1957).

[6]. Goldsmith, D. L., "Motions of Links in the 3-Sphere." Bulletin of the A.M.S. (1974).

[7]. González-Acuña, F., "Dehn's Construction on Knots." Boletin de La Sociedad Matematica Mexicana 15, no. 2 (1970).

[8]. Kauffman, L., and Taylor, L., "Signature of Links." To appear.

[9]. Lickorish, W. B. R., "A Representation of Orientable Combinatorial 3-Manifolds." Ann. of Math. 78, no. 3 (1962).

[10]. Murasugi, K., "On a Certain Numerical Invariant of Link Types." Trans. A.M.S. 117, 387-422 (1965).

[11]. Seifert, H., "Die Verochlingungsinvarianten der Zyklischen Knotenüberlagerungen." Hamburg. Math. Abh, 84-101 (1936).

[12]. ———, "Uber das Geschlecht von Knoten." Math. Ann. 110, 571-592 (1934).

[13]. Waldhausen, "Uber Involutionen der 3-Sphäre." Topology 8, 81-91 (1969).

KNOT MODULES

Jerome Levine

Among the more interesting invariants of a locally flat knot of codimension two are those derived from the homology (with local coefficients) of the complement X. Since, by Alexander duality, X is a homology circle, one can consider the universal abelian covering $\tilde{X} \to X$ and the homology groups $H_q(\tilde{X})$, which we denote by A_q, are modules over $\Lambda = Z[t, t^{-1}]$. There is also product structure which will be brought in later.

The modules $\{A_q\}$ have been the subject of much study. In the classical case of one-dimensional knots the Alexander matrix (see [F]) gives a presentation of A_1. The knot polynomials and elementary ideals are then derived from the Alexander matrix but depend only on A_1. These considerations generalize to higher dimensions (see [L 1]). The $Q[t, t^{-1}]$-modules $\{A_q \otimes_Z Q\}$ are completely characterized in [L 1] – this is a relatively simple task since $Q[t, t^{-1}]$ is a principal ideal domain. We will be concerned here with the integral problem.

There is quite a bit already known; we refer the reader to [K], [S], [G], [Ke], [T 1]. It is the purpose of this note to announce an almost complete algebraic characterization of the $\{A_q\}$ – except for the case $q = 1$. In addition we will derive a large array of invariants of a more tractable nature from the $\{A_q\}$ and try to give an exact description of their range. Some of these invariants are already known, but many are new. Finally, we will be able to show that these invariants completely determine A_q, under certain restrictions. In this case the invariants consist of ideals, ideal classes and Hermitian forms over certain rings of algebraic integers.

§1. *Module properties of* $\{A_q\}$

It is well-known that A_q is finitely generated, as Λ-module, and multiplication by the element $t-1 \in \Lambda$ defines an automorphism of A_q (see e.g. [K]). But the deepest property is that of duality. This has been observed in many ways, but I would like to present a new formulation which seems like the most suitable.

The duality theorem of [Mi] yields the isomorphism:

$$\bar{H}_q(\tilde{X}) \approx H_\pi^{n+2-q}(\tilde{X}, \partial\tilde{X}) . \tag{1}$$

In this equation, n is the dimension of the knot (a homotopy sphere which is a smooth submanifold of S^{n+2}) and $H_\pi^*(\tilde{X}, \partial\tilde{X})$ is the homology of the cochain complex $\mathrm{Hom}_\Lambda(C_*(\tilde{X}, \partial\tilde{X}), \Lambda)$. $C_*(\tilde{X})$ and $C_*(\tilde{X}, \partial\tilde{X})$ are considered as *left* Λ-modules and the *right* action of Λ on Λ puts the structure of a right Λ-module on $H_\pi^*(\tilde{X}, \partial\tilde{X})$. $\bar{H}_q(\tilde{X})$ denotes the *right* Λ-module defined from the original left Λ structure by the usual means: $a\lambda = \bar{\lambda}a$, where $\lambda \in \Lambda$, $a \in H_q(\tilde{X})$ and $\lambda \to \bar{\lambda}$ is the anti-automorphism of Λ defined by $f(t) \to f(t^{-1})$. Now (1) represents an isomorphism of right Λ-modules.

We now use the universal coefficient spectral sequence (see e.g. [M; p. 323]) to reduce $H_\pi^*(\tilde{X}, \partial\tilde{X})$ to information about $H_*(\tilde{X}, \partial\tilde{X})$. Since Λ has global dimension 2 and the $\{A_q\}$ are Λ-torsion modules, the spectral sequence collapses to a set of short exact sequences. Using (1) and the trivial nature of ∂X, we derive the following exact sequences for $0 < q \leq n$:

$$0 \to \mathrm{Ext}^2_\Lambda(A_{n-q}, \Lambda) \to \bar{A}_q \to \mathrm{Ext}^1_\Lambda(A_{n+1-q}, \Lambda) \to 0 \tag{2}$$

and

$$A_q = 0, \text{ for } q > n .$$

To properly interpret (2) we define T_q to be the Z-torsion submodule of A_q, and $F_q = A_q/T_q$. It is not hard to show that T_q is finite (see [K]). It can then be shown that $\mathrm{Ext}^2_\Lambda(A_i, \Lambda)$ is a Z-torsion module and depends only on T_i, while $\mathrm{Ext}^1_\Lambda(A_i, \Lambda)$ is Z-torsion free and depends on F_i. As a result, (2) can be rewritten:

(3) $\quad \bar{T}_q \approx \mathrm{Ext}^2_\Lambda(T_{n-q}, \Lambda)$ for $0 < q < n$, $T_q = 0$ for $q \geq n$.

(4) $\quad \bar{F}_q \approx \mathrm{Ext}^1_\Lambda(F_{n+1-q}, \Lambda)$ for $0 < q \leq n$, $F_q = 0$ for $q > n$.

§2. *Product structure on* $\{A_q\}$

The chains of $\tilde{X}$ admit an intersection pairing with values in Λ (see [Mi], [B]) which satisfies the Hermitian property: $\alpha \cdot \beta = (-1)^{q(n+2-q)} \overline{\beta \cdot \alpha}$, when $\alpha \in C_q(\tilde{X})$, $\beta \in C_{n+2-q}(\tilde{X})$. This induces a Hermitian pairing in the usual way on $H_*(\tilde{X})$, but, since A_q is Λ-torsion, this pairing is trivial. One can then define a linking pairing: $A_q \times A_{n+1-q} \to Q(\Lambda)/\Lambda$ where $Q(\Lambda)$ is the quotient field of Λ, in a manner entirely analogous to the usual linking pairing in the Z-torsion part of the homology of a manifold. This is just the *Blanchfield pairing* (see [B], [Ke], [T 2]). Under the canonical isomorphism $\mathrm{Hom}_\Lambda(A, Q(\Lambda)/\Lambda) \approx \mathrm{Ext}^1_\Lambda(A, \Lambda)$, for any Λ-torsion module A, the isomorphism (4) is adjoint to the Blanchfield pairing (which vanishes on Z-torsion). The Hermitian property of this pairing yields the following strengthening of (4);

(4)′ If $n = 2q-1$, the isomorphism of (4) corresponds to a pairing $\langle\, , \rangle : F_q \times F_q \to Q(\Lambda)/\Lambda$ satisfying the Hermitian property: $\langle \alpha, \beta \rangle = (-1)^{q+1} \overline{\langle \beta, \alpha \rangle}$.

One can define a more obscure linking pairing on the Z-torsion: $[\, ,] : T_q \times T_{n-q} \to Q/Z$, which is Z-linear, $(-1)^{q(n-q)}$ symmetric and admits t as an isometry i.e. $[t\alpha, t\beta] = [\alpha, \beta]$. In the case of a fibered knot (see [S]) T_q is the Z-torsion subgroup of $H_q(F)$, where F is the fiber, and $[\, ,]$ coincides with the usual linking pairing on $H_*(F)$. This pairing relates to (3) as follows. It can be shown that $\mathrm{Ext}^2_\Lambda(T, \Lambda) \approx \mathrm{Hom}_Z(T, Q/Z)$, canonically, as Λ-modules, for any finite Λ-module T. It turns out that, under this isomorphism, the isomorphism of (3) is adjoint to $[\, ,]$. The symmetry of $[\, ,]$ yields a strengthening of (3):

(3)′ If $n = 2q$, the isomorphism of (3) corresponds to a Z-linear pairing $[\, ,] : T_q \times T_q \to Q/Z$ satisfying the symmetry property $[\alpha, \beta] = (-1)^q [\beta, \alpha]$.

§3. *Obstructions to smoothness of 3-dimensional knots*

If $<,> A_q \times A_q \to Q(\Lambda)/\Lambda$ is the pairing of (4)′, when q is even, Trotter defines an associated unimodular, even, integral quadratic form λ (see [T 2]). The signature $\sigma(\lambda)$ is a multiple of 8. A smooth, or even PL locally flat, knot bounds a submanifold M of S^{n+2} and it is not hard to see that $\sigma(\lambda)$ is the signature of M. We conclude from Rohlin's theorem:

(5) If $n = 3$, the quadratic form associated to the pairing $<,>$ of (4)′ has signature $\equiv 0 \bmod 16$, when the knot is smooth or PL locally flat.

There do exist topological locally flat knots for which $\sigma(\lambda) \not\equiv 0 \bmod 16$ (see [CS] or [Ka]).

§4. *Realization Theorem:* We now present our main geometric result.

THEOREM. *Suppose that* $\{F_q, T_q\}$ *is a family of finitely generated* Λ*-modules on each of which* $t-1$ *is an automorphism. Suppose, furthermore, that* F_q *is* Z*-torsion free,* T_q *is finite and they satisfy* (3), (3)′, (4) *and* (4)′, *for a certain* $n \geq 1$, *and* (5) *if* $n = 3$. *We also assume* $T_1 = 0$.

Then there exists a smooth n*-dimensional knot in* $(n+2)$*-space with* F_q, T_q *and the pairings* $<,>$ *of* (3)′ *and* $[,]$ *of* (4)′ *as the associated knot modules and linking pairings.*

REMARKS:

(i) One can realize many $T_1 \neq 0$ using the twist-spinning construction of Zeeman [Z].

(ii) In the case $n = 3$, I do not know which $(F_2, <,>)$ not satisfying (5) can be realized by topological knots.

(iii) In the case $n = 2q-1$, $q \geq 2$, the isotopy class of the knot is completely determined by $(F_q, <,>)$ when $\tilde{X}$ is $(q-1)$-connected (see [Ke] or [L 2] and [T 2]).

(iv) This theorem includes previous results of [K], [G], [Ke]. In particular, it is interesting to compare the middle-dimensional results of [K] and [G], which are stated in terms of presentations of F_q or T_q.

§5. *Algebraic study of* $\{T_q\}$

We now turn to the algebraic consideration of the modules F_q, T_q and pairings $<,>$ $[,]$. We will attempt to extract reasonable invariants, determine the range of these invariants and, in some cases, use the invariants to classify.

Let T be a finite Λ-module. We may, without loss of generality, assume T is p-primary for some prime number p. Consider the associated modules:

$$T_{(i)} = \frac{\mathrm{Ker}\, p^{i+1}}{\mathrm{Ker}\, p^{i}}, \qquad T^{(i)} = \frac{p^i T}{p^{i+1} T}.$$

These are modules over the principal ideal domain $\Lambda_p = Z/(p)[t, t^{-1}]$.

PROPOSITION.

(i) *There is a natural exact sequence of* Λ_p*-module*:

$$0 \longrightarrow T_{(i+1)} \xrightarrow{p} T_{(i)} \longrightarrow T^{(i)} \xrightarrow{p} T^{(i+1)} \longrightarrow 0 .$$

(ii) *Given any finite collection* $\{T_i, T^i\}$ *of* Λ_p*-modules together with exact sequences*: $0 \to T_{i+1} \to T_i \to T^i \to T^{i+1} \to 0$, *there exists a finite* p*-primary* Λ*-module* T *such that* $T_{(i)} \approx T_i$, $T^{(i)} \approx T^i$ *and the exact sequence of* (i) *corresponds to the given one.*

The modules $T_{(i)}$, $T^{(i)}$ are described entirely by polynomial invariants in Λ_p. These include the *local Alexander polynomials* considered in [K] and [G]. The proposition makes it a straightforward matter to write down the range of these invariants for T_q if $q < \frac{1}{2}n$.

When $n = 2q$, there is more to be said. For example, let $\Delta_i = T_{(i)}/p\, T_{(i+1)}$. The pairing $[\,,\,]$ of (3)′ yields a non-singular

$(-1)^q$-symmetric pairing $\Delta_i \times \Delta_i \to Z/(p)$, for which the action of t is an isometry. Conversely, given $\{\Delta_i\}$ with such pairings, there exists T with a pairing $[\,,]$ inducing the given ones. Now it is not too difficult to determine those Λ_p-modules Δ_i which admit such pairings. It is interesting that one obtains *different* answers for q even and odd and, therefore, the possible T_q, for n-dimensional knots where $n = 2q$, are not identical.

Of course, the polynomial invariants derived here do not classify the module T_q, in general.

§6. *Algebraic study of* $\{F_q\}$

Let F be a finitely generated Λ-module which is Z-torsion-free. Let $\phi \in \Lambda$ be an irreducible polynomial and define:

$$F(\phi, i) = \operatorname{Ker} \phi^i / \operatorname{Ker} \phi^{i-1}; \text{ then } F(\phi, i) \text{ is a } \Lambda/(\phi)\text{-module .}$$

Multiplication by ϕ induces a monomorphism: $\phi : F(\phi, i) \to F(\phi, i-1)$. Suppose $R = \Lambda/(\phi)$ is a Dedekind domain (for example, if ϕ is quadratic this will happen when the discriminant of ϕ is square-free) (see also [T 1]). Then the $\{F(\phi, i)\}$ or, even better, the quotients $F(\phi, i-1)/\phi F(\phi, i)$ yield invariants of F in the form of ideals in R, ideal classes, and ranks. These include all the rational invariants [L 1] and the ideal class invariants of [FS], and the ideals are certainly related to the elementary ideals of F in Λ (see [F]). Furthermore, it is not difficult to determine the range of these invariants, for $q < \frac{1}{2}(n+1)$, by constructing F to realize any collection of $\{F(\phi, i)\}$.

The effects of the duality relations (4), (4)$'$ on these invariants seems complicated, in general. This is also true of the question of classification. Both of these problems are made manageable by imposing a "homogeneity" restriction on F.

Suppose F is ϕ-primary i.e. $\phi^r F = 0$, for some r. Then we may consider F as a module over $\Lambda/(\phi^r) = S$. Let S_0 be the localization of S at the prime (ϕ) – S_0 is a principal ideal domain. Then

$F \subset F \otimes_S S_0$ (because F is Z-torsion free). Now $F \otimes S_0 \approx \sum_{i \leq r} F_i$, where F_i is a free $S_0/(\phi^i)$-module. We say F is *homogeneous of degree* d if $F_i = 0$ for all $i \neq d$. (I'd like to thank David Eisenbud for this formulation of homogeneity.)

PROPOSITION.

(i) *If* F *is homogeneous of degree* d, *then the isomorphism class of* F *is determined by the isomorphism class of the nested sequence of* R-*modules*:

$$F(\phi, d) \to F(\phi, d-1) \to \cdots \to F(\phi, 1) \ .$$

All the $F(\phi, i)$ *are* R-*torsion free modules of the same rank.*

(ii) *Given any sequence*: $B_d \to B_{d-1} \to \cdots \to B_1$ *of* R-*torsion free modules of the same rank, there is a homogeneous* ϕ-*primary* Λ-*module* F *of degree* d, *whose associated sequence* $F(\phi, d) \to \cdots \to F(\phi, 1)$ *is isomorphic to the given one.*

We are still assuming R is Dedekind. If the class number of R is zero, i.e. it is a principal ideal domain, the classification of the nested sequence $\{F(\phi, i)\}$ can be formulated in terms of row-equivalence of matrices over R. If the rank of the $F(\phi, i)$ is *one*, the $\{F(\phi, i)\}$ are just a sequence of ideals in R, determined up to scalar multiplication.

Note that $S_0 = Q[t, t^{-1}]/(\phi^r)$ and so the condition of homogeneity can be formulated in terms of the polynomial invariants of [L 1].

Of course these results extend to sums of homogeneous modules.

Suppose $n = 2q-1$, and F has a pairing $\langle , \rangle$ as in (4)$'$. If ϕ is relatively prime to $\bar{\phi}$, then $\langle , \rangle$ pairs the ϕ-primary component of F to the $\bar{\phi}$-primary component (when F is the sum of its ϕ-primary components, over all ϕ). No further restriction is imposed on the ϕ-primary component by the existence of $\langle , \rangle$.

If $\phi, \bar{\phi}$ are associate elements of Λ we may assume $\phi = \bar{\phi}$ (see [L1]). $F(\phi, 1)$ inherits a $(-1)^{q+1}$-Hermitian non-degenerate pairing from $\langle , \rangle$ which we denote by:

$$\langle , \rangle' : F(\phi, 1) \times F(\phi, 1) \to S_0/(\phi) = Q(R)$$

the quotient field of R.

PROPOSITION. *Suppose* F *is homogeneous* ϕ*-primary of degree* d, *where* $\phi = \bar{\phi}$. *Then*

(i) $F(\phi, i)$ *is dual to* $F(\phi, d-i+1)$ *under* $\langle , \rangle'$, *i.e.* $\langle F(\phi, i), F(\phi, d-i+1) \rangle \subset R$ *and the induced pairing* $\langle , \rangle_i : F(\phi, i) \times F(\phi, d-i+1) \to R$ *is non-singular.*

(ii) *The injections* $\phi : F(\phi, i+1) \to F(\phi, i)$ *and* $F(\phi, d-i+1) \to F(\phi, d-i)$ *are adjoint with respect to* $\langle , \rangle_i$ *and* $\langle , \rangle_{i+1}$.

(iii) *The isomorphism class of* $(F, \langle , \rangle)$ *is determined by that of the system* $(\{F(\phi, i)\}, \langle , \rangle')$.

(iv) *Given* $\langle , \rangle'$ *on* $F(\phi, 1)$ *satisfying* (i), (ii), *there exists* $\langle , \rangle$ *on* F *inducing it.*

Thus the isomorphism classes of such $(F, \langle , \rangle)$ correspond to the isomorphism classes of torsion-free R-modules B equipped with a non-degenerate (non-singular, if d is odd) $(-1)^{q+1}$-Hermitian pairing and a sequence of submodules of equal rank: $B_d \subset B_{d-1} \subset \cdots \subset B_c = B$, where $d = 2c-1$ or $2c-2$, by setting $B_i = F(\phi, i)$.

A solution of the local classification problem i.e. over the completions of R, can be derived from [J].

The simplest case is rank one. The $\{B_i\}$ are fractional ideals of R; the $(-1)^{q+1}$-Hermitian pairing corresponds to a non-zero $\lambda \in Q(R)$ such that $\lambda = (-1)^{q+1} \bar{\lambda}$ and:

$$(*) \qquad \lambda B\bar{B} \begin{cases} = R & \text{if } d \text{ odd}, \\ \subseteq R & \text{if } d \text{ even}. \end{cases}$$

Equivalence becomes $\{B_i,\lambda\} \sim \{\mu B_i, \lambda/\mu\bar{\mu}\}$, for any non-zero $\mu \in Q(R)$ (compare [T 1]). For example, if ϕ is quadratic we may write $\phi = at^2 + (1-2a)t + a$, for an integer a; R is Dedekind if and only if $4a-1$ is square-free. The class number of R is a divisor of the class number of the ring of algebraic integers R_0 in the algebraic number field generated by a root of ϕ. Condition (*) is never satisfied if q is even and d odd. Otherwise such a λ exists for any B.

If $a = p^m$, for some prime p, the computations become reasonable. For example, the class number of R is $1/m$ times the class number of R_0, and for q and d odd, for each B, there are two (for m odd) or four (for m even) inequivalent Hermitian forms. If d is even, there are an infinite number of inequivalent forms. We record here the non-trivial class numbers of R for $p^m \leq 125$:

class number	p^m
2	13, 23, 29, 31, 47, 49, 64, 67, 121
3	53, 71, 83
4	73, 89 .

BRANDEIS UNIVERSITY

REFERENCES

[B.] Blanchfield, R. C.: Intersection theory of manifolds with operators with applications to knot theory, Annals of Math. 65 (1957), 340-356.

[C.S.] Cappell, S., and Shaneson, J.: On topological knots and knot cobordism, Topology 12 (1973), 33-40.

[F.] Fox, R. H.: A quick trip through knot theory, Topology of 3-manifolds, Ed. M. K. Fort, Jr. Prentice-Hall, Englewood, N. J.

[F.S.] Fox, R. H., and Smythe, N.: An ideal class invariant of knots, Proc. A.M.S. 15 (1964), 707-709.

[G.] Gutierrez, M.: On Knot Modules, Inv. Math.

[J.] Jacobowitz, R.: Hermitian forms over local fields, Amer. J. of Math. 84 (1962), 441-465.

[Ka.] Kato, M.: Classification of compact manifolds homotopy equivalent to a sphere, Sci-papers College general Educ., U. of Tokyo 22 (1972), 1-26.

[Ke.] Kearton, C.: Classification of simple knots by Blanchfield duality, Bull. A.M.S. 79 (1973), 952-956.

[K.] Kervaire, M.: Les noeuds de dimensions superieures, Bull. Soc. Math. France 93 (1965), 225-271.

[L 1] Levine, J.: Polynomial invariants of knots of codimension two, Annals of Math. 84 (1966), 537-554.

[L 2] ———— : Algebraic classification of some knots of codimension two, Comm. Math. Helv. 45 (1970), 185-198.

[M.] Mac Lane, S.: Homology, Springer-Verlag, Berlin, 1963.

[Mi.] Milnor, J.: A duality theorem for Reidemeister torsion, Annals of Math. 76 (1962), 137-147.

[S.] Sumners, D.: Polynomial invariants and the integral homology of coverings of knots and links, Inv. Math. 15 (1972), 78-90.

[T 1] Trotter, H. F.: On the algebraic classification of Seifert matrices, Proceedings of the Georgia Topology Conference 1970, University of Georgia, 92-103.

[T 2] ———— : On S-equivalence of Seifert matrices, Inv. Math. 20 (1973), 173-207.

[Z.] Zeeman, E. C.: Twisting spun knots, Trans. A.M.S. 115 (1965), 471-495.

THE THIRD HOMOTOPY GROUP OF SOME HIGHER DIMENSIONAL KNOTS

S. J. Lomonaco, Jr.

0. *Introduction*

In 1962 Fox [1] posed the problem of computing the second homotopy group of the complement $S^4 - k(S^2)$ of a $(4,2)$-knot as a $Z\pi_1$-module. Although Epstein [3] had previously shown that π_2 as an abelian group (without $Z\pi_1$-action) was algebraically uninteresting, Fox pointed out that this might not be the case when the action of π_1 on π_2 is considered. Since then some progress has been made. In [6, 7, 8] a presentation of the second homotopy group of an arbitrary spun knot [5] was calculated as a $Z\pi_1$-module and found to be algebraically non-trivial. In particular,

THEOREM 0. *If* $k(S^2) \subset S^4$ *is a* 2*-sphere formed by spinning an arc* a *about the standard* 2*-sphere* S^2 *and* $(x_1, \cdots, x_n : r_1, \cdots, r_m)$ *is a presentation of* $\pi_1(S^4 - k(S^2))$, *then*

$$\left(X_1, \cdots, X_n : \sum_i (\partial r_j/\partial x_i) X_i = 0 \ (0 \leq j \leq m)\right)$$

is a presentation of $\pi_2(S^4 - k(S^2))$ *as a* $Z\pi_1$*-module, where* $r_0 = r_0(x_1, \cdots, x_n)$ *is the image of the generator of* $\pi_1(S^2 - a)$ *under the inclusion map and the symbols* $\partial r_j/\partial x_i$ *denote the images of Fox's derivatives* [9] *in* $\pi_1(S^4 - k(S^2))$.

Little appears to be known about the higher dimensional homotopy groups. In this paper a procedure is given for computing a presentation of π_3 of a spun knot as a $Z\pi_1$-module. Specifically,

THEOREM 1. *Let* $(S^4, k(S^2))$ *be defined as in Theorem 0 above. Then* $\pi_3(S^4 - k(S^2))$ *is isomorphic as a* $Z\pi_1$*-module to* $\Gamma(\pi_2(S^4 - k(S^2)))$, *where* Γ *denotes a functor defined by J. H. C. Whitehead* [10, 11] *and later generalized by Eilenberg and MacLane* [12, 13]. *Hence,* π_3 *as a* $Z\pi_1$*-module is determined by* π_1 *and* π_2.

COROLLARY 2. *If* $\pi_2 \neq 0$, *then* π_3 *of a spun knot as a group (i.e., without* $Z\pi_1$*-structure) is free abelian of infinite rank. Otherwise,* $\pi_3 = 0$.

THEOREM 3. *Let* $k(S^2) \subset S^4$ *be a 2-sphere formed by spinning an arc* a *about the standard 2-sphere* S^2 *and* $(x_1, \cdots, x_n : r_1, \cdots, r_m)$ *be a presentation of* $\pi_1(S^4 - k(S^2))$. *Let* $r_0 = r_0(x_1, \cdots, x_n)$ *be the image of the generator of* $\pi_1(S^2 - k(S^2))$ *under the inclusion map and* X_i *and* $\partial r_i/\partial x_j$ *be as in Theorem 0. Then as a* $Z\pi_1$*-module,* $\pi_3(S^4 - k(S^2))$ *is generated by the symbols*

$$\gamma(X_i),\ [X_i, gX_j] \qquad (1 \leq i, j \leq n;\ g \in \pi_1)$$

subject to the relations

$$\begin{aligned} 2\gamma(X_i) &= [X_i, X_i] & & 1 \leq i, j \leq n \\ \gamma\left(\sum_{j=1}^{n} (\partial r_k/\partial x_j) X_j\right) &= 0 & & 0 \leq k \leq m \\ \left[X_i, g \sum_j (\partial r_k/\partial x_j) X_j\right] &= 0 & & g \in \pi_1 \\ [X_i, gX_j] &= g[X_j, g^{-1}X_i] & & \end{aligned}$$

where $[X_i, gX_i]$ *is the Whitehead product of* X_i *and* gX_j *and* $\gamma(X_i)$ *is represented by the composition of the Hopf map* $S^3 \to S^2$ *with a representative of* X_i.

Applications of the above theorem to specific examples can be found in the last section of this paper.

I would like to thank Richard Goldstein for his helpful comments during the preparation of this paper and also Peter Kahn for suggesting the above more general formulation of Theorem 0.

REMARK. The methods of this paper may easily be extended to p-spun knots.

I. Definition of a Spun Knot

Let S^2 be a standard 2-sphere in the 3-sphere S^3 and let α be a polyhedral arc with endpoints lying on S^2 and with interior lying entirely within one of the two components of $S^3 - S^2$. (See Figure 1.)

If α is spun about S^2 holding S^2 fixed, a knotted 2-sphere $k(S^2)$ in S^4 is generated [5]. If one would like to think of the spinning as taking place in time, then at time 0, the arc α would appear on the right of the 2-sphere as indicated in the figure. It would then immediately vanish into another 3-dimensional hyperplane and after rotating through 180° suddenly reappear inside S^2 as indicated by the dotted arc on the left of Figure 1. Again it would disappear into another 3-dimensional hyperplane and rotate through the remaining 180° until it suddenly reappeared on the right closing up the knotted 2-sphere $k(S^2)$.

II. $\pi_3 = \Gamma(\pi_2)$

The complement $X = S^4 - k(S^2)$ of an arbitrary spun knot $(S^4, k(S^2))$ will not be examined in more detail. Let $X_0 = S^3 - k(S^2)$ be the 3-dimensional cross-section shown in Figure 1, and X_+ and X_- denote the closures of the two components of $X - X_0$. Let $p : \tilde{X} \to X$ be the universal covering of X and $\tilde{X}_i = p^{-1}(X_i)$ for $i = +,\ 0,$ and $-$.

Since $\pi_1(X_i) \to \pi_1(X)$ are all onto, it follows from the homotopy sequence of the fibration

$$\pi_1(X) \to \tilde{X}_i \to X_i$$

that $\tilde{X}_i$ is connected and

$$1 \to \pi_1(\tilde{X}_i) \to \pi_1(X_i) \to \pi_1(X) \to 1$$

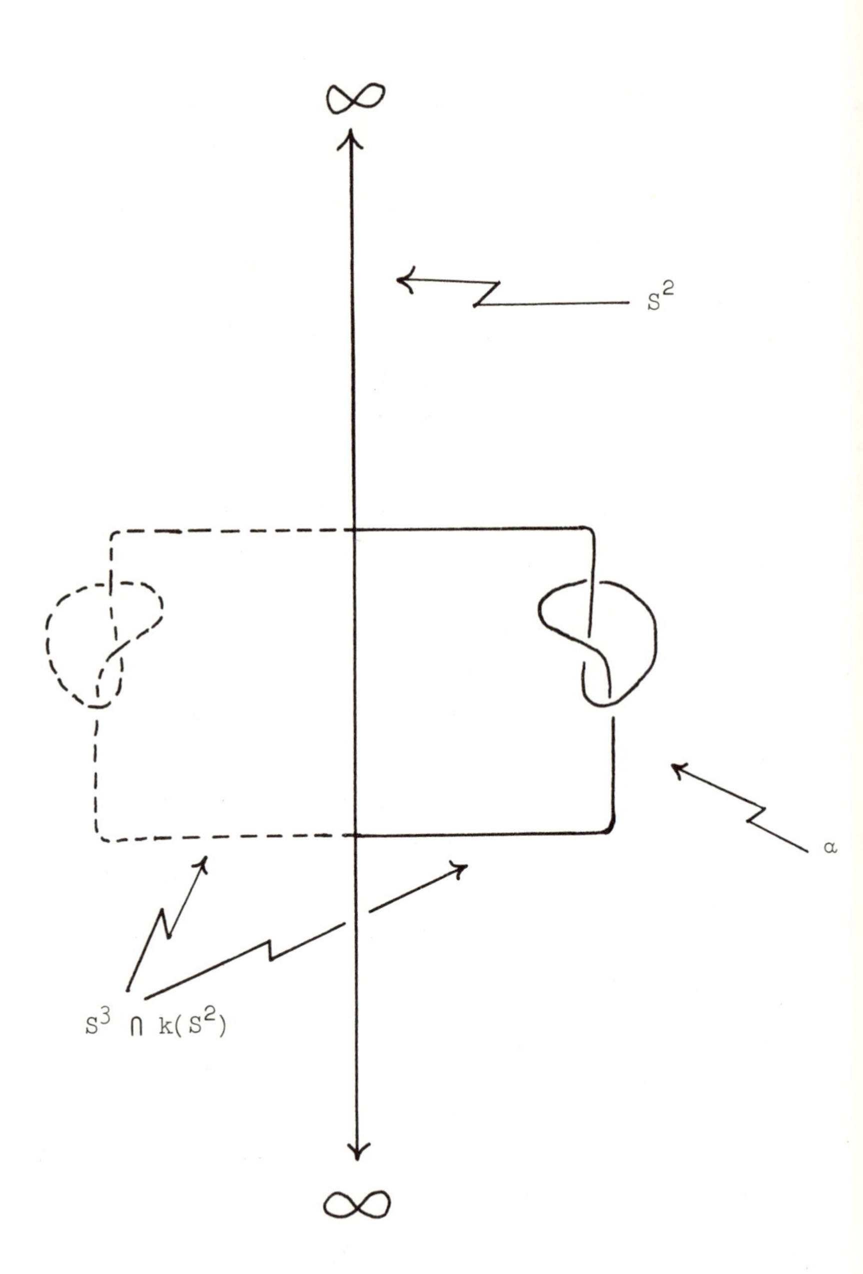

Figure 1. Spun 2-sphere

is exact for $i = +, 0,$ and $-$. Moreover, since $\pi_1(X_\pm) \to \pi_1(X)$ is an isomorphism onto [5], it follows that $\tilde{X}_\pm$ are simply connected, and hence are the universal covers of $X_\pm$. Thus,

LEMMA. *The lift* $\tilde{X}_0$ *of* X_0 *to the universal cover of* X *is connected and* $\pi_1(\tilde{X}_0)$ *is the kernel of* $\pi_1(X_0) \to \pi_1(X)$. *Moreover, the lifts* $\tilde{X}_\pm$ *of* $X_\pm$ *are the universal covers of* $X_\pm$.

Since $\tilde{X}_+$ and $\tilde{X}_-$ both collapse to the right half of $\tilde{X}_0$ via a deformation arising from the spinning, Hurewicz's theorem coupled with the asphericity of knots [4] yields that $H_n(\tilde{X}_\pm) = 0$ for $n \geq 1$. Hence, from the Mayer-Vietoris sequence for the triad $(\tilde{X}; \tilde{X}_+, \tilde{X}_-)$, we have $H_n(\tilde{X}) \simeq H_{n-1}(\tilde{X}_0)$ for $n \geq 2$. Thus,

LEMMA.

$$H_2(\tilde{X}) \simeq H_1(\tilde{X}_0) \quad \textit{and}$$

$$H_n(\tilde{X}) = 0 \quad \textit{for} \quad n > 2 .$$

Proof. Since $\tilde{X}$ collapses to a 3-dimensional CW-complex [8], the last part of this lemma is obviously true for $n > 3$. $H_3(\tilde{X}) \simeq H_2(\tilde{X}_0)$ can be shown to be equal to zero by an analysis of the following decomposition of $\tilde{X}_0$.

Let $X_0^{\pm}$ denote the closure of the two components of $X_0 - S^2$. Then $X_0 = X_0^+ \cup X_0^-$ and $X_{00} = X_0^+ \cap X_0^-$ is S^2 minus the two endpoints of a. Hence, X_{00} is a homotopy 1-sphere and $\pi_1(X_{00})$ is infinite cyclic. Since $\pi_1(X_0^{\pm}) \to \pi_1(X)$ is an isomorphism onto [5], it follows from the homotopy sequence of the fibration

$$\pi_1(X) \to \tilde{X}_0^{\pm} \to X_0^{\pm}$$

that $\tilde{X}_0^{\pm}$ are simply connected. Applying the asphericity of knots [4], we have that $H_2(\tilde{X}_0^{\pm}) = 0$. After inspecting the Mayer-Vietoris sequence for the triad $(\tilde{X}_0; \tilde{X}_0^+, \tilde{X}_0^-)$, we have

$$H_2(\tilde{X}_0) \simeq H_1(\tilde{X}_{00}) .$$

Since the image of a generator of $\pi_1(X_{00})$ in $\pi_1(X)$ has a linking number of ± 1 with respect to $k(S^2)$, $\pi_1(X_{00}) \to \pi_1(X)$ is a monomorphism. Thus, from the homotopy sequence of the fibration

$$\pi_1(X) \to \tilde{X}_{00} \to X_{00} ,$$

we have that $\tilde{X}_{00}$ is simply connected. Hence, $H_3(\tilde{X}) \simeq H_2(\tilde{X}_0) \simeq H_1(\tilde{X}_{00}) = 0$.

With the above lemma and J. H. C. Whitehead's Certain Exact Sequence [10, 11], we have

$$\pi_n(X) \simeq \Gamma_n(X) \qquad \text{for } n \geq 3 .$$

Hence, $\Gamma_3(X) = \Gamma(\pi_2(X))$, where Γ is an algebraic functor defined by J. H. C. Whitehead [10, 11] and later generalized by Eilenberg and MacLane [12, 13]. This formula gives an effective procedure for computing $\pi_3(X)$. In summary, we have

THEOREM 1. $\pi_3(S^4 - k(S^2)) \simeq \Gamma(\pi_2(X))$. *Hence, the third homotopy group of a spun knot as a* $Z\pi_1$*-module is determined by the first and second homotopy groups. As an abelian group, it is determined solely by* π_2.

From [3], $\pi_2(X)$, if non-zero, is free abelian of infinite rank. Since Γ never decreases the rank of a free abelian group, we have

COROLLARY 2. *The third homotopy group of a spun knot as a group is free abelian of infinite rank if the second homotopy group is non-zero. Otherwise, it is zero.*

III. Whitehead's Functor

A more detailed understanding of J. H. C. Whitehead's functor Γ [10, 11] is needed to compute a presentation of $\pi_3(X)$. Very briefly, Γ is defined as follows. (For more details see [10, 11].)

Let A be an additive abelian group. Then $\Gamma(A)$ is an additive abelian group generated by the symbols

$$\{\gamma(a)\}_{a \in A}$$

subject to the relations

$$\gamma(-a) = \gamma(a) \tag{1}$$

$$\begin{aligned}\gamma(a+b+c) - \gamma(b+c) - \gamma(c+a) - \gamma(a+b) \\ + \gamma(a) + \gamma(b) + \gamma(c) = 0 \ .\end{aligned} \tag{2}$$

Define $[a,b]$ by

$$\gamma(a+b) = \gamma(a) + \gamma(b) + [a,b] \ .$$

Then, $[a,b]$ is a measure of how close γ is to a homomorphism.

The following relations are consequences of (1) and (2).

$$\gamma(0) = 0$$

$$2\gamma(a) = [a,a]$$

$$[a,b+c] = [a,b] + [a,c]$$

$$[a,b] = [b,a]$$

$$\gamma\left(\sum_i a_i\right) = \sum_i \gamma(a_i) + \sum_{i<j} [a_i,a_j]$$

$$\gamma(na) = n^2\gamma(a) \ .$$

A proof of the following theorem can be found in [10, 11].

THEOREM. *If* A *is an additive abelian group with generators* a_i *and relations* b_j, *then* $\Gamma(A)$ *is an additive abelian group with generators*

$$\{\gamma(a_i)\} \cup \{[a_i,a_j]\}_{i<j}$$

and relations

$$\{\gamma(b_j) = 0\} \cup \{[a_i,b_j] = 0\} \ .$$

Finally, if A admits a group of operators W, then so does $\Gamma(A)$, according to the rule

$$w\gamma(a) = \gamma(wa)$$

for $w \in W$ and $a \in A$.

IV. Computation of $\pi_3(S^4 - k(S^2))$

From Section II, $\pi_3(X) \simeq \Gamma(\pi_2(X))$, and from Section III, $\pi_3(X)$ is generated by

$$\{\gamma(\xi)\}_{\xi \in \pi_2(X)} \quad \text{and} \quad \{[\xi, \xi']\}_{\xi, \xi' \in \pi_2(X)} .$$

In [10, 11] J. H. C. Whitehead demonstrates that $[\xi, \xi']$ is the Whitehead product of ξ and ξ' and that $\gamma(\xi)$ is represented by the composition of the Hopf map $S^3 \to S^2$ with a representative of ξ. Hence, we have

THEOREM 3. *Let* $k(S^2) \subset S^4$ *be a 2-sphere formed by spinning an arc* a *about the standard 2-sphere* S^2 *and* $(x_1, \cdots, x_n : r_1, \cdots, r_m)$ *a presentation of* $\pi_1(S^4 - k(S^2))$. *Let* $r_0 = r_0(x_1, \cdots, x_n)$ *be the image of the generator of* $\pi_1(S^2 - k(S^2))$ *under the inclusion map and* X_i *and* $\partial r_i/\partial x_j$ *be as in Theorem 0. Then as a* $Z\pi_1$*-module,* $\pi_3(S^4 - k(S^2))$ *is generated by the symbols*

$$\gamma(X_i), \ [X_i, gX_i] \qquad (1 \leq i, j \leq n; \ g \in \pi_1)$$

subject to the relations

$$\begin{aligned}
&2\gamma(X_i) = [X_i, X_i] && 1 \leq i, j \leq n \\
&\gamma\left(\sum_{j=1}^{n} (\partial r_k/\partial x_j) X_j\right) = 0 && 0 \leq k \leq m \\
&\left[X_i, g \sum_j (\partial r_k/\partial x_i) X_j\right] = 0 && g \in \pi_1 \\
&[X_i, gX_j] = g[X_i, g^{-1}X_j] ,
\end{aligned}$$

where $[X_i, gX_i]$ *is the Whitehead product of* X_i *and* gX_j *and* $\gamma(X_i)$ *is represented by the composition of the Hopf map* $S^3 \to S^2$ *with a representative of* X_i.

V. Examples

EXAMPLE 1. If the trefoil is spun about S^2, then

$$\pi_1(S^4 - k(S^2)) = |a,b : baba^{-1}b^{-1}a^{-1}|$$

$$\pi_2(S^4 - k(S^2)) = |B : (1-a+ba)B = 0 \quad |$$

$$\pi_3(S^4 - k(S^2)) = \left| \gamma(B), [B,gB], (g \in \pi_1) : \begin{array}{l} 2\gamma(B) = [B,B] \\ (1-a+ba)\gamma(B) = -[B,baB] \\ [B,gB] - [B,gaB] + [B,gbaB] = 0 \\ [B,gB] = g[B,g^{-1}B] \end{array} \right|$$

where $[B,gB]$ is the Whitehead product of B and gB and $\gamma(B) =$ (Hopf map) o B. (See Figure 2.)

EXAMPLE 2. If the square knot is spun about S^2, then

$$\pi_1(S^4 - k(S^2)) = |a,b,c : baba^{-1}b^{-1}a^{-1}, caca^{-1}c^{-1}a^{-1}|$$

$$\pi_2(S^4 - k(S^2)) = |B, C : (1-a+ba)B = 0 = (1-a+ca)C|$$

$$\pi_3(S^4 - k(S^2)) = \left| \begin{array}{l} \gamma(B) \\ \gamma(C) \\ [B,gB] \\ [C,gC] \\ [B,gC] \\ [C,gB] \end{array} : \begin{array}{l} 2\gamma(B) = [B,B],\ 2\gamma(C) = [C,C] \\ (1-a+ba)\gamma(B) = -[B,baB] \\ (1-a+ca)\gamma(C) = -[C,caC] \\ [B,gB] - [B,gaB] + [B,gbaB] = 0 \\ [C,gB] - [C,gaB] + [C,gbaB] = 0 \\ [B,gC] - [B,gaC] + [B,gcaC] = 0 \\ [C,gC] - [C,gaC] + [C,gcaC] = 0 \\ [B,gB] = g[B,g^{-1}B],\ [C,gC] = g[C,g^{-1}C] \\ [B,gC] = g[C,g^{-1}B] \end{array} \right|$$

where g ranges over π_1.

COMMUNICATIONS RESEARCH DIVISION
INSTITUTE FOR DEFENSE ANALYSES
PRINCETON, NEW JERSEY
AND
STATE UNIVERSITY OF NEW YORK
ALBANY, NEW YORK

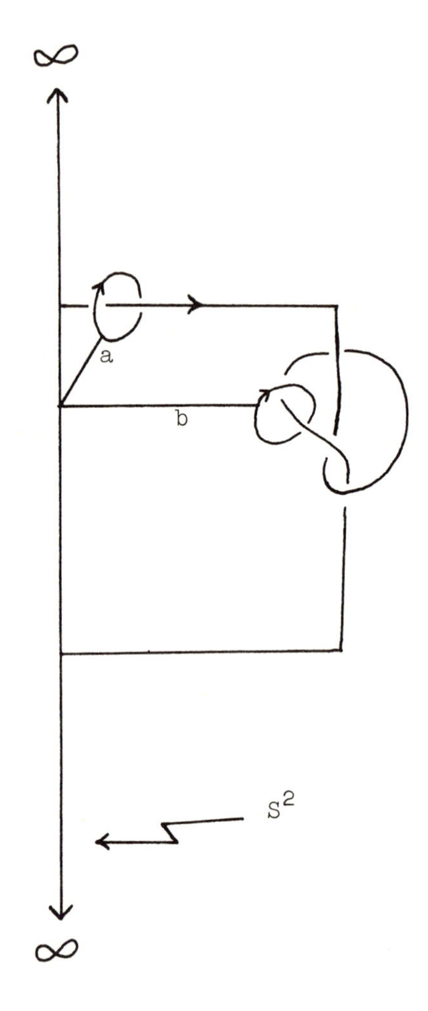

Figure 2. Spun Trefoil

REFERENCES

[1] Fox, R. H., *Some problems in knot theory*, in Topology of 3-Manifolds Symposium, Prentice-Hall, Englewood Cliffs, N. J. 1962.

[2] Crowell, R. H., and Fox, R. H., Knot Theory, Ginn, 1963.

[3] Epstein, D. B. A., *Linking spheres*, Proc. Camb. Phil. Soc. math. phys. sci. 56 (1960), 215-219.

[4] Papakyriakopoulos, C. D., *On Dehn's lemma and the asphericity of knots*, Ann. of Math. 66 (1957), 1-26.

[5] Artin, Emil, *Zur isotopie zweidimensional Flachen im* R^4, Hamburg Abh. 4 (1925), 174-177.

[6] Andrews, J. J., and Lomonaco, S. J., *The second homotopy group of 2-spheres in 4-space*, A.M.S. Bull., 75 (1969), 169-171.

[7] ______________________, *The second homotopy group of 2-spheres in 4-space*, Ann. of Math. 90 (1969), 199-204.

[8] Lomonaco, S. J., Jr., *The second homotopy group of a spun knot*, Topology, 8 (1969), 95-98.

[9] Fox, R. H., *Free differential calculus I. Derivations in the free group ring*, Ann. of Math. 57 (1953), 547-560.

[10] Whitehead, J. H. C., *The secondary boundary operator*, Proc. Nat. Acad. Sci. 36 (1950), 55-60.

[11] ______________, A certain exact sequence, Ann. of Math. 52 (1950), 51-110.

[12] Eilenberg, Samuel and Saunders MacLane, *On the Groups* $H(\pi, n)$, I, Ann. of Math. 58 (1953), 55-106; II, Ann. of Math. 70 (1954), 49-139; III, Ann. of Math. 60 (1954), 513-557.

[13] MacLane, Saunders, Homology, Academic Press, 1963.

[14] Andrews, J. J., and Curtis, M. L., *Knotted 2-spheres in the 4-sphere*, Ann. of Math. 70 (1959), 565-571.

[15] Hu,Sze-tsen, Homotopy Theory, Academic Press, 1959.

[16] Cappell, Sylvain, *Superspinning and knot complements*, in Topology of Manifolds Symposium, Markham Publishing Co., Chicago. 1970.

OCTAHEDRAL KNOT COVERS

Kenneth A. Perko, Jr.

One of Fox's continuing interests was the investigation of noncyclic covering spaces of knots, *i.e.*, those which belong to a homomorphism of the knot group onto a noncyclic group of permutations [2]. Unfortunately, criteria for the existence of such coverings are still rather rare [7, Ch. VI; 10]. With the help of his suggestions on Lemma 1, we derive (geometrically) a necessary and sufficient condition for a knot group to have a representation on the symmetric group of degree four.

THEOREM. *A knot group admits a homomorphism onto* S_4 *if and only if it admits one onto* S_3.

"Only if" follows trivially from the homomorphism ζ of S_4 on S_3 obtained by factoring the former over its normal subgroup isomorphic to the four group [4]. To prove sufficiency, we show that any homomorphism h of a knot group on S_3 may be lifted to an H on S_4 such that $H_\zeta = h$. There are two types of H: those which send all meridians to elements of period 2 (simple H), and those which send them to elements of period 4 (locally cyclic H). Let M_3 be the branched 3-fold (dihedral) covering space of S^3 associated with an arbitrary h [2, §§4-5].

LEMMA 1. h *lifts to a simple* H *if and only if* $H_1(M_3; Z)$ *maps homomorphically onto* Z_2.

LEMMA 2. h *lifts to a locally cyclic* H *whenever some odd multiple of the branch curve of index* 2 *in* M_3 *is strongly homologous to zero.*

If the condition of Lemma 2 is not satisfied, then that of Lemma 1 is satisfied by mapping this branch curve to the generator of Z_2. This proves the theorem, modulo the lemmas.

Proof of Lemma 1. Let i symbolize (jk) and $-i$ symbolize (i4) where $\{i, j, k\} = \{1, 2, 3\}$. Then ζ maps $\pm i$ to i. h may be thought of as an assignment f of symbols $i = 1, 2, 3$ to segments of a knot diagram (x, y, z at each crossing) such that $f(x) + f(z) \equiv 2f(y) \bmod 3$, where y is the overpass [2, §1]. Consider, at each crossing, the cellular decomposition of M_3 discussed in [8]. (*Cf.* [7, Ch. III].) Let x, y, z represent also the 2-cells which lie beneath corresponding segments and are visible on the right from the (i+1)th (mod 3) copy of S^3. Branch relations for $H_1(M_3; Z)$ insure that the other 2-cells adjoining the branch curve of index 2 are homologous to $-x, -y, -z$. At a crossing where $f(x) = f(y)$ the Wirtinger-like homology relation is $x-2y+z \sim 0$, while at a crossing where $f(x) \neq f(y)$ it is $x+y+z \sim 0$. This may be verified by examining the various possibilities. Clearly $H_1(M_3; Z)$ maps homomorphically onto Z_2 precisely when there exists a mapping m of all x, y, z on integers $\{0, 1\}$ such that all these relations are congruent to 0 mod 2 (*i.e.*, there are either none or two 1's at the second type of crossing and $m(x) = m(z)$ at the first). If we interpret $m(x) = 1$ as placing a minus sign before $f(x)$, we see that these conditions are identical with those for the existence of a lifted, simple H. Again, this may be verified by examining the various possibilities.

Proof of Lemma 2. Now let i symbolize (ij4k) and $-i$ symbolize (ik4j) where $i \equiv j-1 \equiv k-2 \bmod 3$. Again, ζ maps $\pm i$ to i. Here, however, it is necessary to distinguish between two different types of crossing where $f(x) \neq f(y)$, depending on whether the segment x for which $f(x) \equiv f(y)-1 \bmod 3$ lies to the right or left of y. At a crossing of the first type, the associated equation for constructing a hypothetical 2-chain which bounds t times the curve of index 2 is $x+y+z = t$. (*Cf.* [8, §2].) For the second, it is $x+y+z = 2t$. At a crossing where $f(x) = f(y)$, it is

$x-2y+z=0$. Here, of course, we let x, y, z represent also the dummy coefficients assigned to the 2-cells x, y, z. Coefficients of the other 2-cells adjoining this branch curve are then $t-x$, $t-y$, $t-z$ by the branch equations. If all of these equations have a solution (in integers) for some odd t, then we may assign signs $+$ or $-$ to each $f(x)$ according as the congruence class mod 2 of the coefficient x is 1 or 0, and such an assignment will yield a lifted, locally cyclic H. Again, this latter assertion may be verified by examining the various possibilities to see that the behavior of the sign of the symbols $\pm i$ is reflected by these equations (interpreted as congruences mod 2) at each type of crossing.

It may be conjectured that every h lifts to a locally cyclic H.

From the coset representations of (abstract) S_4 which belong to its nonconjugate subgroups [1] we may construct, for each H, a partially ordered set of connected covering spaces of S^3, branched along the knot, which cover each other as indicated below:

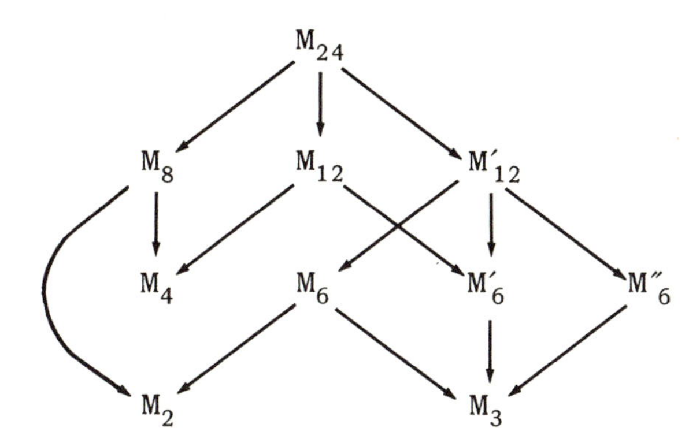

The covering maps may be thought of as the identification of corresponding points in various copies of S^3. *Cf.* [9].

In view of the recent result of Hilden [3] and/or Montesinos [6], these coverings may be relevant to the classification problem for 3-manifolds. Note that any H is consistent with the D-operations of [5], discovered independently by Fox and adverted to in [2, §4].

NEW YORK, NEW YORK

REFERENCES

[1] Burnside, W., Theory of groups of finite order, Ch. XII (Cambridge Univ. Press, 1897) (2nd ed., Dover, N. Y., 1955).

[2] Fox, R. H., Metacyclic invariants of knots and links, Canad. J. Math. 22 (1970), 193-201; *cf.* same J. 25 (1973), 1000-1001.

[3] Hilden, H. M., Every closed orientable 3-manifold is a 3-fold branched covering space of S^3. Bull. Amer. Math. Soc. 80 (1974), 1243-1244.

[4] Huppert, B., Endliche Gruppen I, p. 174 (Springer, Berlin, 1967).

[5] Montesinos, J. M., Reducción de la Conjetura de Poincaré a otras conjeturas geométricas, Revista Mat. Hisp.-Amer. 32 (1972), 33-51.

[6] ———, A representation of closed, orientable 3-manifolds as 3-fold branched coverings of S^3, Bull. Amer. Math. Soc. 80 (1974), 845-846.

[7] Neuwirth, L. P., Knot Groups (Princeton Univ. Press, 1965).

[8] Perko, K. A., Jr., On covering spaces of knots, Glasnik Mat. 9 (29) (1974), 141-145.

[9] ———, On dihedral linking numbers of knots, Notices Amer. Math. Soc. 21 (1974), A-327.

[10] Riley, R., Homomorphisms of knot groups on finite groups, Math. Comp. 25 (1971), 603-619.

SOME KNOTS SPANNED BY MORE THAN ONE UNKNOTTED SURFACE OF MINIMAL GENUS

H. F. Trotter

§1. *Introduction*

A *spanning surface* of a tame knot K in S^3 is a tame orientable surface F embedded in S^3 with K as the unique component of its boundary. We call two such surfaces *directly equivalent* if there is an orientation-preserving homeomorphism of S^3 onto itself that carries one surface onto the other and preserves the orientation of K. They are said to be *inversely equivalent* if there exists such a homeomorphism reversing the orientation of K (but still preserving that of S^3), and are *equivalent* if they are either directly or inversely equivalent. (We shall not be concerned here with the stronger notion of equivalence under isotopy leaving K fixed.)

In this paper we give some examples of knots with spanning surfaces of minimal genus that fall into more than one (direct) equivalence class. Examples of knots of this kind have been given by Alford, Schaufele, and Lyon [1, 2, 6]. The inequivalent surfaces exhibited in these examples have complements which are not homeomorphic. The contrary is true in our examples. In fact, all the surfaces that we consider are "unknotted" in the sense of having complements which are handlebodies.

We prove inequivalence by showing that the Seifert matrices of the relevant surfaces are not congruent. (Thus we have some "natural" examples of matrices which are S-equivalent [11] but not congruent.) This matrix condition is of course only sufficient for inequivalence, not necessary. It can be shown to hold for infinitely many knots of genus one. Although I am sure that it holds for infinitely many knots of every genus, I have no proof of the fact.

In Section 2 we describe the (generalized) pretzel knots which furnish our examples, and in Section 3 discuss their Seifert matrices and related algebraic invariants. Section 4 describes the method used to prove non-congruence of the matrices, and Section 5 summarizes the arithmetic involved in our examples.

§2. *Pretzel knots and surfaces*

For $(p_1, \cdots, p_n)$ an n-tuple of integers, let $F(p_1, \cdots, p_n)$ be the surface consisting of two horizontal disks (lying one above the other like the top and bottom of a vertical cylinder) joined by n twisted but unknotted vertical bands, where the ith band in order has $|p_i|$ half-twists, right or left-handed according to the sign of p_i. Figure 1 shows an equivalent surface in a form that is easier to draw. Let $K(p_1, \cdots, p_n)$ be the knot or

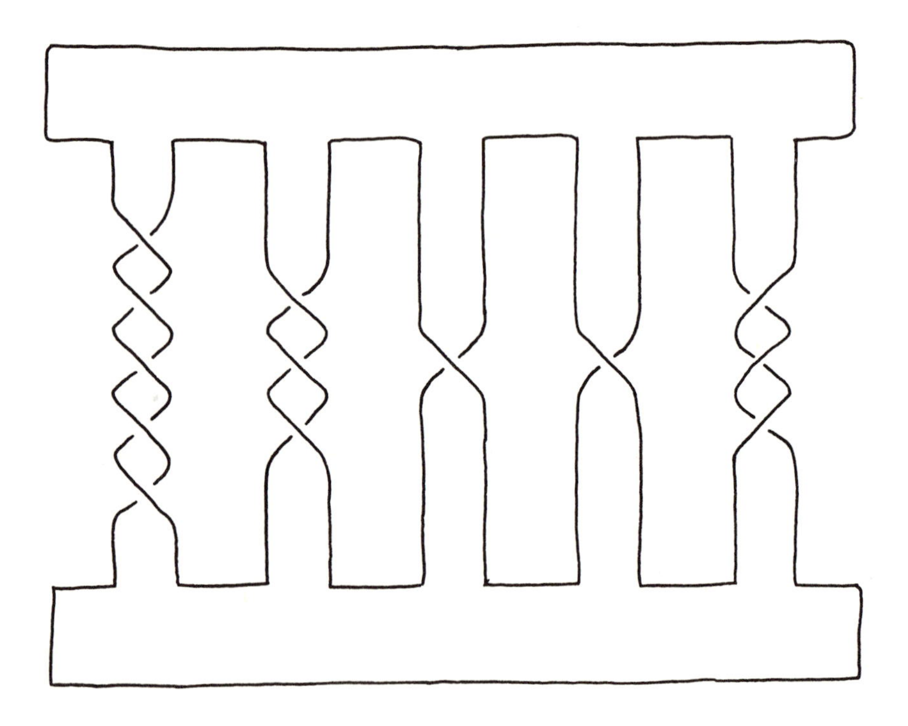

F(5, 3, 1, 1, −3)

Fig. 1.

link formed by the boundary of $F(p_1, \cdots, p_n)$. Reidemeister [8] called knots of the form $K(p_1, p_2, p_3)$ "pretzel knots" and it seems appropriate to extend the terminology.

F is orientable if and only if all the p_i have the same parity, and the boundary of F has more than one component if they are all even, or if they are all odd and n is even. We therefore assume from now on that n and all the p_i are odd. $F(p_1, \cdots, p_n)$ is then an orientable surface of genus h, where $n = 2h + 1$.

Let us say that $(q_1, \cdots, q_n)$ is a *cyclic rearrangement* of $(p_1, \cdots, p_n)$ if there is some k such that $q_i = p_{i+k}$ for all i, interpreting the subscripts modulo n. (This is not quite the same thing as a cyclic permutation, since the p_i need not be distinct.) The following statement is obvious from the construction of F.

(2.1) *If* $(q_1, \cdots, q_n)$ *is a cyclic rearrangement of* $(p_1, \cdots, p_n)$ *then* $F(q_1, \cdots, q_n)$ *and* $F(p_1, \cdots, p_n)$ *are directly equivalent.*

Contemplating the effect on Figure 1 of a $180°$ rotation about a vertical axis lying in the plane of the paper makes the following clear.

(2.2) $F(p_1, \cdots, p_n)$ *and* $F(p_n, \cdots, p_1)$ *are inversely equivalent.*

Let us call an n-tuple $(p_1, \cdots, p_n)$ *fully asymmetric* if the *only* rearrangements of it that yield a directly equivalent surface are the cyclic rearrangements. In later sections we shall prove:

(2.3) *The* n-*tuples* $(5, 3, 1)$ *and* $(5, 3, 1, 1, 1)$ *are fully asymmetric.*

The same method of proof can presumably yield many more examples, but individual calculations are required in each case, and it is difficult to draw general conclusions. I conjecture that all n-tuples (with n and all the p_i odd) are fully asymmetric, unless both $+1$ and -1 occur in the n-tuple. (In the latter case the surface has an "unknotted" handle that can be moved around freely.) I have, however, no solid supporting evidence.

It follows from (2.2) that if $(p_1,\cdots,p_n)$ is fully asymmetric, and $(q_1,\cdots,q_n)$ is a rearrangement of it, then $F(q_1,\cdots,q_n)$ is inversely equivalent to $F(p_1,\cdots,p_n)$ if and only if $(q_1,\cdots,q_n)$ is a cyclic rearrangement of $(p_n,\cdots,p_1)$. Thus (2.3) amounts to the following assertions.

(2.4) $F(5,3,1)$ *and* $F(5,1,3)$ *are not directly equivalent but are inversely equivalent.*

(2.5) *No two of* $F(5,3,1,1,1)$, $F(5,1,3,1,1)$, $F(5,1,1,3,1)$, *and* $F(5,1,1,1,3)$ *are directly equivalent, but the first and fourth are inversely equivalent, and so are the second and third.*

Figure 2 illustrates an obvious equivalence between the knots $K(\cdots,p,1,\cdots)$ and $K(\cdots,1,p,\cdots)$. More generally, any p_i equal to 1 or -1 can be permuted freely in the n-tuple without changing the equivalence class of the associated knot. (See Conway's remarks on "flyping" in [5], and the operation of type $\Omega.5$ of Reidemeister [7].) As immediate consequences of (2.4) and (2.5) we have:

(2.6) *The knot* $K(5,3,1)$ *has unknotted spanning surfaces of minimal genus falling into at least two distinct classes under direct equivalence.*

(2.7) *The knot* $K(5,3,1,1,1)$ *has unknotted spanning surfaces of minimal genus falling into at least four classes under direct equivalence, and into at least two classes under equivalence.*

Similar examples obviously arise from any fully asymmetric n-tuples that contain 1 or -1 (with the exception of trivial cases like $(3,1,1,1,1)$ for which all rearrangements are cyclic).

§3. *Seifert matrices of pretzel surfaces*

Let F be a spanning surface of genus h for the oriented knot K. The *Seifert form* of F is a bilinear form S_F defined on the homology group $H_1(F)$ by taking $S_F(u,v)$ to be the linking number in S^3 of a

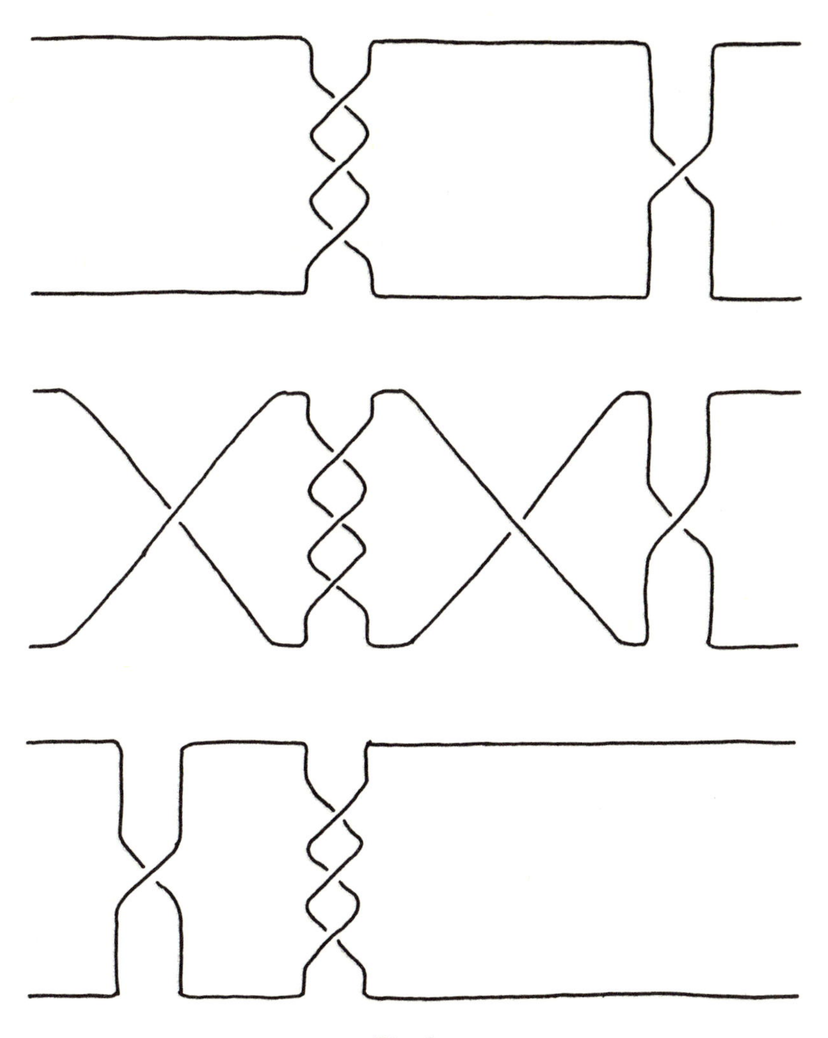

Fig. 2.

cycle representing u and a translate in the positive normal direction to F of a cycle representing v. (The positive direction is to be determined by some convention from the orientations of S^3 and K.) A *Seifert matrix* for F is obtained by choosing a basis $u_1, \cdots, u_{2h}$ for $H_1(F)$ and setting $v_{ij} = S_F(u_i, u_j)$ for $1 \leq i, j \leq 2h$. A different choice of basis gives a matrix W such that $W = PVP'$ with P an integral unimodular matrix.

The integral congruence class of its Seifert matrices is thus an invariant of F.

Let $n = 2h + 1$ and consider the surface $F(p_1, \cdots, p_n)$ where $p_i = 2k_i + 1$. The cycles $u_1, \cdots, u_{2h}$, where u_i runs up the (i+1)st band and down the ith, form a basis for $H_1(F)$. Define $V(p_1, \cdots, p_n)$ to be the Seifert matrix for F with respect to the basis $(-1)^{i+1} u_i$. (Putting the alternating sign in here keeps minus signs out of the matrix.) Following around u_i one encounters $p_i + p_{i+1}$ half twists or $k_i + k_{i+1} + 1$ full twists. Thus (with appropriate choice of sign convention for linking numbers) the diagonal entry v_{ii} is $k_i + k_{i+1} + 1$. Cycles representing u_i and u_{i+1} run in the same direction along the (i+1)st band and intersect once on F. When one of them is pushed off in the positive direction, the linking number is $k_{i+1} + 1$; when the other is pushed off, the linking number is k_{i+1}. We assume conventions to have been chosen so that $v_{i,i+1} = k_{i+1} + 1$, and $v_{i+1,i} = k_{i+1}$. When i and j differ by more than 1, u_i and u_j do not meet or link and $v_{ij} = v_{ji} = 0$.

The following observations are not relevant to the rest of the paper, but seem to be worth commenting on.

(3.1) *The determinant of* $V(p_1, \cdots, p_n)$ *is*

$$\prod_{i=1}^{n} (k_i + 1) - \prod_{i=1}^{n} k_i \ .$$

The proof is a straightforward induction on n. Note that the formula in (3.1) is a symmetric function of the p_i. There is even more symmetry in the situation.

(3.2) *The Alexander polynomial, the signature, and the Minkowski units of* $K(p_1, \cdots, p_n)$ *are independent of the order of the* p_i.

Since the type of K itself is unchanged under cyclic permutation, so are these invariants. The n-cycle $(1\ 2 \cdots n)$ and the transposition $(1\ 2)$

generate the whole symmetric group, so we need only examine what happens when p_1 and p_2 are exchanged. We temporarily adopt the notation $p_1 = 2a+1$, $p_2 = 2b+1$, $p_3 = 2c+1$. The upper left corner of $V(p_1, p_2, \cdots)$ is then

$$\begin{bmatrix} a+b+1 & b+1 \\ b & b+c+1 \end{bmatrix}.$$

Now subtract the first row from the second, change the sign of the first row, and perform the corresponding column operations. The resulting matrix is integrally congruent to the original and has the 2×2 matrix

$$\begin{bmatrix} a+b+1 & a \\ a+1 & a+c+1 \end{bmatrix}$$

in its upper left corner, and is otherwise unchanged. It is the same as $V(p_2, p_1, \cdots)$ except that the entries a and $a+1$ are reversed in position. Now the signature and Minkowski units depend only on $V + V'$, so they are unchanged [10]. The Alexander polynomial is $\det(tV - V')$. Every non-zero term in the determinant of any tridiagonal matrix M must contain $m_{i,i+1}$ if it contains $m_{i+1,i}$, so the determinant is not affected if the two elements are exchanged. Hence the Alexander polynomials of $K(p_1, p_2, \cdots)$ and $K(p_2, p_1, \cdots)$ are the same.

Proposition (3.2) gives an easy way of constructing presumably distinct knots which cannot be distinguished by the "classical" invariants.

§4. *A necessary condition for congruence of Seifert matrices*

If V is a Seifert matrix then $(V - V')^{-1}$ exists and has integer entries. Define

$$\Gamma_V = V(V - V')^{-1} .$$

Then if $W = PVP'$, with P and P^{-1} integral, $\Gamma_W = PVP'(P(V-V')P')^{-1} = P\Gamma_V P^{-1}$, so for V and W to be integrally congruent it is necessary that Γ_V and Γ_W be integrally similar.

Let $\phi(z) = \det(\Gamma - zI)$ be the characteristic polynomial of Γ. Similar matrices of course have the same characteristic polynomial. The theory of integral similarity is fairly simple when ϕ is irreducible (as happens in our examples), and a brief self-contained account of this case can be found in [9], which has references to further literature. The rest of this section is taken almost directly from [9].

We consider matrices with a given irreducible characteristic polynomial ϕ. Let L be the field $Q(\zeta)$ obtained by adjoining a root ζ of ϕ to the rationals, and let R be ring $Z[\zeta]$ generated by ζ.

The *row class* of Γ can then be defined as the class of the ideal of R generated by the determinantal cofactors of the elements of the first row of $\Gamma - \zeta I$. (Two ideals A, B of R are in the same class if $aB = bA$ for some non-zero a, b in R. When R is the full ring of algebraic integers in L, this coincides with the usual definition of ideal classes in a Dedekind ring.) The theorem that we shall use states that two matrices having ϕ for characteristic polynomial are similar if and only if they determine the same row class.

§5. *Calculations*

The theory and methods of calculation used here can all be found in [3]. Let $X = V(5,3,1)$ and $Y = V(5,1,3)$. Then

$$X = \begin{bmatrix} 4 & 2 \\ 1 & 2 \end{bmatrix}, \qquad \Gamma_X - \zeta I = \begin{bmatrix} 2-\zeta & -4 \\ 2 & -1-\zeta \end{bmatrix}$$

$$Y = \begin{bmatrix} 3 & 1 \\ 0 & 2 \end{bmatrix} \qquad \Gamma_Y - \zeta I = \begin{bmatrix} 1-\zeta & -3 \\ 2 & -\zeta \end{bmatrix}$$

where ζ is a root of $\phi(z) = z^2 - z + 6$. The discriminant of ϕ is -23, which is square-free, so $R = Z[\zeta]$ is the full ring of integers in $L = Q(\zeta)$, and the ideal classes of R form a group. The cofactors of the first row in $\Gamma_X - \zeta I$ are $-1-\zeta$ and 2, so the ideal $A = [2, 1+\zeta]$ represents the row class of Γ_X. Similarly, $B = [2, \zeta]$ represents the row class of Γ_Y.

It is easy to check that AB, A^3, and B^3 are the respective principal ideals (2), $(1+\zeta)$, and $(2-\zeta)$. Writing "$\simeq$" for the relation of class equivalence, we therefore have $A^3 \simeq B^3 \simeq AB \simeq 1$, so $B \simeq A^2$. Thus $A \simeq B$ only if A is principal. The norm of A is 2, and if it were principal, its generator would have to be an element of norm 2. This is impossible because the norm of $x + y\zeta$ is $x^2 + xy + 6y^2$, which does not represent 2. Hence the row classes A and B are not the same, X is not congruent to Y, and we have proved (2.4), which is equivalent to the first half of (2.3).

Now let W, X, Y, Z be $V(5,3,1,1,1)$, $V(5,1,3,1,1)$, $V(5,1,1,3,1)$ and $V(5,1,1,1,3)$ respectively. Then, for example,

$$W = \begin{bmatrix} 4 & 2 & 0 & 0 \\ 1 & 2 & 1 & 0 \\ 0 & 0 & 1 & 1 \\ 0 & 0 & 0 & 1 \end{bmatrix} \text{ and } \Gamma_W - \zeta I = \begin{bmatrix} 2-\zeta & -4 & 0 & -4 \\ 2 & -1-\zeta & 0 & -2 \\ 1 & 0 & 1-\zeta & -1 \\ 1 & 0 & 1 & -\zeta \end{bmatrix}$$

where ζ is a root of $\phi(z) = z^4 - 2z^3 + 12z^2 - 11z + 6$. Generators for the ideal representing the row class of Γ_W are $-\zeta^3 - 1$, $2\zeta^2 - 4\zeta + 2$, $-\zeta^2 + 1$, and $\zeta^2 + \zeta$. It turns out to be an ideal of norm 8.

The polynomial $\phi(z)$ can be written as $\psi(w) = w^2 - 11w + 6$, where $w = z(1-z)$. Then ζ is a root of $\zeta^2 + \zeta + \omega$ where ω is a root of ψ, and $K = Q(\omega)$ is a quadratic subfield of $L = Q(\zeta)$. The discriminant of K over Q is 97 and of L over K is $1-4\omega$, which has norm 53. Since both discriminants are square-free, $R = Z[\zeta]$ is the full ring of integers in L and its ideal classes form a group.

The prime 2 factors into the ideals $A = [2, 1+\zeta]$, $B = [2, \zeta]$, (both of norm 2), and $C = [2, \zeta^2 + \zeta + 1]$ (of norm 4). $AB = (7\zeta^2 - 7\zeta + 4)$ and $A^5 = (1+\zeta)$ are principal ideals, so $A \simeq B^{-1}$ and $A^5 \simeq 1$. The ideal given above, generated by the cofactors of the first row of $\Gamma_W - \zeta I$ is equal to A^3, and similar calculations give A^2B, AB^2, and B^3 as the prime factorizations of the ideals obtained from X, Y, and Z. Consequently the row classes associated with W, X, Y, and Z are the classes represented by A^3, A, A^4, and A^2. If these are all distinct then no two of W, X, Y, and Z are congruent, and (2.5) and the second half of (2.3) follow.

Otherwise A is a principal ideal generated by some element α. The ideals (α^5) and $(1+\zeta)$ are equal, so $\alpha^5 = u(1+\zeta)$ for some unit u. L is a totally imaginary field of degree 4, so its units are of the form $\pm e^n$ for some fundamental unit e. The fundamental unit of $K = Q(\omega) = Q(\sqrt{97})$ can be found in tables (e.g. in [4]); expressed in terms of ω it is $\eta = -655 + 1138\omega$. (This can in fact be shown to be a fundamental unit for L, but we do not need that fact.) There are two homomorphisms f, g of R onto Z/31Z, characterized by $f(\zeta) = 7$ and $g(\zeta) = -6$. Under both f and g, ω maps to -11 and η to 2. The fifth powers modulo 31 are ± 1, ± 5, and ± 6. Since 2 is not a fifth power modulo 31, η is not a fifth power in R. Hence every unit is some power of η times a fifth power, and if $u(1+\zeta)$ has a fifth root for any unit u, then so has one of $\eta^i(1+\zeta)$, $0 \leq i \leq 4$. Under g, $1+\zeta$ maps to -5 which is a fifth power, so $\eta^i(1+\zeta)$ cannot be a fifth power unless 5 divides i. Under f, $1+\zeta$ maps into 8, so $1+\zeta$ is not itself a fifth power. This exhausts the possibilities, and we conclude that A cannot be principal.

There is perhaps some interest in indicating the result of an example in which the Seifert matrices turned out to be congruent. For the surfaces $F(1,5,7,-3,-3)$ and $F(1,7,-3,-3,5)$ the Seifert matrices are

$$W = \begin{bmatrix} 3 & 3 & 0 & 0 \\ 2 & 6 & 4 & 0 \\ 0 & 3 & 2 & -1 \\ 0 & 0 & -2 & -3 \end{bmatrix} \quad \text{and} \quad V = \begin{bmatrix} 4 & 4 & 0 & 0 \\ 3 & 2 & -1 & 0 \\ 0 & -2 & -3 & -1 \\ 0 & 0 & -2 & 1 \end{bmatrix} .$$

Then $W = PVP'$, with

$$P = \begin{bmatrix} 39 & -151 & -19 & -22 \\ 37 & -106 & -2 & -16 \\ 22 & -37 & 10 & -6 \\ 39 & -144 & -16 & -21 \end{bmatrix}.$$

P is not the only matrix which will transform V into W, but it can be shown to be the "smallest" one that will do so. One may conclude that congruence of Seifert matrices is not always determinable by inspection.

The calculations reported here were first done while the author held a visiting appointment in the Mathematical Sciences department at the T. J. Watson Research Laboratories of the IBM Corporation. Extensive use was made of the APL interactive programming system, which is very well adapted to calculation with small matrices. It is feasible to verify the assertions made in this section by hand calculation, but the partly trial and error process of arriving at them could hardly have been carried out without mechanical assistance. It is also a pleasure to acknowledge the pleasant and stimulating atmosphere of the Laboratories.

PRINCETON UNIVERSITY

REFERENCES

[1] Alford, W. R., Complements of minimal spanning surfaces of knots are not unique. Ann. of Math. 91 (1970), 419-424.

[2] Alford, W. R., and Schaufele, C. B., Complements of minimal spanning surfaces of knots are not unique II, Topology of Manifolds (Proc. Inst. Univ. Georgia 1969), Markham, Chicago 1970.

[3] Artin, E., Theory of Algebraic Numbers. Math. Inst. Göttingen, 1959.

[4] Cohn, H., A Second Course in Number Theory. Wiley, New York, 1962.

[5] Conway, J. H., An enumeration of knots and links, and some of their algebraic properties, 329-358 in Computational Problems in Abstract Algebra, ed. J. Leech, Pergamon, Oxford, 1969.

[6] Lyon, H. C., Simple knots without unique minimal surfaces. Proc. Amer. Math. Soc., to appear.

[7] Reidemeister, K., Knotentheorie, Ergebnisse Math. 1 (1932) (Chelsea reprint, New York, 1948).

[8] ———, Knoten und Gruppen. Abh. Hamburg Univ. Math. Sem. 5 (1926), 7-23.

[9] Taussky, O., On a theorem of Latimer and MacDuffee. Can. J. Math. 1 (1949), 300-302.

[10] Trotter, H. F., Homology of group systems with applications to knot theory. Ann. of Math. 76 (1962), 464-498.

[11] ———, On S-equivalence of Seifert matrices. Inv. Math. 20 (1973), 173-207.

GROUPS AND MANIFOLDS CHARACTERIZING LINKS

Wilbur Whitten

Let L denote the tame link $K_1 \cup \cdots \cup K_\mu$ in the oriented three-sphere S^3, and let ρ and η be fixed integers; ρ, arbitrary; $\eta = \pm 2$. For each of $i = 1, \cdots, \mu$, let V_i be a closed, second-regular neighborhood of K_i, and let $\mathcal{K}_i$ be a tame knot in Int V_i. For $i \neq j$, we assume that $V_i \cap V_j = \emptyset$. We also assume that V_i has order greater than zero with respect to $\mathcal{K}_i (i = 1, \cdots, \mu)$; that is, each meridional disk of V_i meets $\mathcal{K}_i$. We set $R(L) = \mathcal{K}_1 \cup \cdots \cup \mathcal{K}_\mu$, and we call R(L) a revision of L. If, for each of $i = 1, \cdots, \mu$, the knot $\mathcal{K}_i$ bounds a disk D_i that lies in Int V_i, that has exactly one clasping singularity, and that has K_i as its diagonal, ρ as its twisting number, and η as its self-intersection number [13, §20, p. 232], then we shall denote R(L) by $D(L; \rho, \eta)$, which we call the (ρ, η)-double of L; we call D_i a clasping disk. In this paper, we prove that the group of $D(L; \rho, \eta)$ characterizes the (ambient) isotopy type $\{L\}$ of L when $\mu > 1$; see the announcement [25] for an outline of the proof.

I recently proved the same result for knots in S^3 [24]. J. Simon had previously characterized a knot's type by the free product of two, suitably chosen, cable-knot groups [18]. The "doubled-link" characterizations, presented here and in [24] and [25], are, however, more direct, cover links as well as knots, and yield characterizations of amphicheiral knots [24 and 25, Corollary 2.3]. Moreover, because $\pi_1(S^3 - D(L; \rho, \eta))$ characterizes $\{L\}$, so does $S^3 - D(L; \rho, \eta)$; see [24, Theorem 2.1, p. 263] and Corollary 2.2.

§1. *Preliminaries*

Throughout this work, the three-sphere has a fixed orientation; all mappings are piecewise linear; all submanifolds, subpolyhedra; and all regular neighborhoods, at least second regular. If L is a link in S^3, then $\{L\}$ denotes the (ambient) isotopy type of L; L^*, the mirror image of L. The complement C of an open solid torus in S^3 is a *toral solid*; if a core K of the solid torus is knotted, the manifold C is a K-*knot manifold*. The complement of $\mu(>1)$ mutually disjoint, open, solid tori in S^3 is a μ-*link manifold*.

Let U_1 and U_2 denote solid tori in S^3. The orientation of S^3 induces an orientation in each of U_1 and U_2. A homeomorphism $U_1 \to U_2$ that preserves these orientations and that maps a longitude of U_1 onto a longitude of U_2 is *faithful*.

All links are to be oriented, but the orientation of a link has no bearing on either the link's type or the link's isotopy type. For a knot K, a meridian-longitude pair (m, λ) of oriented, simple, closed curves is always *oriented with respect to* K; that is, m has linking number $+1$ with K, and λ and K are homologous in some second-regular neighborhood of K.

LEMMA 1.1. *Let* L *be a link in* S^3, *and let* $R(L)$ *be any revision of* L. *Then* L *is splittable if and only if* $R(L)$ *is splittable.*

Proof. If L is splittable, there is a polyhedral 2-sphere S in S^3 with disjoint complementary regions C_1 and C_2 such that $S^3 = C_1 \cup S \cup C_2$, such that $L \cap S = \emptyset$, and such that $L \cap C_j \neq \emptyset$ $(j = 1, 2)$. There is an autohomeomorphism h of S^3 that is isotopic to the identity, that leaves each knot K_i fixed point for point, and that moves each solid torus V_i away from S; that is, $S \cap h(V_i) = \emptyset$ $(i = 1, \cdots, \mu)$. Evidently, $R(L) \cap h^{-1}(S) = \emptyset$ and $R(L) \cap h^{-1}(C_j) \neq \emptyset$ $(j = 1, 2)$; hence, $R(L)$ is splittable.

Now suppose that L is unsplittable; we can assume that $\mu > 1$. Let $G = \pi_1(S^3 - L)$, and let $Q_i = \pi_1(V_i - \mathcal{K}_i)$ $(i = 1, \cdots, \mu)$. Because L is unsplittable, the group G is indecomposable; that is, G is not the free product of two nontrivial groups [12, Theorem (27.1), p. 19]. If μ_i is a meridian of V_i, then $\mu_i \cup \mathcal{K}_i$ is unsplittable because V_i has order greater than zero with respect to $\mathcal{K}_i$. Evidently, $Q_i \approx \pi_1(S^3 - (\mu_i \cup \mathcal{K}_i))$; hence, Q_i is indecomposable; see [12] (loc. cit.).

Set $\mathcal{G}_i = \pi_1(S^3 - (\mathcal{K}_1 \cup \cdots \cup \mathcal{K}_i \cup K_{i+1} \cup \cdots \cup K_\mu))$ $(i = 1, \cdots, \mu-1)$, set $\mathcal{G}_0 = G$, and set $\mathcal{G}_\mu = \pi_1(S^3 - R(L))$. Because each of the links L and $\mu_1 \cup \mathcal{K}_1$ is unsplittable, one can easily prove, with the loop theorem and the Dehn lemma, that the inclusion $\partial V_1 \to (S^3 - \mathrm{Int}(V_1 \cup \cdots \cup V_\mu))$ induces a monomorphism $\pi_1(\partial V_1) \to \pi_1(S^3 - \mathrm{Int}(V_1 \cup \cdots \cup V_\mu))$; the inclusion $\partial V_1 \to V_1 - \mathcal{K}_1$, a monomorphism $\pi_1(\partial V_1) \to Q_1$. Because $\pi_1(S^3 - \mathrm{Int}(V_1 \cup \cdots \cup V_\mu)) \approx G$, the Seifert-van Kampen theorem implies that $\mathcal{G}_1 \approx G \underset{\pi_1(\partial V_1)}{*} Q_1$. But the free produce of two indecomposable groups amalgamated over a nontrivial group is itself indecomposable [8, p. 246]. Thus, $\mathcal{G}_1$ is indecomposable.

Suppose, for some $i = 1, \cdots, \mu-1$, that $\mathcal{G}_i$ is indecomposable. Then, clearly, $\mathcal{K}_1 \cup \cdots \cup \mathcal{K}_i \cup K_{i+1} \cup \cdots \cup K_\mu$ is unsplittable. Because $\mu_{i+1} \cup \mathcal{K}_{i+1}$ is also unsplittable, we have $\mathcal{G}_{i+1} \approx \mathcal{G}_i \underset{\pi_1(\partial V_{i+1})}{*} Q_{i+1}$. But $\mathcal{G}_i$ and Q_{i+1} are indecomposable, hence, so is $\mathcal{G}_{i+1}$. Induction now implies that $\mathcal{G}_\mu$ is indecomposable; hence, $R(L)$ is unsplittable [12] (loc. cit.), concluding the lemma's proof.

REMARK. In the foregoing proof, we saw that, if L is unsplittable, then so is $\mathcal{K}_1 \cup \cdots \cup \mathcal{K}_{\mu-1} \cup K_\mu$. Permuting indices, one can, therefore, show, for $i = 1, \cdots, \mu$ and for $\mathcal{K}_0 = \emptyset = \mathcal{K}_{\mu+1}$, that

$$\mathcal{K}_1 \cup \cdots \cup \mathcal{K}_{i-1} \cup K_i \cup \mathcal{K}_{i+1} \cup \cdots \cup \mathcal{K}_\mu$$

is unsplittable, if L is unsplittable.

LEMMA 1.2. *Let* L *and* L' *be links in* S^3, *and let* (ρ, η) *and* (ρ', η') *be pairs of integers;* ρ *and* ρ', *arbitrary;* η *and* η', *in* $\{2, -2\}$. *If* $\{L\} = \{L'\}$, *if* $\rho = \rho'$, *and if* $\eta = \eta'$, *then* $\{D(L; \rho, \eta)\} = \{D(L'; \rho', \eta')\}$. *Conversely, if* $\{D(L; \rho, \eta)\} = \{D(L'; \rho', \eta')\}$, *then* $\{L\} = \{L'\}$; *furthermore,* $\rho = \rho'$ *and* $\eta = \eta'$ *unless*

(1) *some component* $\mathcal{K}_i$ *of* $D(L; \rho, \eta)$ *is a maximal unsplittable sublink of* $D(L; \rho, \eta)$ *and*

(2) $\mathcal{K}_i$ *is either the trivial or the figure-eight knot.*
In particular, $\{D(L;\rho,\eta)\} = \{D(L';\rho',\eta')\}$ *if and only if* $(\{L\},\rho,\eta) = (\{L'\},\rho',\eta')$, *provided that the number of components of* L *is* ≥ 2 *and that* $D(L;\rho,\eta)$ *is unsplittable.*

Proof. Assume that $\{L\} = \{L'\}$, that $\rho = \rho'$, and that $\eta = \eta'$. The links L and L′ then have the same number of components; we have $L = K_1 \cup \cdots \cup K_\mu$ and $L' = K'_1 \cup \cdots \cup K'_\mu$, say. For each of $i = 1, \cdots, \mu$, let V_i be a closed regular neighborhood of K_i; V'_i, a closed regular neighborhood of K'_i. We assume that $V_i \cap V_j = \emptyset = V'_i \cap V'_j$, if $i \neq j$.

Because $\{L\} = \{L'\}$, there is an orientation-preserving autohomeomorphism h_1 of S^3 and there is a permutation p of $\{1, \cdots, \mu\}$ such that $h_1(V_i) = V'_{p(i)}$ $(i = 1, \cdots, \mu)$. The knots $h_1(\mathcal{K}_i)$ and $\mathcal{K}'_{p(i)}$ are (ρ,η)-doubles of $K'_{p(i)}$ $(i = 1, \cdots, \mu)$. W. Graeub has shown that, for any knot K, the system $(\{K\},\rho,\eta)$ determines $\{D(K;\rho,\eta)\}$ [6, p. 47]. Hence, for each of $i = 1, \cdots, \mu$, there is an orientation-preserving autohomeomorphism ϕ_i of S^3 taking $h_1(\mathcal{K}_i)$ onto $\mathcal{K}'_{p(i)}$.

We examine the map ϕ_i. According to our definition of a doubled knot, there are clasping disks Y_i and Y'_i in Int $V'_{p(i)}$ such that $h_1(\mathcal{K}_i) = \partial Y_i$ and $\mathcal{K}'_{p(i)} = \partial Y'_i$. Each of Y_i and Y'_i has a line segment as its set of singularities. On Y_i lays a core k_i of $V'_{p(i)}$ meeting $h_1(\mathcal{K}_i)$ in exactly two points and containing the set s_i of singularities of Y_i; similarly, there is a core k'_i of $V'_{p(i)}$ on Y'_i containing the singularities s'_i of Y'_i. There is an orientation-preserving autohomeomorphism ϕ_{i1} of S^3 acting as the identity on $S^3 - \text{Int}\, V'_{p(i)}$ and taking (k_i, s_i) onto (k'_i, s'_i) [13, Lemma 1, p. 158]. Beginning at Step 2 on p. 47 of [6], one can see how Graeub constructs an autohomeomorphism ϕ_{i2} of S^3 leaving k'_i fixed point for point and mapping $\phi_{i1}(Y_i)$ onto Y'_i. We set $\phi_i = \phi_{i2}\phi_{i1}$.

Choose a simplicial decomposition for $V'_{p(i)}$ containing some triangulation of Y'_i. If the natural number n is sufficiently large, the closure N_i of the *nth*-regular neighborhood of Y'_i belongs to both Int $V'_{p(i)}$ and

$\mathrm{Int}(\phi_{i2}(V'_{p(i)}))$ $(=\mathrm{Int}(\phi_i(V'_{p(i)})))$. The polyhedron N_i is a handlebody, the singular disk Y'_i is a strong deformation retract of N_i, and the group $\pi_1(Y'_1) \approx Z$. Hence, N_i is a solid torus.

The core k'_i of $V'_{p(i)}$ and of $\phi_i(V'_{p(i)})$ belongs to $\mathrm{Int}\, N_i$. I claim that k'_i is also a core of N_i. Let c_i be a core of N_i. If k is a knot in a solid torus V, let $O_V(k)$ denote the order of V with respect to k. We have $O_{V'_{p(i)}}(k'_i) = O_{V'_{p(i)}}(c_i)O_{N_i}(k'_i)$ [13, Theorem 3, p. 175]. Because $O_{V'_{p(i)}}(k'_i) = 1$, we also have $O_{V'_{p(i)}}(c_i) = 1 = O_{N_i}(k'_i)$. Hence, there exist knots d_i and e_i such that $c_i = d_i \# k'_i$ and $k'_i = e_i \# c_i$ [13, Theorem 2, p. 171]. Therefore, $c_i = (d_i \# e_i) \# c_i$. Because factorization is unique in the semigroup of oriented-knot types, $d_i \# e_i$ is trivial. But this implies that each of d_i and e_i is trivial [5, p. 142]. Consequently, k'_i is a core of N_i [13, Theorem 2, p. 171].

We now construct an autohomeomorphism ψ_i of S^3 that takes $\phi_i(V'_{p(i)})$ onto $V'_{p(i)}$ and that acts as the identity on N_i. Define $\psi_i|(S^3 - \mathrm{Int}(\phi_i(V'_{p(i)}))) = \phi_{i2}^{-1}|(S^3 - \mathrm{Int}(\phi_i(V'_{p(i)})))$ and $\psi_i|N_i = 1$. The inclusion $\partial(\phi_i(V'_{p(i)})) \to \phi_i(V'_{p(i)}) - \mathrm{Int}\, N_i$ induces an isomorphism $\pi_1(\partial(\phi_i(V'_{p(i)}))) \to \pi_1(\phi_i(V'_{p(i)}) - \mathrm{Int}\, N_i)$; the inclusion $\partial(V'_{p(i)}) \to V'_{p(i)} - \mathrm{Int}\, N_i$, an isomorphism $\pi_1(\partial(V'_{p(i)})) \to \pi_1(V'_{p(i)} - \mathrm{Int}\, N_i)$. Hence, the isomorphism $(\psi_i|\partial(\phi_i(V'_{p(i)})))_*$ induces an isomorphism $\pi_1(\phi_i(V'_{p(i)}) - \mathrm{Int}\, N_i) \to \pi_1(V'_{p(i)} - \mathrm{Int}\, N_i)$. There is a homeomorphism $\psi'_i : (\phi_i(V'_{p(i)}) - \mathrm{Int}\, N_i) \to (V'_{p(i)} - \mathrm{Int}\, N_i)$ inducing the latter isomorphism [21, Corollary 6.5, p. 80]. Because ϕ_{i2}^{-1} leaves k'_i pointwise fixed, $\psi_i|\partial(\phi_i(V'_{p(i)}))$ takes a meridian-longitude pair onto a meridian-longitude pair. Therefore, ψ'_i takes meridian-longitude pairs for each of $\phi_i(V'_{p(i)})$ and N_i onto meridian-longitude pairs for each of $V'_{p(i)}$ and N_i, respectively.

But this means that $\psi'_i|\partial(\phi_i(V'_{p(i)}) - \mathrm{Int}\, N_i)$ and $\psi_i|\partial(\phi_i(V'_{p(i)}) - \mathrm{Int}\, N_i)$ differ on $\partial V'_{p(i)} \cup \partial N_i$ by a map σ_i such that each of $\sigma_i|\partial V'_{p(i)}$ and $\sigma_i|\partial N_i$ is isotopic to the identity [11]; that is, there is an autohomeomorphism σ_i of $\partial V'_{p(i)} \cup \partial N_i$ such that σ_i is isotopic to the identity and such that $\psi_i = \sigma_i \psi'_i$ on $\partial(\phi_i(V'_{p(i)})) \cup \partial N_i$. Because each of $\partial V'_{p(i)}$ and ∂N_i is collared in $V'_{p(i)} - \text{Int } N_i$, the map σ_i can be extended to an autohomeomorphism of $V'_{p(i)} - \text{Int } N_i$. We set $\psi_i|(\phi_i(V'_{p(i)}) - \text{Int } N_i) = \sigma_i \psi'_i$ to complete the definition of ψ_i.

We now define $h_2|(S^3 - \text{Int}(V'_{p(1)} \cup \cdots \cup V'_{p(\mu)})) = 1$ and $(h_2|V'_{p(i)}) = ((\psi_i \phi_i)|V'_{p(i)})$ $(i = 1, \cdots, \mu)$. Then the autohomeomorphism $h = h_2 h_1$ of S^3 takes $D(L; \rho, \eta)$ onto $D(L'; \rho, \eta)$, thereby finishing the proof of the lemma's first conclusion, $(\{L\}, \rho, \eta)$ determines $\{D(L; \rho, \eta)\}$.

Now, the "converse." H. Seifert showed in [17; §§9, 10, 11; pp. 77-79] that, when L is a knot, $\{D(L; \rho, \eta)\}$ determines $\{L\}$; that, when L is a nontrivial knot, $\{D(L; \rho, \eta)\}$ also determines each of ρ and η; and that, when L is unknotted, $D(L; \rho, \eta)$ is amphicheiral, if $\{D(L; \rho, \eta)\}$ does not determine (ρ, η). Finally, H. Schubert proved that the only amphicheiral, doubled knots are the trivial knot and the figure-eight knot [14, Theorem 5, p. 145]; this completes the lemma's proof when $\mu = 1$.

Assume now that $\mu \geq 2$, and suppose that $D(L; \rho, \eta)$ is unsplittable. For each of $i = 1, \cdots, \mu$, let Y_i and Y'_i be clasping disks that K_i bounds; assume that $Y_i \subset \text{Int } V_i$ and that $Y'_k \cap Y'_j = \emptyset$ when $k \neq j$. Lemma 1.1 and the unsplittability of $D(L; \rho, \eta)$ imply that L is also unsplittable. Hence, $S^3 - \text{Int}(V_1 \cup \cdots \cup V_\mu)$ is boundary irreducible, and there is an orientation-preserving autohomeomorphism of S^3 moving Y'_i onto $Y_i (i = 1, \cdots, \mu)$ and leaving each point of $D(L; \rho, \eta)$ fixed. Seifert essentially constructed such a homeomorphism in his proof in [17] of Lemmas 5 and 7; for our proof, one need make only minor changes in Seifert's work. Therefore, if $\mu \geq 2$ and if $D(L; \rho, \eta)$ is unsplittable, then $\{D(L; \rho, \eta)\}$ determines the triple $(\{L\}, \rho, \eta)$; cf. [17, §9, p. 77].

Finally, suppose that $\{D(L_1; \rho, \eta), \cdots, D(L_m; \rho, \eta)\}$ is the set of maximal unsplittable sublinks of $D(L; \rho, \eta)$. We have seen that $\{D(L_j; \rho, \eta)\}$ determines $\{L_j\}$ $(j = 1, \cdots, m)$; consequently, $\{D(L; \rho, \eta)\}$ determines $\{L\}$, because $L = L_1 \cup \cdots \cup L_m$. If, furthermore, no $D(L_j; \rho, \eta)$ satisfies simultaneously the conditions (1) and (2) of the hypothesis, then $\{D(L_j; \rho, \eta)\}$ determines (ρ, η) as well as $\{L_j\}$ $(j = 1, \cdots, m)$; therefore, $\{D(L; \rho, \eta)\}$ determines $(\{L\}, \rho, \eta)$, as claimed, completing the lemma's proof.

§2. *The characterizations*

Let L denote the link $K_1 \cup \cdots \cup K_\mu$ in S^3, let $D(L;\rho,\eta)$ be the (ρ,η)-double of L, and, for each of $i = 1,\cdots,\mu$, let W_i be a closed regular neighborhood of K_i. We assume that $W_i \subset \mathrm{Int}\, V_i$ $(i=1,\cdots,\mu)$, and we set $C^3(L;\rho,\eta) = S^3 - \mathrm{Int}\,(W_1 \cup \cdots \cup W_\mu)$.

THEOREM 2.1. *Let* L *and* L' *be links in* S^3, *and let* ρ *and* η *be fixed integers;* ρ, *arbitrary;* $\eta = \pm 2$. *Then* L *and* L' *are of the same (ambient) isotopy type if and only if* $\pi_1(C^3(L;\rho,\eta)) \approx \pi_1(C^3(L';\rho,\eta))$.

COROLLARY 2.2. *Let* L *and* L' *be links in* S^3, *and let* ρ *and* η *be fixed integers;* ρ, *arbitrary;* $\eta = \pm 2$. *Then* L *and* L' *belong to the same (ambient) isotopy type if and only if* $C^3(L;\rho,\eta) \cong C^3(L';\rho,\eta)$.

Proof. The necessity follows from Lemma 1.2; the sufficiency, from Theorem 2.1.

Proof of Theorem 2.1. Lemma 1.2 immediately establishes the necessity. To prove the sufficiency, we assume, henceforth, that $\pi_1(C^3(L;\rho,\eta)) \approx \pi_1(C^3(L';\rho,\eta))$.

Let $L_1,\cdots,L_m$ be the maximal unsplittable sublinks of L, and suppose $m > 1$. Applying Lemma 1.1, one can easily show that $D(L_1;\rho,\eta),\cdots,D(L_m;\rho,\eta)$ are the maximal unsplittable sublinks of $D(L;\rho,\eta)$. Hence, $\pi_1(C^3(L;\rho,\eta)) = \pi_1(C^3(L_1;\rho,\eta)) * \cdots * \pi_1(C^3(L_m;\rho,\eta))$; furthermore, each factor is indecomposable [12, Theorem (27.1), p. 19]. Since $\pi_1(C^3(L;\rho,\eta)) \approx \pi_1(C^3(L';\rho,\eta))$, we have $\pi_1(C^3(L';\rho,\eta)) = G_1 * \cdots * G_m$, with $G_j \approx \pi_1(C^3(L_j;\rho,\eta))$ $(j=1,\cdots,m)$. Therefore, $D(L';\rho,\eta)$ is splittable [12] (loc. cit.). Let $D(L'_1;\rho,\eta),\cdots,D(L'_{m'};\rho,\eta)$ be the maximal unsplittable sublinks of $D(L';\rho,\eta)$; we have $m' > 1$, and $L'_1,\cdots,L'_{m'}$ are the maximal unsplittable sublinks of L'. Setting $G'_k = \pi_1(C^3(L'_k;\rho,\eta))$ $(k=1,\cdots,m')$, we have $\pi_1(C^3(L';\rho,\eta)) = G'_1 * \cdots * G'_{m'}$. Each factor G'_k is indecomposable because of [12] (loc. cit.). Therefore, $m = m'$, and

$G'_{k_j} \approx G_j$ $(j=1,\cdots,m)$ for some rearrangement $(G'_{k_1},\cdots,G'_{k_m})$ of $(G'_1,\cdots,G'_m)$ [8, p. 245]. Consequently, $\pi_1(C^3(L'_{k_j};\rho,\eta)) \approx \pi_1(C^3(L_j;\rho,\eta))$ $(j=1,\cdots,m)$; thus, if the sufficiency of our condition holds for each pair (L_j, L'_{k_j}) $(j=1,\cdots,m)$ of *unsplittable* links, then it holds for the pair (L, L') of splittable links, because, for any link L, if the maximal unsplittable sublinks are $L_1,\cdots,L_m$, then the collection $\{\{L_1\},\cdots,\{L_m\}\}$ determines $\{L\}$, as one can easily prove.

We shall, therefore, assume not only that $\pi_1(C^3(L;\rho,\eta)) \approx \pi_1(C^3(L';\rho,\eta))$, but also that each of L and L' and, hence, each of $D(L;\rho,\eta)$ and $D(L';\rho,\eta)$ is unsplittable. Finally, note that the number of components in each of $L, L', D(L;\rho,\eta)$, and $D(L';\rho,\eta)$ is μ; we shall assume that $\mu > 1$, because the theorem is true when $\mu = 1$ [24].

Now, some notation. We have $L' = K'_1 \cup \cdots \cup K'_\mu$ and $D(L';\rho,\eta) = \mathcal{K}'_1 \cup \cdots \cup \mathcal{K}'_\mu$. For each of $i = 1,\cdots,\mu$, set $T_i = \partial V_i$, and let W'_i denote a closed regular neighborhood of $\mathcal{K}'_i(= D(K'_i;\rho,\eta))$; assume that $W'_k \cap W'_j = \emptyset$ when $k \neq j$. Set $C' = C^3(L';\rho,\eta) = S^3 - \operatorname{Int}(W'_1 \cup \cdots \cup W'_\mu)$ and set $C = C^3(L;\rho,\eta)$. Also, set $M = S^3 - \operatorname{Int}(V_1 \cup \cdots \cup V_\mu)$ and set $\Lambda_i = V_i - \operatorname{Int} W_i$ $(i=1,\cdots,\mu)$.

The space M is an aspherical μ-link manifold and Λ_i is an aspherical 2-link manifold [12] (loc. cit.). We have $C = (\cdots((M \cup_{T_1} \Lambda_1) \cup_{T_2} \Lambda_2) \cdots) \cup_{T_\mu} \Lambda_\mu$. For each of $i = 1,\cdots,\mu$, set $M_i = C - \operatorname{Int} V_i$. The link $\mathcal{K}_1 \cup \cdots \cup \mathcal{K}_{i-1} \cup K_i \cup \mathcal{K}_{i+1} \cup \cdots \cup \mathcal{K}_\mu$ is unsplittable; see the remark preceding Lemma 1.2, p. 65. Furthermore, each Λ_i is a deformation retract of an unsplittable 2-link's complement. Thus, because $\mu > 1$, each of M_i and Λ_i $(i=1,\cdots,\mu)$ is boundary irreducible. Moreover, it is not hard to show that none of the inclusion-induced monomorphisms $\pi_1(T_i) \to \pi_1(M_i)$ and $\pi_1(T_i) \to \pi_1(\Lambda_i)$ $(i=1,\cdots,\mu)$ is surjective, because there is only one link whose group is free abelian of rank two [9]. Therefore, for each of $i = 1,\cdots,\mu$, it follows that $\pi_1(C) \approx \pi_1(M_i) \underset{\pi_1(T_i)}{*} \pi_1(\Lambda_i)$ and that this group is a nontrivial free-product with amalgamated subgroup.

Because $\pi_1(C') \approx \pi_1(C)$ and because each of C' and C is aspherical, there exists a homotopy equivalence $f: C' \to C$; cf. [10, p. 93]. Set $T = T_1 \cup \cdots \cup T_\mu$. A result of J. R. Stallings and F. Waldhausen [20, Lemma 1.1, p. 506] guarantees the existence of a mapping $g: C' \to C$ with the following properties:

(1) $g \simeq f$;

(2) g *is transverse with respect to* T; *that is, there exist product neighborhoods* $U(g^{-1}(T))$ *and* $U(T)$ *such that* g *maps each fiber of* $U(g^{-1}(T))$ *homeomorphically onto a fiber of* $U(T)$;

(3) $g^{-1}(T)$ *is a compact, orientable, and, as we shall see, disconnected surface properly imbedded in* C';

(4) *if* F *is any component of* $g^{-1}(T)$, *then* $\ker(\pi_j(F) \to \pi_j(C')) = 1\ (j = 1, 2)$.

We divide the remainder of the proof into seven parts.

1. *For each of* $i = 1, \cdots, \mu$, *the space* $g^{-1}(T_i)$ *is not empty.*

If $g^{-1}(T_i) = \emptyset$, then either $g(C') \subset M_i$ or $g(C') \subset \Lambda_i$, because T_i separates C. Let x be a point in C', and let y be a point in T_i. When a suitable path is chosen from y to $g(x)$, then either $g_*(\pi_1(C',x)) \subseteq \pi_1(M_i,y)$ or $g_*(\pi_1(C',x)) \subseteq \pi_1(\Lambda_i, y)$, depending on whether $g(C') \subset M_i$ or $g(C') \subset \Lambda_i$. Thus, $g_*(\pi_1(C', x))$ is a *proper* subgroup of $\pi_1(C, y)$, because $\pi_1(C, y)$ is a nontrivial free-product with amalgamation. But g is a homotopy equivalence (by (1)); therefore, $g_*(\pi_1(C', x)) = \pi_1(C, y)$, yielding a contradiction.

We digress to prove a lemma needed in part 2.

LEMMA 2.3. *Any properly imbedded, incompressible annulus* A *in* C *is boundary parallel.*

Proof. Let α and β denote the components of ∂A, and suppose that $\alpha \subset \partial W_{j_1}$ and $\beta \subset \partial W_{j_2}$. Assume that A is in general position with respect to $T_{j_1} = \partial V_{j_1}$; the components of $A \cap T_{j_1}$ are mutually disjoint, simple, closed curves. Because T_{j_1} is incompressible in C and because $D(L;\rho,\eta)$ is unsplittable, we can remove those curves in $A \cap T_{j_1}$ bounding disks in A. Assuming this has been done, choose the curve α' in $A \cap T_{j_1}$ cobounding with α a subannulus A' of A properly imbedded in $V_{j_1} - \mathrm{Int}\, W_{j_1}$; we are assuming, of course, that $A \cap T_{j_1}$ still has curves in it. The winding number of α in V_{j_1} is 0 [13, Theorem 4, p. 175]. Thus, because $\alpha' \sim \pm\alpha$ in V_{j_1}, the winding number of α' in V_{j_1} is 0. But if a curve is on the boundary of a solid torus, then the curve's winding number and order coincide [13, Lemma 1, p. 170]. Therefore, the order of V_{j_1} with respect to α' is 0, and so α' either bounds a disk in T_{j_1} or is a meridian of V_{j_1}.

Clearly, the incompressibility of A prevents α' from bounding in T_{j_1}; thus, α' is a meridian. Because the winding number of α in V_{j_1} is 0, the linking number of $\alpha' \cup \alpha$ is also 0 [23, p. 374]. Thus, two trivial knots with linking number 0 bound the annulus A'; hence, A' is planar [7, p. 136], implying that $\alpha' \cup \alpha$ is splittable. Hence, the order of W_{j_1} with respect to α must be 0, because, otherwise, the order of V_{j_1} with respect to α would be > 0 because the order of V_{j_1} with respect to $\mathcal{K}_{j_1}$ is 2 [13, Theorem 3, p. 175; Lemma 2, p. 238], and $O_{V_{j_1}}(\alpha) > 0$ implies that $\alpha' \cup \alpha$ is unsplittable. Therefore, either α bounds a disk on ∂W_{j_1} or α is a meridian of W_{j_1}. Certainly, α does not bound on ∂W_{j_1}. But if α is a meridian of W_{j_1}, then $\alpha' \cup \mathcal{K}_{j_1}$ has linking number ± 1, which is a contradiction. Consequently, $A \cap T_{j_1}$ must now be empty; hence, $j_1 = j_2$ and $A \subset \mathrm{Int}\, V_j\,(j = j_i = j_2)$, because T_{j_1} separates W_{j_1} and W_{j_2} when $j_1 \neq j_2$.

We now have $\partial A = \alpha \cup \beta \subset \partial W_j$ as well as $A \subset \mathrm{Int}\, V_j$. Neither α nor β bounds a disk on ∂W_j; therefore, $\alpha \cup \beta$ bounds two annuli, A_1 and A_2, on ∂W_j.

I claim that there exists a toral solid X_1 and a 2-link manifold X_2 such that $V_j - \mathrm{Int}\, W_j = X_1 \cup_A X_2$. To see this, consider the torus $A_1 \cup A$. There are toral solids, U_1 and U_2, such that $S^3 = U_1 \cup_{A_1 \cup A} U_2$. If both W_j and $S^3 - \text{Int } V_j$ belong to the same solid, say U_2, then, obviously, the claim holds: take $X_1 = U_1$ and $X_2 = V_j - \text{Int}(U_1 \cup W_j)$. On the other hand, suppose $W_j \subset U_1$, say, and suppose $S^3 - \text{Int } V_j \subset U_2$. Then A_2 separates U_1, and so there is a toral solid X_1 such that $U_1 = X_1 \cup_{A_2} W_j$; note that $\partial X_1 = A_2 \cup A$. But $S^3 - \text{Int } X_1$ is a toral solid containing both W_j and $S^3 - \text{Int } V_j$. Because this situation is analogous to the first case with both W_j and $S^3 - \text{Int } V_j$ in U_1, the claim's proof follows.

Suppose X_1 is a knot manifold. Then X_1 belongs to a polyhedral 3-cell in V_j [1, Lemma 1, p. 226]. Consequently, $O_{V_j}(\alpha) = 0 = O_{V_j}(\beta)$ [13, Theorem 1, p. 171]. Because $O_{V_j}(\alpha) = O_{V_j}(\mathcal{K}_j)\, O_{W_j}(\alpha)$ and because $O_{V_j}(\mathcal{K}_j) = 2$, we see that $O_{W_j}(\alpha) = 0$. Thus, because α does not bound on ∂W_j, each of α and β must be a meridian of W_j. But if $V_{(0,2)}$ is a second-regular neighborhood of a clasping disk D in S^3 with trivial diagonal, with $\rho = 0$, and with $\eta = +2$, then there is a homeomorphism $e : (V_j, \mathcal{K}_j) \to (V_{(0,2)}, \partial D)$; each of $e(\alpha)$ and $e(\beta)$ is a meridian of ∂D, the knot ∂D is trivial, and $e(X_1)$ is a knot manifold. Therefore, the trivial knot is the composite of two knots one of which is nontrivial. Because this is impossible [5, pp. 141-142], X_1 is a solid torus.

Suppose now that the inclusion-induced homomorphism $\pi_1(A) \to \pi_1(X_1)$ is not surjective; we shall deduce a contradiction. Assume that $\partial X_1 = A_1 \cup A$. The space $X_1 \cup_{A_1} W_j$ is, evidently, a toral solid. If $X_1 \cup_{A_1} W_j$ were a knot manifold, then we could find a 3-cell in V_j containing it [1] (loc. cit.), implying that $O_{V_j}(\mathcal{K}_j) = 0$ [13, Theorem 1, p. 171]. Because $O_{V_j}(\mathcal{K}_j) = 2$, however, the space $X_1 \cup_{A_1} W_j$ is a solid torus. If x is a generator of $\pi_1(X_1)$, and if y is a generator of $\pi_1(W_j)$, we have $\pi_1(X_1 \cup_{A_1} W_j) = |x,y : x^p = y^q|$. Because A is incompressible, we have $p \neq 0$; because the inclusion-induced homomorphism $\pi_1(A) \to \pi_1(X_1)$ is not surjective, we have $p \neq \pm 1$. Therefore, because $\pi_1(X_1 \cup_{A_1} W_j) \approx Z$, we have $q = \pm 1$; first, q is certainly not 0; second, if $q \notin \{0, 1, -1\}$, then $|x,y : x^p = y^q|$ is the group of a torus link different from a trivial knot [5, p. 144]. Because $q = \pm 1$, the inclusion-induced homomorphism

$\pi_1(A_1) \to \pi_1(W_j)$ is surjective, and so $O_{W_j}(\alpha) = 1$. Thus, $O_{V_j}(\alpha) = O_{V_j}(\mathcal{K}_j)\, O_{W_j}(\alpha) = 2$.

Now let k be a core of X_1. We have $O_{V_i}(\alpha) = O_{X_1}(\alpha)\, O_{V_i}(k)$, the order $O_{X_1}(\alpha)$ is clearly $|p|$, and $|p| \neq 1$; therefore, $2 = |p|\, O_{V_j}(k)$, the order $O_{X_1}(\alpha) = |p| = 2$, and $O_{V_j}(k) = 1$. If $W_{V_j}(k)$ denotes the winding number of k in V_j, then $W_{V_j}(k) = 1$, because $O_{V_j}(k) = 1$ [13, Lemma 1, p. 170]. Thus, because $W_{V_j}(\alpha) = W_{X_1}(\alpha)\, W_{V_j}(k)$ [13, Theorem 4, p. 175] and because $W_{V_j}(\alpha) = W_{V_j}(\mathcal{K}_j)\, W_{W_j}(\alpha) = 0$, we have $W_{X_1}(\alpha) = 0$. But $\alpha \subset \partial X_1$; thus, $W_{X_1}(\alpha) = O_{X_1}(\alpha) = 2$ and $W_{X_1}(\alpha) = 0$ – an absurdity. Consequently, the inclusion-induced homomorphism $\pi_1(A) \to \pi_1(X_1)$ is surjective. Thus, the inclusion $A \to X$ is a homotopy equivalence [22]; therefore, A is boundary parallel in C [15, Theorem 3.1, p. 168].

2. *We can assume that each component* F *of* $g^{-1}(T_i)$ *and, hence, each component of* $g^{-1}(T)$ *is a torus that is not boundary parallel.*

Because $g \simeq f$, the mapping g is a homotopy equivalence; thus, $g_*: \pi_1(C') \to \pi_1(C)$ is an isomorphism. Moreover, because T_i is incompressible in C, property (4) implies that $\pi_1(F)$ is isomorphic to a subgroup of $\pi_1(T_i)$. Therefore, because F is orientable, it is either a 2-sphere, a disk, an annulus, or a torus. Property (4) implies that $\pi_2(F) = 0$, so that F is not a 2-sphere. If F is a disk, one can construct a map $g': C' \to C$ satisfying the properties (1) through (4) and the property that $g'^{-1}(T_i)$ has fewer components than $g^{-1}(T_i)$; see the second paragraph in [24, 2, p. 265]. Therefore, we can assume that F is either an annulus or a torus.

Suppose F is an annulus. Because $D(L'; \rho, \eta)$ and L' are unsplittable, and because $\mu > 1$, Lemma 2.3 applies to C', and so F is boundary parallel. As when F was a disk, we now easily replace g by a map g' satisfying the properties (1) through (4) and the further property that $g'^{-1}(T_i)$ has fewer components than $g^{-1}(T_i)$. Thus, we can assume that F is not an annulus, and that, therefore, each component of $g^{-1}(T_i)$ is a torus. Finally, we can assume that the torus F is not boundary parallel, because, otherwise, we could "remove" it in the obvious way; cf. [24, 2].

3. *For each of* $i = 1, \cdots, \mu$ *and for each component* F *of* $g^{-1}(T_i)$, *we can assume that* $g|F$ *is a homeomorphism.*

Let x be a point on F. The homomorphism $(g|F)_*: \pi_1(F,x) \to \pi_1(T_i, g(x))$ is a monomorphism, because g has properties (4) and (1) and because T_i is incompressible in C. Hence, $g|F$ is homotopic to a covering map $k: F \to T_i$ ([16], [21, Lemma 1.4.3, p. 61]). Because g is transverse with respect to T_i, this homotopy extends to a homotopy $\{h_t\}$ $(0 \leq t \leq 1)$ of g that is constant off a small product neighborhood of F. Note that h_{1*} is an isomorphism. Now $\pi_1(F,x) \subseteq h_{1*}^{-1}(\pi_1(T_i, h_1(x)))$; therefore, $h_{1*}(\pi_1(F,x)) = \pi_1(T_i, h_1(x))$ ([3, Theorem 1, p. 575] or [15, Theorem 1.3, p. 161]). Thus, $k = h_1|F$ is a homeomorphism, verifying 3.

4. *For each of* $i = 1, \cdots, \mu$, *we can assume that* $g^{-1}(T_i)$ *is connected.*

The proof is inductive. If $g^{-1}(T_1)$ is connected, the first inductive step, for $i = 1$, is complete. Otherwise, suppose that $g^{-1}(T_1)$ is disconnected. The construction of a map $g'': C' \to C$ with the same properties, (1) through (4), as g, but with $g''^{-1}(T_1)$ having fewer components than $g^{-1}(T_1)$, involves essentially three steps. First, construct a path α in C' so that α and $g^{-1}(T_1)$ meet only in the endpoints of α – each endpoint on a different component of $g^{-1}(T_1)$ – and so that $g\alpha$ is a null-homotopic loop in C. Second, split C' along $g^{-1}(T_1)$; the path α belongs to some component X resulting from this splitting. Using α, construct a homotopy from g to g' taking X onto T_1. Third, using g', construct a map $g'': C' \to C$ such that g'' satisfies properties (1) through (4), such that $g''^{-1}(\dot{T}) \subset g^{-1}(T)$, and such that $g''^{-1}(T_1)$ contains exactly one component less than $g^{-1}(T_1)$. We shall omit the details of this construction, because these details are in [4, §6, p. 155].

If necessary, we continue the constructions inductively – an induction within the original one – and we obtain a map $h_1: C' \to C$ such that h_1

satisfies properties (1) through (4), such that $h_1^{-1}(T) \subset g^{-1}(T)$, and such that $h_1^{-1}(T_1)$ contains exactly one component. Note that $h_1^{-1}(T_i) \neq \emptyset$, for any i, for the same reason that $g^{-1}(T_i) \neq \emptyset$ (see 1). The remaining steps in the original induction are now clear, completing the proof of 4, and we can assume that $g^{-1}(T)$ has exactly μ components, $T'_1, \cdots, T'_\mu$, with $g^{-1}(T_i) = T'_i$ $(i = 1, \cdots, \mu)$.

LEMMA 2.4. *Any properly imbedded, incompressible annulus* A *in* Λ_i *is boundary parallel.*

Proof. Because the inclusion-induced homomorphism $\pi_1(\Lambda_i) \to \pi_1(C)$ is a monomorphism and because A is incompressible in Λ_i, it follows that A is also incompressible in C. The proof of Lemma 2.3, therefore, shows two things: (1) we cannot have one component of ∂A in T_i and the other component of ∂A in ∂W_i; (2) if $\partial A \subset \partial W_i$, then A is boundary parallel. Thus, we shall assume that $\partial A \subset T_i$.

Now Λ_i is homeomorphic to a link manifold of a Whitehead link (Figure 1 depicts a Whitehead link) each of whose components is trivial. As one can readily see, such a link is interchangeable. Consequently, there is an autohomeomorphism ϕ of Λ_i taking T_i onto ∂W_i and taking ∂W_i onto T_i. The foregoing paragraph implies that $\phi(A)$ is boundary parallel; therefore, A itself is boundary parallel, concluding the proof of Lemma 2.4.

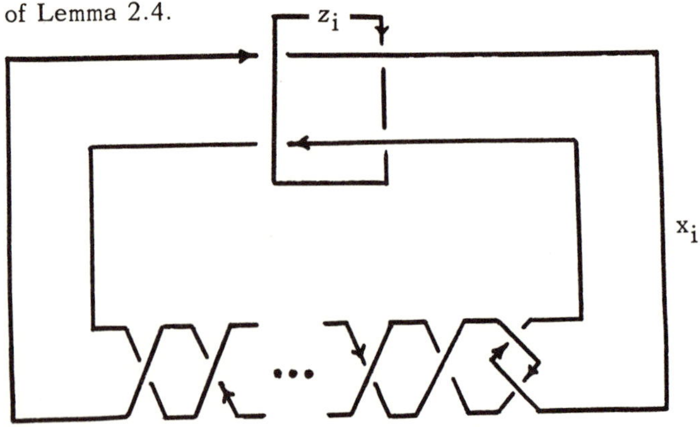

Fig. 1.

LEMMA 2.5. *Every properly imbedded, incompressible torus in* Λ_i $(i=1,\cdots,\mu)$ *is boundary parallel.*

Proof. Let S be a properly imbedded, incompressible torus in Λ_i not parallel to ∂W_i, let V'' be a knotted solid torus in S^3, and let $h: V_i \to V''$ be a faithful homeomorphism. The torus $h(S)$ is obviously incompressible in $S^3 - \mathrm{Int}(h(W_i))$ as well as in $h(\Lambda_i)$. Hence, there is a knot manifold Ω and a solid torus B such that $S^3 = \Omega \cup_{h(S)} B$. The torus $\partial B (= h(S))$ is not parallel to $h(\partial(W_i))$; thus, a core of B must be a companion of $h(D(K_i;\rho,\eta))$. But this implies that ∂B is parallel to $h(T_i)$ [13, Lemma 3, p. 238] and, therefore, that S is parallel to T_i *which establishes the lemma.*

5. (a) *There are mutually disjoint, solid tori,* $V'_1,\cdots,V'_\mu$, *such that* $\partial V'_i = T'_i$ *and such that* $W'_i \subset \mathrm{Int}\, V'_i$ $(i=1,\cdots,\mu)$ *for a suitable change in the subscripts of* $W'_1,\cdots,W'_\mu$.

(b) *Setting* $M' = S^3 - \mathrm{Int}(V'_1 \cup \cdots \cup V'_\mu)$, *we have* $M' \cong M$, *and we can assume that* $g|M'$ *is a homeomorphism.*

Let x_i be a point in T'_i. There are submanifolds M'_i and Λ'_i of C' such that $C' = M'_i \cup_{T'_i} \Lambda'_i$; because T'_i is incompressible in C', $\pi_1(C',x_i) = \pi_1(M'_i,x_i) \underset{\pi_1(T'_i,x_i)}{*} \pi_1(\Lambda'_i,x_i)$. Now $g^{-1}(T_i) = T'_i$, g_* is an isomorphism, and $\pi_1(M_i, g(x_i)) \underset{\pi_1(T_i,g(x_i))}{*} \pi_1(\Lambda_i, g(x_i))$ is a *nontrivial* free-product with amalgamation; therefore, one of the sets $g(M'_i)$ and $g(\Lambda'_i)$ belongs to M_i, and the other belongs to Λ_i. Assume that $g(M'_i) \subseteq M_i$ and $g(\Lambda'_i) \subseteq \Lambda_i$. Then $g_*(\pi_1(M'_i,x_i)) \subseteq \pi_1(M_i,g(x_i))$ and $g_*(\pi_1(\Lambda'_i,x_i)) \subseteq \pi_1(\Lambda_i,g(x_i))$. Because g_* is an isomorphism and because $g_*(\pi_1(T'_i,x_i)) = \pi_1(T_i,g(x_i))$ (because of 3), we have $g_*(\pi_1(M'_i,x_i)) = \pi_1(M_i,g(x_i))$ and $g_*(\pi_1(\Lambda'_i,x_i)) = \pi_1(\Lambda_i,g(x_i))$ [2, Proposition 2.5, p. 485]. But Λ'_i is either a toral solid or a link manifold; therefore, Λ'_i is a 2-link manifold, because its group is that of a link with two components. Consequently, $\partial\Lambda'_i = T'_i \cup \partial W'_i$ for some W'_j with i replacing j. Set $V'_i = \Lambda'_i \cup W'_i$.

To prove that V'_i is a solid torus, we first show that the following supposition leads to an absurdity. Suppose Λ'_i contains a properly imbedded, incompressible torus Q'_i that is not boundary parallel. We have $\Lambda'_i = X_i \cup_{Q'_i} Y_i$, and $\pi_1(\Lambda'_i) = \pi_1(X_i) \underset{\pi_1(Q'_i)}{*} \pi_1(Y_i)$. Each of X_i and Y_i is clearly either a knot manifold or a link manifold. As we have seen, $\pi_1(\Lambda'_i) \approx \pi_1(\Lambda_i)$. Hence, [12, Theorem (27.1), p. 19] implies that Λ'_i is a deformation retract of an unsplittable 2-link's complement, because Λ_i has this property (recall that Λ_i is a deformation retract of the complement of a Whitehead link). Therefore, Λ'_i is irreducible; consequently, each of X_i and Y_i is irreducible, because Q'_i is incompressible in Λ'_i. Thus, each of X_i and Y_i is aspherical.

If the inclusion-induced monomorphism $\pi_1(Q'_i) \to \pi_1(X_i)$ is surjective, then the asphericity of X_i implies that the inclusion map $Q'_i \to X_i$ is a homotopy equivalence [22]. Therefore, $X_i \cong Q'_i \times I$ [15, Theorem 3.1, p. 168]. But then Q'_i is boundary parallel, contrary to our supposition. Thus, the inclusion-induced monomorphism $\pi_1(Q'_i) \to \pi_1(X_i)$ is not surjective; similarly, the inclusion-induced monomorphism $\pi_1(Q'_i) \to \pi_1(Y_i)$ is not surjective; therefore, $\pi_1(X_i) \underset{\pi_1(Q'_i)}{*} \pi_1(Y_i)$ is a *nontrivial* free-product with amalgamation.

Because the 2-link manifold Λ'_i is irreducible, it is aspherical. As we have seen, $(g|\Lambda'_i)_* : \pi_1(\Lambda'_i) \to \pi_1(\Lambda_i)$ is an isomorphism. Consequently, $g|\Lambda'_i$ is a homotopy equivalence [22]. Let $\tau' : \Lambda_i \to \Lambda'_i$ be a homotopy inverse of $g|\Lambda'_i$. There is a mapping $\tau : \Lambda_i \to \Lambda'_i$ satisfying the following properties: (1) $\tau \simeq \tau'$; (2) τ is transverse with respect to Q'_i; (3) $\tau^{-1}(Q'_i)$ is a compact orientable surface properly imbedded in Λ_i; (4) if S is any component of $\tau^{-1}(Q'_i)$, then $\ker(\pi_j(S) \to \pi_j(\Lambda_i)) = 1$ $(j = 1,2)$.

Because $\pi_1(X_i) \underset{\pi_1(Q'_i)}{*} \pi_1(Y_i)$ is a nontrivial free-product with amalgamation, $\tau^{-1}(Q'_i)$ is not empty; the proof is similar to that in step 1. Property (4) implies that $\pi_1(S)$ is isomorphic to a subgroup of $\pi_1(Q'_i)$, because Q'_i is incompressible in Λ'_i. Hence, S is either a 2-sphere, a disk, an annulus, or a torus. It is neither a 2-sphere nor a disk (cf. the proof of step 2).

If S is an annulus, Lemma 2.4 says that it is boundary parallel. But then, as in the proof of 2, we could "remove" it. Thus, we can assume that S is not an annulus; therefore, it is a torus. Moreover, we can assume that this torus is not boundary parallel, because otherwise, we could "remove" it; cf. [24, 2]. Thus, if Λ'_i contains a properly imbedded, incompressible torus that is not boundary parallel, then so does Λ_i, contradicting Lemma 2.5.

We interrupt the proof of 5 to prove the following lemma.

LEMMA 2.6. *The linking number of* $\mathcal{K}'_i$ *with any closed curve on* $T'_i(=\partial V'_i)$ *is zero.*

Proof. Let c be any closed oriented curve on T'_i, and let $\mathfrak{L}(c, \mathcal{K}'_i)$ denote the linking number of c and $\mathcal{K}'_i$. Let (α, β) denote a meridian-longitude pair for the toral solid V'_i. Orient each of α and β, and assume that $\alpha \sim 0$ in $S^3 - \operatorname{Int} V'_i$ and that $\beta \sim 0$ in V'_i. Furthermore, if V'_i is a knot manifold, assume that β is homologous to a core of $S^3 - \operatorname{Int} V'_i$. There are integers a and b such that $c \sim a\alpha + b\beta$ on T'_i. Now $\pi_1(S^3 - (\beta \cup \mathcal{K}'_i)) \approx \pi_1(\Lambda'_i) \approx \pi_1(\Lambda_i)$, and $\pi_1(\Lambda_i)$ is the group of a 2-link with linking number 0; moreover, a 2-link's Alexander polynomial determines (independent of the link) the absolute value of the linking number [19]. Thus, $\mathfrak{L}(\beta, \mathcal{K}'_i) = 0$. Furthermore, $\mathfrak{L}(\alpha, \mathcal{K}'_i) = 0$, because α bounds in $S^3 - \operatorname{Int} V'_i$. But $\mathfrak{L}(c, \mathcal{K}'_i) = a\mathfrak{L}(\alpha, \mathcal{K}'_i) + b\mathfrak{L}(\beta, \mathcal{K}'_i)$, and the lemma follows.

Continuing the proof of 5, we let D'_i be a clasping disk spanning $\mathcal{K}'_i$ and missing $\mathcal{K}'_j$ when $j \neq i$. Seifert's proof of Lemma 7 in [17, p. 75] yields an ambient isotopy holding $D(L'; \rho, \eta)$ fixed and moving D'_i into $\operatorname{Int} V'_i$. To insure that Seifert's work can be applied here, we need only note that $T'_i(=\partial V'_i)$ is incompressible in C' and that the linking number of $\mathcal{K}'_i$ with any closed curve on T'_i is zero; see Lemma 2.6.

Assume now that $D'_i \subset \operatorname{Int} V'_i$. There exists a solid torus N'_i such that $D'_i \subset \operatorname{Int} N'_i$ and such that the diagonal k'_i of D'_i is a core of N'_i

[13, Proof of Theorem 1, p. 235]. Because D'_i is obviously a strong deformation retract of N'_i (see [13] (loc. cit.)), we can assume that $N'_i \subset \text{Int } V'_i$; evidently, we can also arrange for W'_i to be in $\text{Int } N'_i$. Because N'_i has order 2 with respect to $\mathcal{K}'_i$ [13, Lemma 2, p. 238] and because $D(L'; \rho, \eta)$ is unsplittable, $\partial N'_i$ is incompressible in C' and, hence, incompressible in Λ'_i. Because $\partial N'_i$ is not parallel to $\partial W'_i$ and because, as we have proved, every properly imbedded, incompressible torus in Λ'_i is boundary parallel, $\partial N'_i$ is parallel to T'_i. But this implies that $T'_i (= \partial V'_i)$ is compressible in V'_i. Therefore, V'_i is a solid torus. Consequently, $\Lambda'_i \cong N'_i - \text{Int } W'_i$. Furthermore, there is a faithful homeomorphism $(N'_i, \mathcal{K}'_i) \to (V_i, \mathcal{K}_i)$, which one can easily construct with results in [6, pp. 47-54]. Therefore, $N'_i - \text{Int } W'_i \cong \Lambda_i$, and so $\Lambda'_i \cong \Lambda_i$.

Because $g(\Lambda'_i) \subseteq \Lambda_i \subseteq V_i$ $(i = 1, \cdots, \mu)$ and because $V_i \cap V_j = \emptyset$ when $i \neq j$, we obviously have $V'_i \cap V'_j = \emptyset$ when $i \neq j$. This concludes the proof of 5 (a).

To prove 5 (b), notice in the first paragraph of this step 5 that $g(M'_1) \subseteq M_1$ and that $(g|M'_1)_* : \pi_1(M'_1, x_1) \to \pi_1(M_1, g(x_1))$ is an isomorphism. Notice also that $M'_1 = (M'_1 - \text{Int } V'_2) \cup_{T'_2} \Lambda'_2$ and that $M_1 = (M_1 - \text{Int } V_2) \cup_{T_2} \Lambda_2$. Now $(g|M'_1)_*$ is an isomorphism, $(g|M'_1)_*(\pi_1(M'_1 - \text{Int } V'_2, x_2)) \subseteq \pi_1(M_1 - \text{Int } V_2, g(x_2))$, and $(g|M'_1)_*(\pi_1(\Lambda'_2, x_2)) = \pi_1(\Lambda_2, g(x_2))$; thus $(g|M'_1)_*(\pi_1(M'_1 - \text{Int } V'_2, x_2)) = \pi_1(M_1 - \text{Int } V_2, g(x_2))$; that is, $\pi_1(M'_1 - \text{Int } V'_2) \approx \pi_1(M_1 - \text{Int } V_2)$ [2, Proposition 2.5, p. 485].

Arguing inductively, we see that $\pi_1(M'_1 - \text{Int } (V'_2 \cup \cdots \cup V'_\mu)) \approx \pi_1(M_1 - \text{Int } (V_2 \cup \cdots \cup V_\mu))$; that is, $(g|M')_* : \pi_1(M') \to \pi_1(M)$ is an isomorphism. The mappings $g|T'_i$ $(i = 1, \cdots, \mu)$ are homeomorphisms. Therefore, there exists a homotopy from g to a map $g' : C' \to C$ such that $(g'|M') : M' \to M$ is a homeomorphism and such that the homotopy is constant on $C' - \text{Int } M'$ [21, Theorem 6.1, p. 77]. Thus, $M' \cong M$, and we can assume that $g|M'$ is a homeomorphism, concluding 5 (b)'s proof.

6. (a) *For each of* $i = 1, \cdots, \mu$, *we have* $\Lambda'_i \cong \Lambda_i$.

(b) *If* k_i *is a core of* V'_i, *then* $\{k_1 \cup \cdots \cup k_\mu\} = \{L'\}$.

From the proof of 5(a), we know that $\Lambda'_i \cong \Lambda_i$; this proves 6(a). To prove 6(b), recall that the diagonal k'_i of the clasping disk D'_i, used in the proof of 5(a), is a core of N'_i and that $N'_i \cap N'_j = \emptyset$ whenever $i \neq j$. The definition of a doubled link (see the introduction) implies that $D(L';\rho,\eta) = D(k'_1 \cup \cdots \cup k'_\mu;\rho,\eta)$. Lemma 1.2 now implies that $\{k'_1 \cup \cdots \cup k'_\mu\} = \{L'\}$. Assuming that D'_i has been moved into Int V'_i $(i=1,\cdots,\mu)$, we see that k'_i is also a core of V'_i. There is an autohomeomorphism of S^3 acting as the identity on $S^3 - \text{Int}(V'_1 \cup \cdots \cup V'_\mu)$ and taking k'_i onto k_i $(i=1,\cdots,\mu)$ [13, Lemma 1, p. 158]. Therefore, $\{k_1 \cup \cdots \cup k_\mu\} = \{k'_1 \cup \cdots \cup k'_\mu\} = \{L'\}$, proving 6(b).

7. *The links* L' *and* L *belong to the same (ambient) isotopy type.*

The proof of 7 is similar to the proof of 6 in [24]. Recall that ρ and η are fixed integers; ρ, arbitrary; $\eta = \pm 2$. We shall prove 7 for ρ arbitrary and $\eta = +2$. The claim then holds for the pair $(\rho,-2)$: first notice that $D(L^*;-\rho,-\eta)$ is the mirror image of $D(L;\rho,\eta)$; hence, if $\pi_1(C^3(L;\rho,-2)) \approx \pi_1(C^3(L';\rho,-2))$, then $\pi_1(C^3(L^*;-\rho,+2)) \approx \pi_1(C^3(L'^*;-\rho,+2))$, whence $\{L^*\} = \{L'^*\}$ and, therefore, $\{L\} = \{L'\}$. Thus, to prove 7, it suffices, by 6(b), to prove that $\{k_1 \cup \cdots \cup k_\mu\} = \{L\}$ when $\eta = +2$.

Set $G_1 = \pi_1(\Lambda'_i)$ and set $G_2 = \pi_1(\Lambda_i)$. Choose a basepoint of G_1 on T'_i; choose a basepoint of G_2 on T_i. Now read a presentation for G_j $(j=1,2)$ from Figure 1.

REMARK. We shall assume that the orientation of S^3 has been chosen so that the twist knot of Figure 1 has $\eta = +2$.

We have

$$(*) \quad G_j = |u_j,z_j,x_j : z_j u_j z_j^{-1} u_j^{-1},\ u_j = x_j u_j^\rho z_j^{-1} x_j^{-1} z_j u_j^{-\rho} x_j^{-1} u_j^\rho z_j^{-1} x_j z_j u_j^{-\rho}| .$$

The pair (u_1,z_1) is a meridian-longitude pair in the link manifold M'; the pair (u_2,z_2), a meridian-longitude pair in M. Now abolish the subscripts in G_2. Because $g|M'$ is a homeomorphism, we can assume that $g_*(z_1) = u^r z^v$ and that $g_*(u_1) = u^{\pm 1} z^q$. Notice, however, that if

$\alpha : G \to G/G'$ is the canonical abelianization of G, then the second relation for G_1 in (*) becomes $[\alpha(z)]^q = 1$ under the homomorphism $\alpha g_* : G_1 \to G/G'$. The integer q is, therefore, zero, because $\alpha(z)$ is a free generator of G/G'. Consequently, $g|M'$ matches meridians with meridians or their inverses; hence, $k_1 \cup \cdots \cup k_\mu$ and L and, therefore, L' and L (by 6 (b)) are equivalent.

The orientation of S^3 induces an orientation in each of V'_i and V_i; in turn, the oriented solid torus V'_i induces an orientation in T'_i; the solid torus V_i, an orientation in T_i. A pair of transverse, simple, closed curves on T'_i oriented with respect to k_i represents the meridian-longitude pair (u_1, z_1); a similar pair of curves on T_i oriented with respect to K_i represents the pair (u, z). Each of these pairs of curves has intersection number $+1$ or each has intersection number -1, because there is an orientation-preserving homeomorphism $e : V'_i \to V_i$ satisfying $e_*(u_1) = u$ and $e_*(z_1) = z$. Therefore, to prove that $\{k_1 \cup \cdots \cup k_\mu\} = \{L\}$, we suppose that g does not preserve the intersection number; that is, we suppose that one of the following holds: (a) $g_*(z_1) = z$, $g_*(u_1) = u^{-1}$; (b) $g_*(z_1) = z^{-1}$, $g_*(u_1) = u$.

If S is a subset of a group H, then $\langle S\rangle$ denotes the consequence (or normal closure) of S in H. Set $\Gamma = G_1/\langle z_1 u_1^{-(\rho+1)}\rangle$ and set $\Omega = G/\langle zu^{(\rho+1)}\rangle$. A straightforward argument shows that $g_*(\langle z_1 u_1^{-(\rho+1)}\rangle) = \langle zu^{(\rho+1)}\rangle$ in either case (a) or case (b). Hence, g_* induces an isomorphism $\Gamma \to \Omega$.

If H is a knot group, let $\Delta_H(t)$ denote its Alexander polynomial. We have $\Delta_\Gamma(t) = t^2 - t + 1$ and $\Delta_\Omega(t) = -(2\rho+1)t^2 + (4\rho+3)t - (2\rho+1)$. Obviously, $\Delta_\Gamma(t) \neq \Delta_\Omega(t)$, unless $\rho = -1$. Therefore, neither (a) nor (b) can hold when $\rho \neq -1$.

To prove that neither (a) nor (b) holds when $\rho = -1$, set $\rho = -1$ in the presentation (*), set $\Sigma = G_1/\langle z_1 u_1^{-1}\rangle$ $(= |u_1, x_1 : u_1 = x_1 u_1^{-2} x_1^{-1} u_1^2 x_1^{-1} u_1^{-2} x_1 u_1^2|)$, and set $\theta = G/\langle zu\rangle$ $(\approx Z)$. The isomorphism g_*, in either case, induces an isomorphism $\Sigma \to \theta$. But $\Delta_\Sigma(t) = 2t^2 - 3t + 2$ and $\Delta_\theta(t) = 1$, showing that $\Sigma \not\approx \theta$. Consequently, neither

(a) nor (b) can occur; thus, $\{k_1 \cup \cdots \cup k_\mu\} = \{L\}$ and, therefore, $\{L'\} = \{L\}$, completing the theorem's proof.

PRINCETON UNIVERSITY
and
UNIVERSITY OF SOUTHWESTERN LOUISIANA

REFERENCES

[1] Bing, R. H., and Martin, J. M., *Cubes with knotted holes*. Trans. Amer. Math. Soc. 155 (1971), 217-231.

[2] Brown, E. M., *Unknotting in* $M^2 \times I$. Trans. Amer. Math. Soc. 123 (1966), 480-505.

[3] Feustel, C. D., *Some applications of Waldhausen's results on irreducible surfaces*. Trans. Amer. Math. Soc. 149 (1970), 575-583.

[4] ______________, *A splitting theorem for closed orientable 3-manifolds*, Topology 11 (1972), 151-158.

[5] Fox, R. H., *A quick trip through knot theory*, Topology of 3-Manifolds and Related Topics (Proc. the Univ. of Georgia Inst., 1961). Prentice-Hall, Englewood Cliffs, N. J., 1962, 120-167.

[6] Graeub, W., *Die semilinearen Abbildungen*. S.-B. Heidelberger Akad. Wiss. Math.-Natur. K1. (1950), 205-272.

[7] Kyle, R. H., *Embeddings of Möbius bands in 3-dimensional space*. Proc. Roy. Irish Acad. Sect. A 57 (1955), 131-136.

[8] Magnus, W., Karrass, A., and Solitar, D., *Combinatorial Group Theory*, Pure and Applied Mathematics 13, Interscience Pub. (John Wiley and Sons, Inc.), New York, 1966.

[9] Neuwirth, L. P., *A note on torus knots and links determined by their groups*. Duke Math. J. 28 (1961), 545-551.

[10] ______________, *Knot Groups*. Annals of Mathematics 56, Princeton Univ. Press, Princeton, N. J., 1965.

[11] Nielsen, J., *Untersuchungen zur Topologie der geschlossenen zweiseitigen Flächen*, I. Acta Math. 50 (1927), 189-358.

[12] Papakyriakopoulos, C. D., *On Dehn's lemma and the asphericity of knots*. Ann. of Math. 66 (1957), 1-26.

[13] Schubert, H., *Knoten und Vollringe*. Acta Math. 90 (1953), 131-286.

[14] ______________, *Knoten mit zwei Brücken*. Math. Z. 65 (1956), 133-170.

[15] Scott, G. P., *On sufficiently large 3-manifolds*. Quart. J. Math. Oxford Ser. 23 (1972), 159-172.

[16] Seifert, H., *Bemerkung zur stetigen Abbildung von Flächen*. Abh. Math. Sem. Univ. Hamburg 12 (1938), 29-37.

[17] Seifert, H., *Schlingknoten.* Math. Z. 52 (1949), 62-80.

[18] Simon, J., *An algebraic classification of knots in* S^3. Ann. of Math. 97 (1973), 1-13.

[19] Torres, G., *On the Alexander polynomial.* Ann. of Math. 57 (1953), 57-89.

[20] Waldhausen, F., *Gruppen mit Zentrum und 3-dimensionale Mannigfaltigkeiten.* Topology 6 (1967), 505-517.

[21] ————, *On irreducible 3-manifolds which are sufficiently large.* Ann. of Math. 87 (1968), 56-88.

[22] Whitehead, J. H. C., *Combinatorial homotopy,* I. Bull. Amer. Math. Soc. 55 (1949), 213-245.

[23] Whitten, W., *Isotopy types of knot spanning surfaces.* Topology 12 (1973), 373-380.

[24] ————, *Algebraic and geometric characterizations of knots.* Invent. Math. 26 (1974), 259-270.

[25] ————, *Characterizations of knots and links.* Bull. Amer. Math. Soc. 80 (1974), 1265-1270.

Group Theory

HNN GROUPS AND GROUPS WITH CENTER

John Cossey and N. Smythe

We shall show that if H is a subgroup with nontrivial center of a group in a certain class G, then H is an extension of a free group by a subgroup of the rationals. The class $\mathcal{G}$ is described in Section 1; essentially it consists of groups which can be constructed by a sequence of free products with amalgamation and HNN-constructions with free amalgamated and associated subgroups, starting with free groups. The class $\mathcal{G}$ contains all torsion-free 1-relator groups, and also fundamental groups of 3-manifolds with incompressible boundary, in particular knot groups and link groups. The proof is modelled on [8], but note that we do not assume H to be finitely generated.

§1. For our own convenience in the definition of the class $\mathcal{G}$ we shall follow the development of the subgroup theorem for HNN-groups given in [13], summarized below. The reader familiar with this theory as developed by Karrass, Pietrowski and Solitar [8] or Cohen [4] should have little difficulty with translation.

A *diagram of groups* (D, A) consists of

(i) a (connected) directed graph D

(ii) for each vertex v of D, a group A_v

(iii) for each directed edge e of D leading from the vertex λe to the vertex ρe, a homomorphism $A_e : A_{\lambda e} \to A_{\rho e}$.

The *mapping cylinder* of (D, A) is a group m(D, A) given by

$$\text{generators} : \bigcup_{v \in D} A_v \cup \{t_e : e \text{ an edge of } D\}$$

relations : relations of A_v, $t_e A_e(a) t_e^{-1} = a$ for $a \epsilon A_{\lambda e}$

$t_e = 1$ for $e \epsilon T$

where T is a maximal tree in D.

The isomorphism type of $m(D, A)$ is independent of the choice of T. If each A_e is a monomorphism then each vertex group A_v is embedded in the mapping cylinder in the obvious manner; in this case the mapping cylinder is called the *graph product* of (D, A).

If D is a tree, the mapping cylinder is simply the colimit of (D, A); the graph product in this case is called a "tree product" by Karrass and Solitar (although their tree is slightly different). For the diagram

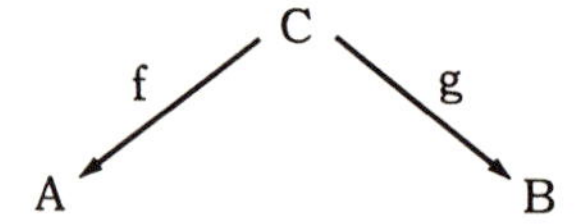

the graph product is $A *_C B$, the free product of A and B amalgamating $f(C)$ with $g(C)$.

For the diagram

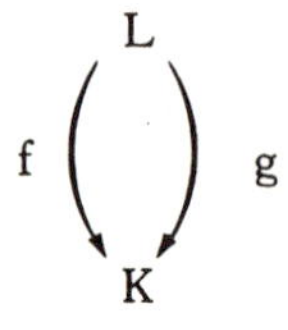

the graph product is the HNN-construction with base K and associated subgroups $f(L)$ and $g(L)$, i.e.

$$< K, t : tf(L)t^{-1} = g(L) > .$$

The mapping cylinder has a universal description in the category of groupoids [13]; its description in the category of groups is complicated by the non-uniqueness of the maximal tree T. However we only need here the fact that there is a homomorphism from the mapping cylinder of (D,A) onto the colimit of (D, A), the kernel of which is normally generated by

$\{t_e : e$ an edge of $D\}$. Thus if we are given a group G and homomorphism $\phi_v : A_v \to G$ for each vertex v, such that each diagram

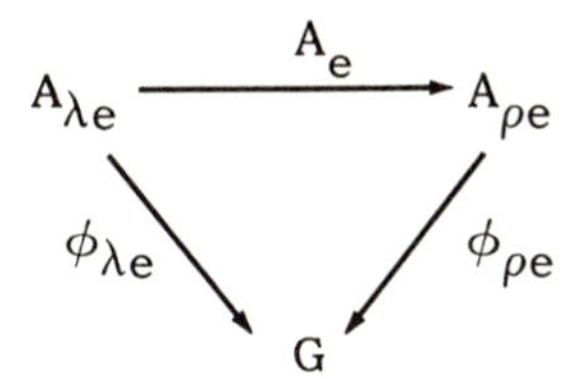

commutes, then there is a uniquely determined homomorphism from the mapping cylinder of (D, A) into G.

As developed in [13], the theory of mapping cylinders allows groupoids at the vertices of a diagram; in particular we need to allow disjoint unions of groups to occur. An example should suffice to illustrate how the mapping cylinder is then obtained. Consider the diagram

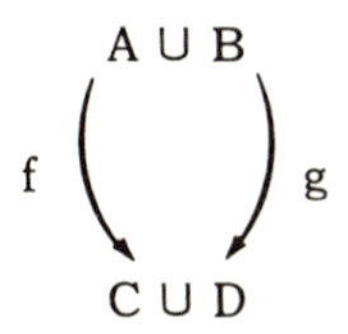

the vertices being disjoint unions of groups A and B, C and D respectively. Suppose $f(A) \subset C$, $g(A) \subset C$, $f(B) \subset C$ and $g(B) \subset D$. The diagram can then be expanded to a diagram in which only groups appear at each vertex

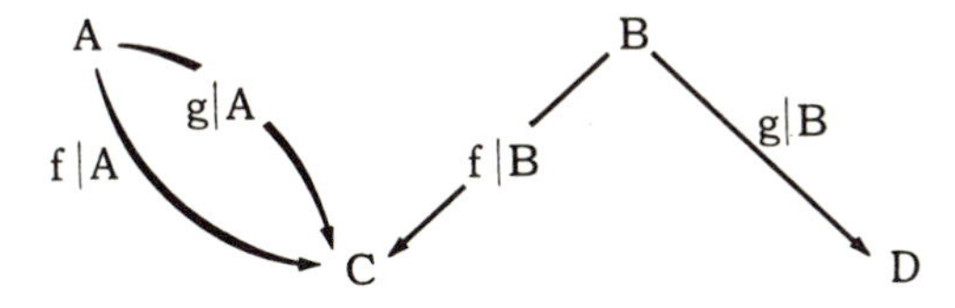

The mapping cylinder (or graph product) of the original diagram is then defined to be the mapping cylinder of this expanded diagram.

The subgroup theorem for graph products can now be stated.

THEOREM 1. *Suppose* M *is the graph product of a diagram* (D, A) *and* H *is a subgroup of* M. *Then* H *is the graph product of a diagram* (D,B), *where the vertex groupoid* B_v *is a disjoint union of groups* $dA_v d^{-1} \cap H$, *with* d *ranging over a set of double coset representatives for* M mod (H,A). *The edge maps* B_e *are induced from the maps* A_e.

(The theorem also holds for mapping cylinders with an appropriate modification to allow for the fact that A_v is not necessarily embedded in M in this case.)

Finally we come to the description of the class $\mathcal{G}$. Let $\mathcal{G}_0$ be the class consisting of the trivial group alone. If $\mathcal{G}_{n-1}$ has been defined, $\mathcal{G}_n$ is to consist of graph products of diagrams

where L is a disjoint union of free groups

K is a disjoint union of members of $\mathcal{G}_{n-1}$.

(Thus $\mathcal{G}_1$ consists of all free groups, $\mathcal{G}_2$ contains free products of free groups with amalgamations, etc.) Let $\mathcal{G} = \cup \mathcal{G}_n$. It is an immediate corollary of the subgroup theorem stated above that $\mathcal{G}$ *is closed with respect to subgroups.*

It is a consequence of work of Magnus (see [11]) that $\mathcal{G}$ contains all torsion-free 1-relator groups. It is a consequence of work of Haken ([6]; see also Waldhausen [14]) that fundamental groups of 3-manifolds with incompressible boundary, in particular knot groups and link groups, are in $\mathcal{G}$.

Note that every group $G \in \mathcal{G}$ has cohomological dimension ≤ 2 since a 2-complex which is a $K(G, 1)$ can be constructed.

§2. THEOREM 2. *Let* H *be a group in* $\mathcal{G}$ *having non-trivial* center. *Then there is a homomorphism* ϕ *from* H *into the rationals* $\mathfrak{Q}$, *whose kernel is a free group. If* H *is non-abelian, then the centre of* H *is infinite cyclic and is mapped monomorphically by* ϕ.

Proof. Note firstly that if H is abelian, then it is either free abelian of rank at most 2 or is locally infinite cyclic (hence isomorphic to a subgroup of $\mathfrak{Q}$) ([5], Theorem 5, p. 149). Thus in the following we may assume H is non-abelian.

We may assume $H \in \mathcal{G}_n$, and that the theorem is true for members of $\mathcal{G}_{n-1}$. Thus H is a graph product of a diagram

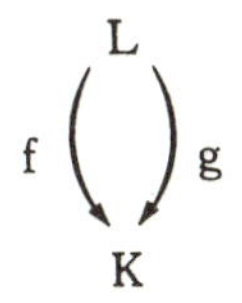

where L is a disjoint union of free groups

K is a disjoint union of members of $\mathcal{G}_{n-1}$.

Expand this diagram to one in which groups occur at each vertex. In this expanded diagram we have

(i) $\mathcal{G}_{n-1}$-groups K_α, for α in a set of vertices A

(ii) free groups L_β, for β in a set of vertices B

(iii) for each $\beta \in B$, two edges from β to vertices in A with corresponding injections $f_\beta : L_\beta \to K_{f\beta}$ and $g_\beta : L_\beta \to K_{g\beta}$.

Within B choose a subset B′ such that the complete subgraph containing $A \cup B'$ is a tree. The graph product S of this subdiagram is a tree product which is embedded in H, and H may be regarded as an HNN-construction with base S, that is, the graph product of the diagram

$$\bigcup_{\beta \in B \setminus B'} L_\beta \rightrightarrows S$$

Case 1: $B = B' = \emptyset$.

Then A must be a singleton $\{\alpha\}$ and $H = K_\alpha \in \mathcal{G}_{n-1}$. The result follows from the induction assumption.

Case 2: $B = B' \neq \emptyset$.

Then $H = S$, a free product of the groups K_α amalgamating the subgroup $f_\beta(L_\beta)$ in $K_{f\beta}$ with the subgroup $g_\beta(L_\beta)$ in $K_{g\beta}$. Then the center of H, ZH, is contained in the intersection $\bigcap_{\beta \in B'} f_\beta(L_\beta)$ of all these amalgamated subgroups.

Hence each L_β has nontrivial center. Since L_β is free, it must be infinite cyclic. Thus ZH is infinite cyclic. Furthermore each K_α contains ZH so has nontrivial center. By the induction assumption, if K_α is non-abelian there is a homomorphism $\phi_\alpha : K_\alpha \to \mathcal{Q}$ which is one to one on the center of K_α. If K_α is abelian, there is a homomorphism $\phi_\alpha : K_\alpha \to \mathcal{Q}$ which is one to one on ZH. Since L_β is cyclic, ϕ_α is one to one on $f_\beta(L_\beta)$, for $\alpha = f\beta$, and on $g_\beta(L_\beta)$, for $\alpha = g\beta$.

The tree product S can be built up inductively vertex by vertex. Then we can inductively construct a map $\phi'_\alpha : K_\alpha \to \mathcal{Q}$ as follows. Suppose we have defined ϕ'_α for α in a subset A' of A, and α^* is a vertex of A joined to A' via a vertex β of B, for example

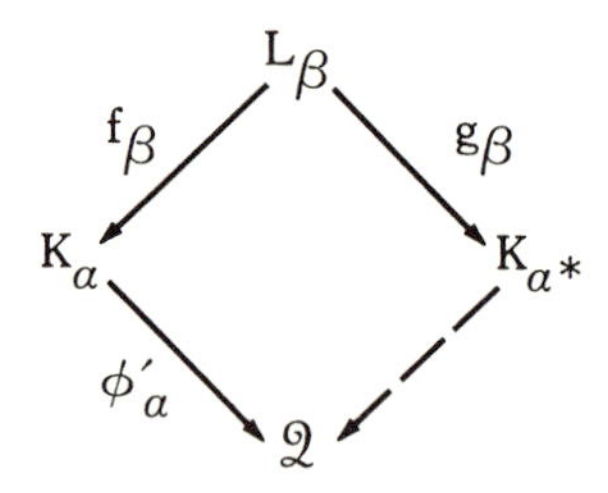

Define $\phi'_{\alpha^*} = q \cdot \phi_{\alpha^*}$ where $q = \phi'_\alpha f_\beta(x)/\phi_{\alpha^*} g_\beta(x)$

for x a generator of L_β .

Thus $\phi'_{f\beta} f_\beta = \phi'_{g\beta} g_\beta$ for all $\beta \in B$. The collection $\{\phi'_\alpha : \alpha \in A\}$ therefore induces a homomorphism $\phi : S \to \mathcal{Q}$ which is one to one on the center.

The kernel of ϕ is free; this can be seen using a cohomological argument or directly, from the subgroup theorem. Thus, since $(\ker \phi) \times ZH$ is a subgroup of H, $\ker \phi$ can have cohomological dimension at most 1; the Swan-Stallings theorem tells us that $\ker \phi$ is free. Alternatively, the subgroup theorem says that $\ker \phi$ is a graph product of a diagram

$$L^* \rightrightarrows K^*$$

where each group in L^* is of the form $(\ker \phi) \cap dL_\beta d^{-1} \cong (\ker \phi) \cap L_\beta = \{1\}$, and each group in K^* is of the form $(\ker \phi) \cap dK_\alpha d^{-1} \cong (\ker \phi) \cap K_\alpha = \ker \phi_\alpha$ which is free; the graph product of such a diagram is free.

Case 3: $B \setminus B' \neq \emptyset$, $ZH \cap S^H = 1$.

Then H/S^H is freely generated by $\{t_\beta : \beta \in B \setminus B'\}$, with nontrivial center. Hence H/S^H is infinite cyclic.

By the cohomological argument given above, S^H is free. Since H is assumed non-abelian, S^H is non-trivial so the cohomological argument also gives us that ZH has cohomological dimension 1. Thus ZH is infinite cyclic. It is mapped monomorphically by the quotient map to H/S^H.

Case 4: $B \setminus B' = \{\beta\}$, $ZH \cap S^H \neq 1$.

Suppose ZH is not contained in S^H. Now H is a split extension of S^H by $\langle t_\beta \rangle$; suppose $t_\beta^m x \in ZH$, with $m \neq 0$, $x \in S^H$. Then $S^H \cap \langle t_\beta^m x \rangle = 1$, so H contains $S^H \times \langle t_\beta^m x \rangle$. Hence S^H has cohomological dimension at most 1, and is therefore free. But $ZH \cap S^H \neq 1$, so S^H has non-trivial center, and so must be infinite cyclic. Now there is only one non-abelian extension of an infinite cyclic S^H $(= \langle a \rangle$, say) by $\langle t_\beta \rangle$, that is

$$\langle t_\beta, a : t_\beta a t_\beta^{-1} = a^{-1} \rangle .$$

But the center of this group is $\langle t_\beta^2 \rangle$, which does not meet $\langle a \rangle = S^H$.

Thus $ZH \subset S^H$. We shall in fact show that $ZH \subset S$. S^H is a graph product of an infinite diagram

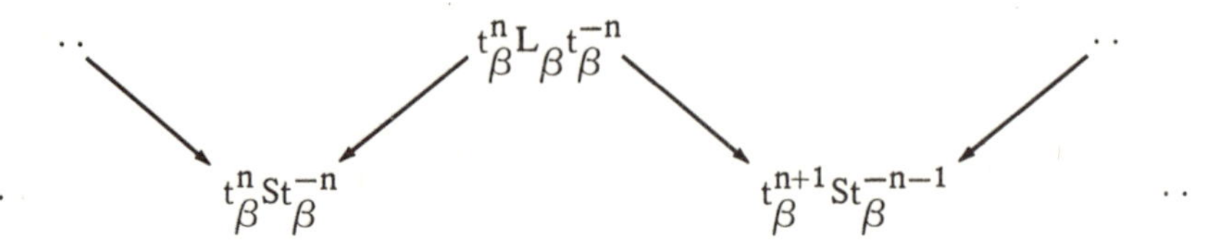

that is to say, a free product of conjugates of S amalgamating conjugates of $f_\beta(L_\beta)$ with conjugates of $g_\beta(L_\beta)$. The center of S^H is therefore contained in the intersection of all these conjugates. In particular L has non-trivial center so must be infinite cyclic. Therefore ZH is infinite cyclic and is contained in S. The arguments of cases 1 and 2 apply to S; hence there is a homomorphism $\phi : S \to \mathcal{Q}$ which is one to one on ZH.

Let $z \in ZH \cap f_\beta(L_\beta)$, $z \neq 1$; then $g_\beta(f_\beta^{-1}(z)) = t_\beta z t_\beta^{-1} = z$. Since L_β is cyclic, t_β must act trivially on $f_\beta(L_\beta) \bmod \ker \phi$ and $\phi f_\beta = \phi g_\beta$. There is therefore an extension of ϕ to a homomorphism $\bar{\phi} : H \to \mathcal{Q}$ which is one to one on $f_\beta(L_\beta)$, therefore on ZH. By the previous argument using cohomological dimension or the subgroup theorem, $\ker \bar{\phi}$ is free.

Case 5: $B \backslash B'$ has more than one element.

Then H is a free product of groups $\langle t_\beta, S \rangle$, $\beta \in B \backslash B'$ amalgamated over S. Hence $ZH \subset S$. Choose $\beta_0 \in B \backslash B'$ and let $S_0 = \langle S, t_\beta : \beta \neq \beta_0 \rangle$. Then H is the graph product of

$$L_{\beta_0} \rightrightarrows S_0$$

The argument for case 4 shows that ZH is cyclic, ZH is contained in $f_{\beta_0}(L_{\beta_0})$ and t_{β_0} commutes with $f_{\beta_0}(L_{\beta_0})$. Hence as in the previous case there is a map $\phi : S \to \mathcal{Q}$ which is one to one on ZH, which may be extended to $\bar{\phi} : H \to \mathcal{Q}$ by setting $\bar{\phi}(t_\beta) = 1$ for all β. This map $\bar{\phi}$ is one to one on ZH and has free kernel. q.e.d.

§3. As an application of Theorem 2, we derive two results about embedding knot groups in other knot groups due to Chang [3].

We start by proving

THEOREM 3. *Let* $G \in \mathcal{G}$, *and suppose that* G *has a non-abelian subgroup* H *with the properties* (i) $ZH \neq 1$ *and* (ii) H *is either normal or of finite index in* G. *Then either* $ZG \neq 1$, *or* G *has a free normal subgroup* F, *with* G/F *isomorphic to a subgroup of the rationals extended by an automorphism of order* 2 *acting invertingly (so that if* G *is finitely generated,* G/F *is isomorphic to the infinite dihedral group).*

Proof. Note first that if H is of finite index, so is $\bigcap_{g \in G} H^g = H_0$.

From Theorem 2, if $ZH \neq 1$, $ZH_0 \neq 1$, and so we may assume that H is normal in G.

Since ZH is normal in G, and infinite cyclic, the centralizer of ZH in G, C say, is also normal, and G/C is isomorphic to a subgroup of $\mathrm{Aut}(ZH)$. Thus $|G/C| = 1$ or 2, and if $G = C$, $ZH \leq ZG$ and we are finished.

Hence suppose $|G/C| = 2$. Now $ZC \neq 1$, and hence C has a free normal subgroup F with C/F isomorphic to a subgroup of the rationals. We claim $F = F \cap xFx^{-1}$, where $x \in G \setminus C$, and hence F is normal in G. Put $E = F \times ZC$; then C/E is periodic, and hence so is $C/E \cap xEx^{-1}$. But $E \cap xEx^{-1} = (F \cap xFx^{-1}) \cap ZC$, and so $C/F \cap xFx^{-1}$ is infinite cyclic-by-periodic. On the other hand $C/F \cap xFx^{-1}$ is isomorphic to a subgroup of $C/F \times C/xFx^{-1}$, and contains a subgroup isomorphic to $Z \times Z$ if $F \neq F \cap xFx^{-1}$. Thus we get $F = F \cap xFx^{-1}$.

We now have C/F locally cyclic and G/F non-abelian. If $x \in G \setminus C$, then it is easy to check that $x^2 \in F$, and that conjugation by xF inverts the elements of C/F.

The next lemma is easy to establish: we include a proof for completeness.

LEMMA 4. *Suppose* H *is normal or of finite index in the group* $G = Z_m * Z_n$, *and* H *is isomorphic to* $Z_r * Z_s$, *where* m, n, r, s *are integers, and neither* m *nor* n *is* 1, *at least one of* m, n *is not* 2, *and at least one of* r, s *is not* 1. *Then* $H = G$.

Proof. By the subgroup theorem, H is the graph product of the diagram

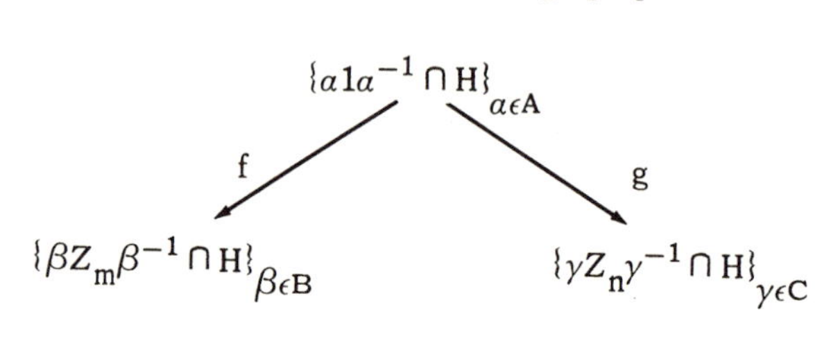

where A is a set of (double) coset representatives for G mod (H, 1), B is a set of double coset representatives for G mod (H, Z_m), and C is a set of double coset representatives for G mod (H, Z_n). The map f sends the component $\alpha 1 \alpha^{-1} \cap H$ corresponding to $\alpha \epsilon A$ into the component $\beta Z_m \beta^{-1} \cap H$ corresponding to the double coset $H\beta Z_m$ containing α. Similarly for g.

Since $H \cong Z_r * Z_s$, there can be no loops in this diagram, for otherwise H would contain a free factor. Furthermore exactly two of the factors $\beta Z_m \beta^{-1} \cap H$, $\gamma Z_n \gamma^{-1} \cap H$ are non-trivial.

Now, if H is normal $\beta Z_m \beta^{-1} \cap H \cong Z_m \cap H$ for all β. Thus there can be at most two elements in B, and hence H has finite index in G.

Thus we need only consider the case H of finite index in G. If $H = \langle x, y : x^s = y^r = 1 \rangle$, then x must be contained in a conjugate of Z_m or of Z_n: say $x \epsilon gZ_m g^{-1}$, and so r divides m. The image of H in the quotient group $Z_{m/r} * Z_n$ is finite and of finite index, and hence $m = r$. Similarly, $s = m$ or $s = n$. Let $|A| = a$, $|B| = b$, $|C| = c$. From Kurosh [10] p. 63, the number of cosets of H contained in the double coset $H\beta Z_m$ is precisely the index of $\beta Z_m \beta^{-1} \cap H$ in $\beta Z_m \beta^{-1}$.

If just one of the $\beta Z_m \beta^{-1} \cap H$ is non-trivial, then precisely one of the $\gamma Z_n \gamma^{-1} \cap H$ is non-trivial and we get

$$(b-1)m+1 = a = (c-1)n+1 .$$

If two of the $\beta Z_m \beta^{-1} \cap H$ are non-trivial, then all the $\gamma Z_n \gamma^{-1} \cap H$ are trivial, and so

$$(b-2)m+2 = a = cn .$$

But the diagram is connected and has no loops, and so its Euler characteristic is 1. Since there are $a+b+c$ vertices and $2a$ edges, we have $b+c-a = 1$. These equations quickly give a contradiction.

We are now in a position to prove

THEOREM 5 (Chang [3]). *Let* H *be the group* $<a,b : a^m = b^n>$, m,n coprime, *and let* G *be a knot group containing* H *either as a normal subgroup or a subgroup of finite index. Then* $G = <A,B : A^m = B^n>$, *and* $a = A^h$, $b = B^h$, *where* h *is prime to* mn.

Proof. Since G is a knot group it follows from Theorem 3 that $ZG \neq 1$, and so by Burde and Zieschang [2] G has generators A, B such that $G = <A,B : A^k = B^\ell>$ with k, ℓ coprime.

Thus $G/ZG \cong Z_k * Z_\ell$, and clearly a non-cyclic subgroup of such a group has trivial center. Thus $ZH = ZG \cap H$, and so $HZG/ZG \cong Z_m * Z_n$.

From Lemma 4 we conclude $HZG = G$, giving H normal in G, and, by a suitable change of generators if necessary, $k = m$, $\ell = n$, and $H = < A^u, B^v : A^{um} = B^{vn}>$, where $(u,m) = (v,n) = 1$. But $A^{um} = B^{un} = B^{vn}$, and so $u = v$, and the result is proved.

§4. REMARKS:

1. Karras, Pietrowski and Solitar prove in [9] that every finitely generated (free-by-infinite cyclic) group with non-trivial center lies in $\mathcal{G}$.

2. If H is a finitely generated group with center in $\mathcal{G}$, then H is the split extension of a finitely generated free group by an infinite cyclic group. It follows that H is finitely presented.

3. There do exist non-abelian groups H with center in $\mathcal{G}$ with H/H' isomorphic to a non-cyclic subgroup of the rationals. The subgroup

generated by $\{y^{a}x^{2}y^{-a} : a$ a positive integer$\}$ in the one relator group $\langle x,y : yx^{2}y^{-1} = x^{4} \rangle$ is an easy example.

4. The subgroups of a group with center in $\mathcal{G}$ are either groups with center or free groups. It follows that all the finitely generated subgroups of such a group are finitely presented, or, groups with center in $\mathcal{G}$ are coherent.

Unfortunately not every group in $\mathcal{G}$ is coherent. For $F_2 \times F_2$, the direct product of two free groups of rank 2, is in $\mathcal{G}$ and it is well known that $F_2 \times F_2$ is not coherent.

It is known that knot groups are coherent (Scott [12]): it is still not known if one-relator groups are coherent. It would be interesting to be able to distinguish the coherent groups in $\mathcal{G}$.

5. It follows from the previous remarks that the groups H with non-trivial center in $\mathcal{G}$ are locally indicable, and hence the integral group ring of H has no zero divisors and only trivial units (see Higman [7]).

6. In defining the class $\mathcal{G}$, we restricted the "amalgamating and associated" subgroups L to be free. We could allow L to belong to any class of groups which (i) is closed under subgroups and (ii) for which the only groups with non-trivial center are infinite cyclic (for instance, choose L as a subgroup of a parafree group [1]), provided we restrict ourselves to groups of cohomological dimension two. We have not however come across any interesting new groups this way.

7. Finally, we would like to pose the question which provoked the results of this paper: namely is Theorem 2 true for groups of cohomological dimension two, or, put in a slightly different way, if G has cohomological dimension 2, and $ZG \neq 1$, is G' free?*

JOHN COSSEY
AUSTRALIAN NATIONAL UNIVERSITY

NEVILLE SMYTHE
AUSTRALIAN NATIONAL UNIVERSITY

---

* Added in proof: R. Bieri has verified this conjecture for a class of groups slightly wider than finitely presented groups.

REFERENCES

[1] Baumslag, Gilbert, Groups with the same lower central series as a relatively free group. II Properties. Trans. Amer. Math. Soc. 142 (1969), 507-538.

[2] Burde, G., and Zieschang, H., Eine Kennzeichnung der Torusknoten. Math. Ann. 167 (1966), 169-176.

[3] Chang, Bai Ching, Some theorems about knot groups. Indiana Univ. Math. J. 22 (1973), 807-812.

[4] Cohen, D. E., Subgroups of HNN groups. J. Austral. Math. Soc. 17 (1974), 394-405.

[5] Gruenberg, K. W., Cohomological Topics in Group Theory. Lecture Notes in Mathematics 143. Springer-Verlag. Berlin, Heidelberg, New York, 1970.

[6] Haken, W., Über das Homöomorphieproblem der 3-Mannigfaltigkeiten I. Math. Z. 80 (1962), 89-120.

[7] Higman, G., The units of group rings. Proc. London. Math. Soc. 46 (1940), 231-248.

[8] Karrass, A., Pietrowski, A., and Solitar, D., An improved subgroup theorem for HNN groups with some applications. Can. J. Math. 26 (1974), 214-224.

[9] ————————————, Finite and infinite cyclic extensions of free groups. J. Austral. Math. Soc. 16 (1973), 458-466.

[10] Kurosh, A. G., Theory of Groups, Volume 1. Chelsea (2nd edition) New York 1955.

[11] Magnus, W., Karrass, A., and Solitar, D., Combinatorial Group Theory. Pure and Applied Mathematics 13, Interscience. New York London Sydney 1966.

[12] Scott, G. P., Finitely generated 3-manifold groups are finitely presented. J. London Math. Soc. 6 (1973), 437-440.

[13] Smythe, N., The subgroup theorem for free products with amalgamation, HNN constructions and colimits. Proceedings of the Second International Conference on the Theory of Groups, Canberra 1973. Edited by M. F. Newman. Lecture Notes in Mathematics 372, Springer Verlag Berlin, Heidelberg, New York 1974.

[14] Waldhausen, F., Whitehead groups of generalized free products. Algebraic K-Theory II. Battelle Institute Conference 1972. Edited by H. Bass. Lecture Notes in Mathematics 342, Springer Verlag. Berlin, Heidelberg, New York 1973.

QUOTIENTS OF THE POWERS OF THE AUGMENTATION IDEAL IN A GROUP RING*

John R. Stallings

Introduction

Let Γ be a commutative ring with 1; G a group, $\varepsilon : \Gamma G \to \Gamma$ the augmentation of the group ring taking $G \to 1$; $J = \operatorname{Ker} \varepsilon$, the augmentation ideal. This paper shows how to compute the quotient groups J^n/J^{n+1} (as well as the multiplicative structure of the graded ring consisting of these quotient groups). This is done in terms of a spectral sequence whose boundary maps are homology operations on groups, with certain functorial properties. We can obtain the spectral sequence either abstractly, in terms of the cobar construction on $K(G, 1)$, or practically (losing a certain amount of homological data) in terms of a presentation of G. We give an application to a group-theoretic problem. In particular, following a suggestion of R. Lyndon, we give an example of a group with n generators and one relation, which cannot be mapped homomorphically onto a free group of rank 2.

Some of this paper was suggested by Brian K. Schmidt's thesis [8], which computed the additive structure of $\Gamma G/J^n$ in terms of a presentation of G.

§1. *The basic idea*

Let $P = \{x : r\}$ be a presentation of G. Let $F = F\{x\}$ be the free group with basis $\{x\}$. Let I be the augmentation ideal in ΓF. For

* This work was done under the auspices of the Miller Institute for Basic Research, the National Science Foundation, and the University of California, Berkeley.

$w \in F$, let $\tilde{w} = 1-w$. The graded ring associated to the filtration of ΓF by powers of I is well-known to be the Γ-tensor algebra on the free Γ-module X generated by the elements $\{\tilde{x}\}$. Let R be the free Γ-module generated by $\{\tilde{r}\}$, which will here be confused with the Γ-sub-module of ΓF generated by $\{\tilde{r}\}$. Then ΓG is the quotient of ΓF by the 2-sided ideal generated by R. Recall that J is the augmentation ideal in ΓG. Then:

$$J/J^2 \approx I/(I^2+R) \; ,$$

which is the cokernel of a certain map $d_0 : R \to I/I^2 \approx X$. This defines additionally a map

$$R \otimes X + X \otimes R \to X \otimes X \approx I^2/I^3 \; .$$

The cokernel of this is

$$I^2/(I^3 + RI + IR) \; .$$

To obtain J^2/J^3 we have to factor the above by something more, namely $R \cap I^2$, which comes from the kernel K of d_0. There is then a map

$$d_1 : K \to \mathrm{Coker}\,(R \otimes X + X \otimes R \to X \otimes X) \; ,$$

whose cokernel is isomorphic to J^2/J^3.

The description of J^3/J^4 is even more complex: First factor $X \otimes X \otimes X$ by the image of $R \otimes X \otimes X + X \otimes R \otimes X + X \otimes X \otimes R$. Then factor this by the image of the kernel of $R \otimes X + X \otimes R \to X \otimes X$. Then factor this by the image of the kernel of $K \to \mathrm{Coker}\,(R \otimes X + X \otimes R \to X \otimes X)$. In the end, J^3/J^4 will be isomorphic to something like this:

$$I^3/(I^4 + (RI^2 + IRI + I^2R) + (RI + IR) \cap I^3 + R \cap (I^3 + RI + IR) \cap I^3) \; .$$

Etc. We shall arrange these successively more complex computations into a spectral sequence.

§2. *Some notational conventions*

$P = \{x : r\}$ will be a presentation of G; Γ a commutative ring with 1; ΓG the group ring; J its augmentation ideal. Modules, tensor products, etc., will be construed to be over Γ. Π means "product," i.e., infinite direct product, while $\amalg$ means "coproduct" or infinite direct sum. Also, $A+B$ usually means direct sum.

X will denote the free module with basis $\{\tilde{x}\}$ in 1-1 correspondence with the generators in P. R will denote the free module with basis $\{\tilde{r}\}$ in 1-1 correspondence with the relators in P.

If M is a module, $M^0 = \Gamma$, $m^{n+1} = M \otimes M^n$, and $T(M) = \coprod_{n \geq 0} M^n$ is the tensor algebra of M, whose multiplication is given by the canonical isomorphism $M^{k+\ell} \approx M^k \otimes M^\ell$. $\hat{T}(M) = \prod_{n \geq 0} M^n$ is the completed tensor algebra, which is the topological algebra obtained from $T(M)$ by completion with respect to the descending sequence of two-sided ideals

$$Z_k = \coprod_{n \geq k} M^n .$$

A basic fact, due to Magnus [5, 4] is that if ΓF is the group ring of the free group on the generators of P, then its completion with respect to the powers I^n of its augmentation ideal is isomorphic to $\hat{T}(X)$, the isomorphism being given by $x \leftrightarrow 1 - \tilde{x}$.

§3. *The spectral sequence from a presentation*

Let A, or $A(P)$, be the completed tensor algebra $\hat{T}(X+R)$.

Define $K_1 = X$, $K_2 = R$, $K_n = 0$ otherwise. Then A is the product of terms called:

$$A^p{}_q = \coprod_{i_1 + \cdots + i_p = q} K_{i_1} \otimes \cdots \otimes K_{i_p} .$$

Here, $A^0{}_0 = \Gamma$. Note that the product of $A^p{}_q$ and $A^r{}_s$ is contained in $A^{p+r}{}_{q+s}$.

We define $A_{-p,q} = A^p{}_q$

$$A_{-p,*} = \coprod_q A_{-p,q} = (K_1 + K_2)^p .$$

Let B denote the uncompleted tensor algebra $T(X+R)$, which is the coproduct of the $A^p{}_q$.

An element $\alpha \in A^p{}_q$ will be said to have *total degree* $n(\alpha) = -p+q$. An element in B has total degree n if it is a sum of finitely many elements of total degree n. An element of A has total degree n if it is in the closure of the elements of B with total degree n; 0 has any total degree. Note that the elements of A of total degree 0 form a sub-algebra isomorphic to $\hat{T}(X)$.

A homomorphism of Γ-modules, $\phi : B \to A$ is said to be a *derivation*, if it is zero on $A_{00} = \Gamma$, and if:

$$\phi(\alpha \otimes \beta) = \phi(\alpha) \otimes \beta + (-1)^{n(\alpha)} \alpha \otimes \phi(\beta) .$$

We define $\partial : A \to A$ as follows:

(1) $\partial(X) = 0$.

(2) $\partial | R$ is defined on the basis $\{\tilde{r}\}$ thus: $\tilde{r} \to 1 - r \in \Gamma F \to A$, by completion. In other words, $\partial \tilde{r}$ is the infinite series, in terms of the Fox derivatives $D_i = \partial/\partial x_i$, and augmentation $\varepsilon : \Gamma F \to \Gamma$.

$$\partial \tilde{r} = \sum (-1)^n (\varepsilon D_{i_1} \cdots D_{i_n}(r)) \tilde{x}_{i_1} \otimes \cdots \otimes \tilde{x}_{i_n} .$$

Cf. [4].

(3) $\partial | B$ is the unique derivation extending the map already defined.

(4) ∂ is the unique continuous extension of this to A.

Now, we note that $\partial^2 = 0$, since ∂ is a derivation on B and $\partial^2 = 0$ on $X + R$ which generates B.

If we define $\Phi^r = \prod_{p \geq r} A_{-p,*}$, then Φ is a decreasing filtration on A, and $\partial \Phi^r \subset \Phi^r$.

Also note that ∂ lowers total degree by 1.

Now, what we have, consisting of A, ∂, Φ, is a filtered chain-complex. It leads therefore to a spectral sequence $E^r{}_{-p,q}$, where:

$$E^r \text{ is an algebra .}$$

$$E^0{}_{-p,q} \approx A_{-p,q}, \text{ and } E^0 \approx B \text{ as an algebra.}$$

The boundary maps:

$$d_r : E^r{}_{-p,q} \to E^r{}_{-p-r,q+r-1}$$

are themselves derivations induced by ∂, and E^{r+1} is the homology of (E^r, d_r).

Furthermore, the shape of the possibly non-zero terms in $E^r{}_{-p,q}$ is of interest: Except for the term $E^r{}_{0,0} = \Gamma$, we have $E^r{}_{-p,q} = 0$ unless $q \geq p$ and $p \geq 1$. We say that such a spectral sequence is an *upper upper left octant spectral sequence*.

3.1. $$E^p{}_{-p,p} \approx E^{p+1}{}_{-p,p} \approx \cdots \approx E^{\infty}{}_{-p,p} .$$

This is true for any upper upper left octant spectral sequence. We refer to this as *convergence on the anti-diagonal*. In general, there will be no convergence phenomenon elsewhere.

The presentation $\{x : r\}$ determines a 2-dimensional cell-complex K. There is a single 0-cell e_0, a 1-cell for each generator x, and a 2-cell for each relator r; the attaching map of the boundary of the 2-cell is determined by the recipe that r gives as a word in the generators. The smash product $\#^n(K, e_0)$ is the n-fold Cartesian product K^n, modulo the subspace consisting of all points having at least one coordinate equal to e_0.

3.2. $$E^1{}_{-p,q} \approx H_q(\#^p(K, e_0), \Gamma) .$$

This can be proved by noting that $E^0{}_{-p,*}$ is just the chain complex of $\#^p(K, e_0)$, because d_0 is a derivation and coincides with the boundary map of the chain-complex of (K, e_0) on $E^0{}_{-1,*}$.

§4. *Functorality on presentations*

Let $P = \{x : r\}$ and $Q = \{y : s\}$ be two presentations. We shall define a notion of map from P to Q that will induce a homomorphism of spectral sequences $E(P)$ to $E(Q)$.

If $\{u\}$ is a set then $F\{u\}$ will denote the free group with that set as basis. A map $f : P \to Q$ consists of two group homomorphisms:

$$\phi_1 : F\{x\} \to F\{y\}$$

$$\phi_2 : F\{x, r\} \to F\{y, s\}$$

satisfying these rules: Define $d_P : F\{x, r\} \to F\{x\}$, by $x \to x$ and r, as basis element of $F\{x, r\}$, $\to r$ as a word in $\{x\}$. Define d_Q similarly. Then we must have:

$$\phi_1 d_P = d_Q \phi_2 \quad \text{and} \quad \phi_1 | \{x\} = \phi_2 | \{x\}$$

$\phi_2(r)$ is a product of terms of the form $ws^{\varepsilon}w^{-1}$

where $w \in F\{y\}$ and $\varepsilon = +1$ or -1.

ϕ_1 defines, by extension to the completed group rings, a homomorphism $f_1 : X \to \hat{T}(Y)$, which is the 0-degree part of $A(Q)$.

ϕ_2 gives one a way to define a map $f_2 : R \to$ the 1-degree part of $A(Q)$, as follows: If

$$\phi_2(r) = \prod_{i=1}^{n} w_i s_i^{\varepsilon_i} w_i^{-1}$$

define

$$\theta_i = (\varepsilon_i - 1)/2 \quad \text{and} \quad \Pi_j = \prod_{i<j} w_i s_i^{\varepsilon_i} w_i^{-1} .$$

Then

$$1 - \phi_2(r) = \sum_{j=1}^{n} \varepsilon_j \Pi_j w_j s_j^{\theta_j} (1 - s_j) w_j^{-1} .$$

So, let $u_j = d_Q(\varepsilon_j \Pi_j w_j s_j^{\theta_j})$ and $v_j = w_j^{-1}$, both being taken as being elements of $\hat{T}(Y)$. Then we define:

$$f_2(\tilde{r}) = \sum_{j=1}^{n} u_j \tilde{s}_j v_j$$

The fact that $\phi_1 d_P = d_Q \phi_2$ now implies that $f_1 + f_2 : X + R \to \hat{T}(Y + S) = A(Q)$ commutes with ∂ in $A(P)$ and $A(Q)$. Extend $f_1 + f_2$ to a continuous algebra homomorphism $f_{\#} : A(P) \to A(Q)$. Since the two boundary maps in $A(P)$ and $A(Q)$ are derivations which commute with $f_{\#}$ on generators of $A(P)$, it follows that $f_{\#}$ is a chain-mapping.

So in the end we have $f_{\#}: A(P) \to A(Q)$, commuting with ∂, preserving total degree and filtration. Thus, $f_{\#}$ defines a map of spectral sequences $f_*: E(P) \to E(Q)$.

4.1. With these definitions, the spectral sequence is a functor from the category of presentations and presentation maps to the category of upper upper left octant spectral sequences.

To show that the map of spectral sequences induced by the composition of two presentation maps is the composition of the two induced maps of spectral sequences is an exercise in keeping your head while the indices proliferate and will be omitted.

4.2. If the map $f: P \to Q$ induces a homology isomorphism of the two-dimensional complexes of the presentations, then

$$f_*: E^r(P) \to E^r(Q)$$

is an isomorphism for $r \geq 1$.

This follows from 4.1 and 3.2.

In particular, the complex of P can always be subdivided into a semi-simplicial two-dimensional complex, giving a presentation Q in which all relators are of the form

$$r: x_{i_1} x_{i_2} x_{i_3}^{-1} = 1 .$$

In such a complex:

4.3. $$\partial \tilde{r} = (-\tilde{x}_{i_1} \otimes \tilde{x}_{i_2} + \tilde{x}_{i_1} + \tilde{x}_{i_2} - \tilde{x}_{i_3}) \otimes (1 - \tilde{x}_{i_3})^{-1} .$$

§5. *Cobar construction* [1] [9]

A *coalgebra* A (over the commutative ring Γ) is:

(1) A graded module with $A_n = 0$ for $n < 0$ and $A_0 = \Gamma$. By $\bar{A}$ we mean $A_1 + A_2 + \cdots$. By $\varepsilon: A \to \Gamma$ is meant $A \to A/\bar{A}$.

(2) A chain-complex, with $\partial: A_n \to A_{n-1}$ such that $\partial^2 = 0$ and $\partial A_1 = 0$.

(3) With a diagonal map $\Delta : A \to A \otimes A$, which is a chain-mapping (relative to the usual ∂ on $A \otimes A$, such that $\partial(\alpha \otimes \beta) = (\partial\alpha) \otimes \beta + (-1)^{\dim \alpha} \alpha \otimes \partial\beta)$, that is strictly associative:

$$(\Delta \otimes 1_A) \circ \Delta = (1_A \otimes \Delta) \circ \Delta ,$$

and for which ε is a 2-sided co-unit:

$$(\varepsilon \otimes 1_A) \circ \Delta = (1_A \otimes \varepsilon) \circ \Delta = 1_A .$$

These equations have to be interpreted in terms of the standard identifications of various tensor products, where 1_A is the identity map on A.

The example of a coalgebra one needs to have in mind is the chain-complex of a semi-simplicial complex K which has only one vertex, and which is provided with the Alexander-Čech diagonal:

$$\Delta(\sigma_{0\cdots n}) = \sum_{i=0}^{n} \sigma_{0\cdots i} \otimes \sigma_{i\cdots n} .$$

Here $\sigma_{j\cdots k}$ is the semisimplicial analogue of that face of the convex simplex with vertices $(x_0, \cdots, x_n)$ which has the vertices $(x_j, \cdots, x_k)$.

We now construct the tensor algebra $T = T(\overline{A})$, giving it the bigradation where $T^p{}_q$ is the coproduct of the terms

$$A_{i_1} \otimes \cdots \otimes A_{i_p}$$

over all positive p-tuples $i_1, \cdots, i_p$ such that $i_1 + \cdots + i_p = q$. As in the earlier construction, $T_{-p,q} = T^p{}_q$ is distributed over the upper upper left octant.

The homomorphism $\partial : A \to A$ restricts to $\partial : \overline{A} \to \overline{A}$. We extend this to a derivation $\partial : T(\overline{A}) \to T(\overline{A})$. As before, a derivation is a homomorphism $D : T(\overline{A}) \to T(\overline{A})$, such that

$$D(\alpha \otimes \beta) = D(\alpha) \otimes \beta + (-1)^{n(\alpha)} \alpha \otimes D(\beta)$$

where $\alpha \in T^p{}_q$ and the total degree $n(\alpha) = -p + q$.

The resultant $\partial : T \to T$ maps $T^p{}_q \to T^p{}_{q-1}$ and satisfies $\partial^2 = 0$. It differs from the "usual ∂" on $\overline{A}^p$ by sign differences; nevertheless, the homology groups of these two boundary maps are isomorphic.

Let $e : \overline{A} \otimes \overline{A} \to \overline{A} \otimes \overline{A}$ be defined by

$$e(\alpha \otimes \beta) = (-1)^q \alpha \otimes \beta, \text{ where } \alpha \in A_q .$$

The diagonal map Δ determines a diagonal map $\overline{A} \to \overline{A} \otimes \overline{A}$, and we define Δ_1 to be the composition

$$\overline{A} \to \overline{A} \otimes \overline{A} \xrightarrow{e} \overline{A} \otimes \overline{A} .$$

Then Δ_1 maps $\overline{A} \to T(\overline{A})$ and so extends to a unique derivation $\Delta_1 : T(\overline{A}) \to T(\overline{A})$. This derivation maps $T^p{}_q$ to $T^{p+1}{}_q$.

The assumption that Δ is a chain-mapping implies now that $\Delta_1 \partial + \partial \Delta_1 = 0$ on elements of $T^1{}_*$, and thence by induction on p, using the fact that Δ_1 and ∂ are derivations, it can be proved in general.

The assumption that Δ is associative now implies that $\Delta_1{}^2 = 0$, first on $T^1{}_*$, and then by induction on p, using the fact that Δ_1 is a derivation, on $T^p{}_*$.

In summary:

$$\partial : T^p{}_q \to T^p{}_{q-1} \quad \text{and} \quad \Delta_1 : T^p{}_q \to T^{p+1}{}_q$$

are derivations.

$$\partial^2 = 0, \ \Delta_1{}^2 = 0, \ \Delta_1 \partial + \partial \Delta_1 = 0 .$$

Therefore, defining $D = \partial + \Delta_1$ we see that $D^2 = 0$, D is a derivation, and D lowers total degree by 1.

We are now in the standard situation of a double complex; so there is a spectral sequence using p as the filtration degree.

E^r is a graded algebra with a derivation

$$d_r : E^r{}_{-p,q} \to E^r{}_{-p-r,q+r-1}$$

induced somehow from D, with respect to which, the homology of E^r is E^{r+1}.

$$E^0_{-p,q} \approx T^p{}_q(\bar{A}), \text{ on which } d_0 = \partial .$$

$$E^1_{-p,q} \approx H_q(\bar{A}^p)$$

E is an upper upper left octant spectral sequence, and so there is convergence on the anti-diagonal.

E is obviously a functor of coalgebras A and appropriately defined coalgebra maps.

§5.1. Suppose A and B are coalgebras and $\phi : A \to B$ is a map of coalgebras such that $\phi_* : H_1(A) \to H_1(B)$ is an isomorphism and $\phi_* : H_2(A) \to H_2(B)$ is onto. Then in the induced map of spectral sequences $\phi_\# : E(A) \to E(B)$, for $r \geq 1$;

$$E^r_{-p,p} \to E^r_{-p,p} \text{ is an isomorphism, and}$$

$$E^r_{-p,p+1} \to E^r_{-p,p+1} \text{ is onto.}$$

I.e., $\phi_\#$ is isomorphic on terms of total degree 0 and epimorphic on terms of total degree 1.

The proof is by induction on r, using the hypothesis for $r = 1$ and the fact that the terms of total degrees 0 and 1 in E^1 can be described (by the Kunneth formula) in terms of $H_1(A)$ and $H_2(A)$. The induction step is an ordinary diagram-chasing argument using the 5-lemma.

5.2. COROLLARY. *If ϕ_* is isomorphic on H_1 and epimorphic on H_2, then $E^\infty_{-p,p}(A) \approx E^\infty_{-p,p}(B)$.*

5.3. COROLLARY. *If A is the coalgebra of a semisimplicial complex K with one vertex, and $\pi_1(K) = G$, and B is the coalgebra of the semisimplicial Eilenberg-MacLane space $K(G, 1)$, then $E^\infty_{-p,p}(A) \approx E^\infty_{-p,p}(B)$.*

This is because there is a map satisfying 5.2 from K to $K(G, 1)$.

Now consider the Eilenberg-MacLane space $K(G,1)$. In a sense, the coalgebra of $K(G,1)$ is the bar-construction of ΓG. Our spectral sequence is the cobar construction of this. The result, if there is any justice, should be equivalent to ΓG. This is a theorem of E. Brown [2].

The key point to note is that on a 2-simplex $(g|h)$, a basis element of ${A^1}_2$, the total boundary D is

$$D(g|h) = -(g)\otimes(h) + (g) + (h) - (gh) .$$

The first term here is derived from the diagonal map, the rest from the boundary map in the semisimplicial complex $K(G,1)$. It follows that the map

$$(g_1)\otimes\cdots\otimes(g_n) \to (1-g_1)\cdots(1-g_n)$$

from the terms of total degree 0 in the cobar construction to the ring ΓG exactly annihilates the image of D. Thus the total 0-dimensional homology of the cobar construction is ΓG, and the filtration on it is that of the powers of the augmentation ideal. It follows that the ${E^\infty}_{-p,p}$ term is then J^p/J^{p+1}. Combined with 5.3, this shows:

5.4. THEOREM. *If* A *is the coalgebra with coefficients* Γ *of a one-vertex semisimplicial complex* K, *and* J *is the kernel* $\Gamma G \to \Gamma$, *where* $G = \pi_1(K)$, *then*

$${E^\infty}_{-p,p}(A) \approx J^p/J^{p+1} .$$

§6. *Comparison of the two spectral sequences*

Given a presentation P, we subdivide it to obtain a homologically equivalent presentation Q whose complex K can be considered to be semi-simplicial. By 4.2, we have $E^r(P) \approx E^r(Q)$ for $r \geq 1$. Comparing the boundary formula 4.3 of $E^r(Q)$ with the boundary formula in the cobar spectral sequence of K, we see there is an isomorphism taking the element $\tilde{r}$ of $E(Q)$ to $\sigma\otimes(1-T)^{-1}$ in $\hat{E}(K)$, where $r = xyz^{-1}$, σ is the 2-simplex corresponding to r and T is the 1-simplex corresponding to z, and $\hat{E}(K)$ is the completion of $E(K)$ in the p-filtration. The composed map defines an isomorphism $E^r(P) \approx E^r(K)$ for $r \geq 1$.

Hence, by 5.4:

6.1. THEOREM. *Let* $E(P)$ *be the spectral sequence defined in* §4 *for the presentation* P *of a group* G. *Then, where* J *is the augmentation ideal in the group ring* ΓG,

$$E^p_{-p,p}(P) \approx J^p/J^{p+1} .$$

§7. *A group-theoretic application*

Suppose G is a group with a presentation with just one relator, $P = \{x_1, \cdots, x_n : r = 1\}$. We want to investigate the question: Onto which free groups can G be mapped?

A free group F of rank m has a presentation $Q = \{y_1, \cdots, y_m\}$ with no relators. If $h : G \to F$ is a homomorphism onto, there are maps of presentations

$$\alpha : P \to Q, \quad \beta : Q \to P$$

such that $\alpha\beta : Q \to Q$ is the identity map, in other words α and h are retractions.

Now, with a given coefficient ring Γ, we can write, when $1 - r \in I^k \setminus I^{k+1}$,

$$1 - r = \sum a_{i_1 \cdots i_k}(1 - x_{i_1}) \cdots (1 - x_{i_k})$$

modulo I^{k+1}. Thus, in the spectral sequence $E(P)$, the derivations d_s are all zero for $s < k - 1$, and

$$(**) \qquad d_{k-1}(\tilde{r}) = \sum a_{i_1 \cdots i_k} \tilde{x}_{i_1} \otimes \cdots \otimes \tilde{x}_{i_k} = \eta(\tilde{x}_1, \cdots, \tilde{x}_n) .$$

We think of η, which is a homogeneous form of degree k in non-commuting variables, as a sort of homology operation.

Since $\alpha : P \to Q$ is forced to be trivial on $\tilde{r}$, and the spectral sequence is functorial, it follows that

$$\eta(\alpha\tilde{x}_1, \cdots, \alpha\tilde{x}_n) = 0 .$$

Now suppose $\alpha\tilde{x}_i = \Sigma b_{ij}\tilde{y}_j$. The fact that α is a retraction implies, if Γ is an integral domain, that the $n \times m$ matrix $(b_{ij}) = B$ has rank m. Furthermore, we could follow α by any automorphism α, $Q \to Q$, and the same would be true of $\gamma\alpha$ as is true of α. The result is to multiply B on the right by an arbitrary invertible $m \times m$ integer matrix. Thus, if the ring Γ is one of the prime fields Z_p we can fix it up so that B is in column echelon form.

7.1. For $\Gamma = Z_p$ a necessary condition that G can be mapped onto a free group of rank m is that for the form η defined by (**), there is a matrix $B = (b_{ij})$ over Γ of size $n \times m$ and rank m, in column echelon form, such that

$$\eta\left(\sum b_{ij}\tilde{y}_j, \cdots, \sum b_{nj}\tilde{y}_j\right) = 0 .$$

I.e.,

$$\sum a_{i_1 \cdots i_k} b_{i_1 j_1} \cdots b_{i_k j_k} \tilde{y}_{j_1} \otimes \cdots \otimes \tilde{y}_{j_k} = 0$$

I.e., for every k-tuple of integers $j_1, \cdots, j_k \in [1, m]$,

$$\sum_{\substack{i_1, \cdots i_k \\ \in [1,n]}} a_{i_1 \cdots i_k} b_{i_1 j_1} \cdots b_{i_k j_k} = 0 .$$

7.2. Suppose the group G has the presentation

$$P = \{x_1, \cdots, x_n : x_1{}^e \cdots x_n{}^e = 1\}, \text{ where } e > 1 .$$

Then if G can be homomorphically mapped onto a free group of rank m, it follows that $m \leq n/2$.

Proof. Select a prime p dividing e. Therefore $e = qf$ where $q = p^k$ and $f \not\equiv 0$ mod p. Then with coefficient ring Z_p, where $\tilde{x} = 1 - x$,

$$\begin{aligned} 1 - r &= 1 - (1-\tilde{x}_1)^{qf} \cdots (1-\tilde{x}_n)^{qf} = 1 - (1-\tilde{x}_1{}^q)^f \cdots (1-\tilde{x}_n{}^q)^f \\ &= f(\tilde{x}_1{}^q + \cdots + \tilde{x}_n{}^q) \text{ modulo } I^{q+1} . \end{aligned}$$

So, in the form η all the coefficients are zero except for those whose indices are all the same, in which case the coefficient is f.

Now if $(b_{ij}) = B$ is an $n \times m$ matrix of rank m in column echelon form and we take the k-tuple $j_1, \cdots, j_k = s, s, \cdots, s, t$ in 7.1, we find that for all $s, t \in [1, m]$,

$$f \sum_{i=1}^{n} b_{is}{}^{q-1}\, b_{it} = 0 \,. \tag{\#}$$

The matrix $C = (c_{si})$ with $c_{si} = b_{is}{}^{q-1}$ is obtained from B by transposing and setting every non-zero entry to 1. Since B is in echelon form, C has the same rank, namely m.

Since $f \neq 0$ in Z_p, the equation (#) says:

$$C \cdot B = 0 \,.$$

Thus the row-space of C is contained in the null space of B. The dimensions of these are m and $n-m$, respectively, and so $m \leq n-m$, or $m \leq n/2$, as asserted.

Note that, conversely, by mapping the odd generators to themselves, the even generators to the inverse of the preceding one, but the last one to 1 if n is odd, we can retract the above group onto a free subgroup of rank $(n-1)/2$ or $n/2$, whichever is an integer.

Another interesting example is this (a similar thing was pointed out to me by R. Lyndon in a letter about 15 years ago): Let $[u, v] = uvu^{-1}v^{-1}$. Given an integer n, for $1 \leq i < j \leq n$, let $\{a_{ij}\}$ be $n(n-1)/2$ pairwise distinct powers of 2. Let c be a power of 2 at least as large as each a_{ij}. Define $e_{ij} = c/a_{ij}$.

Define an expression

$$r(x_1, \cdots, x_n) = \prod_{i<j} [x_i{}^{a_{ij}}, x_j{}^{a_{ij}}]^{e_{ij}} \,. \tag{!}$$

The product has to be taken in some fixed order, which is not going to make any difference for our purposes.

Define a group G now by the presentation:

$$P = \{x_1, \cdots, x_n : r(x_1, \cdots, x_n) = 1\} .$$

7.3. The above-defined group with one relation cannot be mapped onto a free group of rank 2.

Proof. Take $\Gamma = Z_2$. In the spectral sequence $E(P)$ all boundary maps are zero until d_{2c-1} is reached, and then

$$d_{2c-1}(\tilde{r}) = \eta(\tilde{x}_1, \cdots, \tilde{x}_n) = \sum_{i<j} (\tilde{x}_i^{a_{ij}} \tilde{x}_j^{a_{ij}} + \tilde{x}_j^{a_{ij}} \tilde{x}_i^{a_{ij}})^{e_{ij}} .$$

If there were a retraction $G \to F\{u,v\}$, we would have an $m \times 2$ matrix of rank 2 (b_{ij}) with

$$(*) \qquad \eta(b_{11}\tilde{u} + b_{12}\tilde{v}, \cdots, b_{n1}\tilde{u} + b_{n2}\tilde{v}) = 0 .$$

We can analyze each summand of this expression somewhat. Suppose $p, q, r, s \in Z_2$ and a is a power of 2. Look at

$$(p\tilde{u} + q\tilde{v})^a (r\tilde{u} + s\tilde{v})^a + (r\tilde{u} + s\tilde{v})^a (p\tilde{u} + q\tilde{v})^a .$$

We can say the following:

(1) The coefficient of $\tilde{u}^{2a}$ is $p^a r^a + r^a p^a = 0$ and similarly the coefficient of $\tilde{v}^{2a}$ is 0.

(2) The coefficient of $\tilde{u}^a \tilde{v}^a$ is $p^a s^a + r^a q^a = ps + rq =$ Determinant of $(p,q; r,s)$.

(3) If k divides a and $t = a/k > 1$, and $f = t/2$, (of course, k, t, f are powers of 2), then the coefficient of $(\tilde{u}^k \tilde{v}^k)^t$ is:

$$(p^k q^k)^f (r^k s^k)^f + (r^k s^k)^f (p^k q^k)^f = 0 .$$

Therefore, in the expression (*) above, the coefficient of the term

$$(\tilde{u}^{a_{ij}} \tilde{v}^{a_{ij}})^{e_{ij}}$$

comes solely from the ij-th summand of η, and is therefore, by (2) above,

$$b_{i1} b_{j2} + b_{j1} b_{i2} .$$

This must be zero for every pair i, j with $i < j$. Thus, every 2×2 subdeterminant of B is zero, and so B has rank at most one.

7.3 can be rephrased in the following rather curious form:

7.4. If F is a free group, and $x_1, \cdots, x_n \in F$ then $\{x_1, \cdots, x_n\}$ generates a cyclic subgroup if and only if (where r is given by (!) above)

$$r(x_1, \cdots, x_n) = 1 .$$

Proof. Sufficiency is 7.3, since every subgroup of F is free. Necessity follows from the fact that r would be in the commutator subgroup of the cyclic subgroup.

Another remark. If one makes an expression r' by replacing the commutators in r by the expression:

$$\langle u, v \rangle = (uv)^2 u^2 v^2$$

then the resulting form in the one-relator group with this relation $r' = 1$, is identical with that for r, and so the result 7.3 holds for this group also. Every generator of the group occurs in r' with only positive exponents.

§8. *Related questions*

The examples in §7 are simple in the sense that they could have been described without spectral sequences. This is probably true of any spectral sequence argument that only depends on facts about the first non-vanishing d_r. One unsolved problem is therefore to derive interesting group-theoretic results using the machinery derived here in a more essential way.

For another thing, it is my impression that Milnor's isotopy invariants [6] can be described in terms of the spectral sequence of the fundamental group of the complement of a link. But this is not perfectly clear.

Finally, Rips [7] has shown that there is a difference between the "dimension subgroups" and the terms of the lower central series. This paper has discussed powers of the augmentation ideal, which are related to the dimension subgroups. It would be worthwhile to describe some computable spectral sequence which would compute the quotient groups of the lower central series. There should be a map of that spectral sequence into the one defined here, and then homological algebra should be developed sufficiently for one to perceive clearly why it is that when torsion appears in the quotients of the lower central series it is possible for the lower central series to differ from the dimension series. My idea for this is to substitute the Curtis spectral sequence [3] for the cobar construction; this would probably converge on the anti-diagonal to the quotients of the lower central series. The problem would be, how to compute with the Curtis spectral sequence, at least to the point of going through the Rips example in detail? And how to describe a similar spectral sequence in terms of generators and relations?

UNIVERSITY OF CALIFORNIA, BERKELEY

BIBLIOGRAPHY

[1] Adams, J. F., and Hilton, P. J., On the chain algebra of a loop space. Comm. Math. Helv. 30 (1956), 305-330.

[2] Brown, E. H., Jr., Twisted tensor products I. Ann. of Math. 69 (1959), 223-246.

[3] Curtis, E., Lower central series of semi-simplicial complex. Topology 2 (1963), 159-171.

[4] Fox, R. H., Free differential calculus I. Derivation in the free group ring. Ann. of Math. 57 (1953), 547-560.

[5] Magnus, W., Beziehungen zwischen Gruppen und Idealen in einem speziellen Ring. Math. Ann. 111 (1935), 259-280.

[6] Milnor, J., Isotopy of links, Algebraic geometry and topology. Princeton University Press, 280-306.

[7] Rips, E., On the fourth integer dimension subgroup. Israel J. of Math. 12 (1972), 342-346.

[8] Schmidt, B. K., Mappings of degree n from groups to abelian groups. Thesis, Princeton University (1972).

[9] Chen, K.-T., Iterated integrals of differential forms and loop space homology. Ann. of Math. 97 (1973), 217-246.

KNOT-LIKE GROUPS

Elvira Rapaport Strasser

Abstract

If K is a knot, the fundamental group of its complement in the three-sphere is called a knot group. Every knot group having finite presentations, has one consisting of one more generator than the number of defining relations; that is, a presentation of deficiency 1. Abelianized, the knot group is free cyclic. I call a finitely presented group, G, knot-like if it has these two properties.

If G is knot-like, G' its commutator subgroup, and G'/G'' is finitely generated, then G'/G'' is free (Abelian) and its rank is equal to the degree, d, of the Alexander polynomial of G [8]. If G is actually a knot group, and G' is finitely generated, then G' is free of rank d; the proof [7] is topological.

Let P be a presentation of deficiency 1 of the knot-like group G; then P gives rise to a presentation of G' as a product of groups H_i, $i: 0, \pm 1, \cdots$, and a certain presentation of H_0. Let M be the deficiency of the presentation of H_0 so gotten, and suppose P such that M is least possible. The main result of the present paper is that any two of the following conditions imply the third: 1. G' is finitely generated; 2. G' is free; 3. $M = d$. For one-relator presentations, either of the first two conditions implies the rest.

While d is independent of the presentation of a group, M is not: $d \leq M$ and G may have presentations P_1 and P_2 of deficiency 1 such that $M_1 > d$, $M_2 = d$.

§1. *Introduction*

Let $P^* = (x, a_1, \cdots, a_n; r_1, \cdots, r_n)$ present the group G for which the factor commutator subgroup, G/G', is cyclic. Then I will say that G is knot-like. It is no loss of generality to assume that the a-symbols are elements of the commutator subgroup, G'; that r_i has the form $a_i C_i$ for some elements C_i of G' [4]; and that G/G' is generated by the coset containing x. Then the set of all conjugates, $\bar{x}^j a_i s^j$, $i: 1, \cdots, n$, $j: 0, \pm 1, \cdots$, generates G', and the r_i can be expressed as words in these conjugates.

Set $P^* = (x, a; r)$, so that a is an n-tuple of symbols a_i, and r an n-tuple of words in x and a. Set $\bar{x}^j a_i x^j = a_{ij}$; let R_0 be the rewrite of r as an n-tuple of a_{ij}-words, and R_k the rewrite of the n-tuple $\bar{x}^k r x^k$. Then

$$P = (a_{ij}; R_j, i: 1, \cdots, n, j: 0, \pm 1, \cdots)$$

is a presentation of G'; it makes sense to speak of a_{ij} as an element of G'.

All terms used in the sequel without definition may be found in [5].

Take all defining relators reduced and cyclically reduced, and suppose that R_0 contains a_{i,m_i} but not $a_{i,k}$ if $k < m_i$. Suppose further that regardless of first subscripts, t is the smallest second subscript occurring in the rewrite of r_1; then the rewrite of $\bar{x}^t r_1 x^t$ contains $a_{i,0}$ for some i but no negative subscripts. Replacing r_1 by $\bar{x}^t r_1 x^t$ and proceeding similarly with the rest of r makes m_i non-negative for each i. If now $m_1 \neq 0$, then replacing a_1 by $x^{m_1} a_1 x^{-m_1}$ everywhere in r leads to a rewrite containing $a_{1,0}$ so that $m_1 = 0$. Similarly for the remaining m_i. If M is the sum of the numbers M_i, $i: 1, \cdots, n$, then the "spread" of the a_{ij} in R_0 is M.

If the a_{ij} are allowed to commute – that is, if the second commutator subgroup, G'', of G is factored out – then R_0 will consist of words of the form

$$w = (a_{10})^{b_{10}}(a_{11})^{b_{11}} \cdots a_{1,M_1}^{b}(a_{20})^{b_{20}} \cdots$$

for certain integers b_{ij}, b, etc. For a moment write $\bar{x}^j a_i x^j$ as $(a_i)^{x^j}$ in P^*. Then w becomes

$$(a_1)^{b_{10}}(a_1)^{b_{11}x}\cdots(a_1)^{bx^{M_1}}(a_2)^{b_{20}}\cdots$$

which can be written as

$$(a_1)^{b_{10}+b_{11}x+\cdots+bx^{M_1}}(a_2)^{b_{20}}\cdots .$$

Then, modulo G'',

$$r_1 = (a_1)^{P_{11}(x)}(a_2)^{P_{12}(x)}\cdots(a_n)^{P_{1n}(x)}$$

$$r_i = (a_1)^{P_{i1}(x)}\cdots(a_n)^{P_{in}(x)} .$$

Let $P(x) = x^q(c_0+c_1x+\cdots+c_dx^d)$ be the determinant of the $n\times n$ matrix of the $P_{ij}(x)$, so that $d \leq M$ and $c_0+c_1x+\cdots+c_dx^d$ is the Alexander polynomial of G. The constant c_0 is zero only if $P(x)$ is zero. But this cannot happen: since $r_i = a_i$ modulo G', the exponent sum of a_i in r_i is 1, and the exponent sum of a_j in r_i is zero for $j \neq i$, so tha setting $x = 1$ in the expression for r_i given above reduces it to a_i and so $P(1) = 1$. That is, $P(x)$ is not the zero polynomial, and so $c_0 \neq 0$, and $c_d \neq 0$. While M varies with the presentation, d is an invariant of G [1].

From the presentation P of G' one gets a presentation of a certain group

$$H_0 = (a_{1,0}, a_{1,1},\cdots, a_{1,M_1}, a_{2,0},\cdots, a_{2,M_2},\cdots, a_{n,M_n}; R_0) .$$

The deficiency of this is $(M+n)-n = M$.

Adding the integer t to all second subscripts in H_0 gives a group

$$H_t = (a_{1,t},\cdots, a_{1,M_1+t}, a_{2,t},\cdots, a_{n,M_n+t}; R_t) .$$

Clearly, the union of the H_t is the presentation P of G'.

An element of H_0 is 1 in G' if it is consequence (product of conjugates) in G' of the members of the n-tuples of relators $R_0, R_{\pm 1},\cdots$

in G'; it is 1 in H_0 if it is consequence of the relators R_0 in H_0. In general, then, H_0 is not a subgroup of G'. But there is a normal subgroup, call it K_0, in H_0 such that H_0/K_0 is subgroup of G'. Similarly for the H_t. Writing H_t^* for these factorgroups, one gets a presentation of G' as a free product of the H_t^* with amalgamated subgroups as follows.

Let H_{01}^* be the subgroup of H_0^* generated by every symbol in H_0^* except the a_{i,M_i}, $i:1,\cdots,n$; and let H_{02}^* be the subgroup of H_0^* generated by all but the $a_{i,0}$, $i:1,\cdots,n$. Then mapping each a_{ij} into $a_{i,j+1}$ changes H_{01}^* into H_{02}^* and this mapping is an isomorphism since it corresponds to a conjugation by x in G. The subgroups $H_{t,j}^*$ are similarly defined for $j:1,2$ and all integral t.

Let a_{ij} of $H_{t,1}^*$ be matched with $a_{i,j+1}$ of $H_{t+1,1}^*$; this gives an isomorphism. Let a_{ij} of $H_{t,2}^*$ be matched with a_{ij} of $H_{t+1,1}^*$; this also gives an isomorphism and provides an identification of the subgroup $H_{t,2}^*$ of H_t^* with the subgroup $H_{t+1,1}^*$ of H_{t+1}^* along that isomorphism.

§2. *An example*

Let G be a one-relator group and $P^* = (x, a;\ r)$ a presentation of it with $a = a_1$ and $r = r_1 = a^2x^{-1}ax^2a^{-1}x^{-3}a^{-1}x^2$, so that $r = a$ modulo G'. The rewrite of r in the exponential form is $a^{2+x-x^{-1}-x^2}$. The rewrite of $x^{-1}r^{-1}x$ is $a^{x^3+1-x^2-2x}$. Modulo G'', this can be written as $a^{L(x)}$, with $L(x) = x^3-x^2-2x+1$ the Alexander polynomial of G. Replacing r by $x^{-1}r^{-1}x$ in P^*, one gets

$$R_0 = a_3a_0a_2^{-1}a_1^{-2}$$

and

$$H_0 = (a_0, a_1, a_2, a_3;\ R_0)\ .$$

Then

$$H_{01} = (a_0, a_1, a_2;\,)$$

$$H_{02} = (a_1, a_2, a_3;\,)$$

(that is, free groups of rank 3 because of [3]); $d = M = M_1 = 3$, and $m = m_1 = 0$.

Though H_{0i} is free of rank 3 in view of the Freiheitssatz, in this case inspection is sufficient to establish this fact. For example one can argue that R_0 and the symbols a_1, a_2, a_3 freely generate the symbols a_0, a_1, a_2, a_3 so that H_0 is free on a_1, a_2, a_3. But then H_0 is contained in H_{02} and vice versa.

H_{01} and H_{02} are isomorphic under the matching of a_i in H_{01} with a_{i+1} in H_{02} so G' is the free product of the H_i with amalgamated subgroups

$$H_{i,2} = (a_{i+1}, a_{i+2}, a_{i+3};\,) \text{ of } H_i = (a_i, \cdots, a_{i+3}; R_i)$$

and

$$H_{i+1,1} = (a_{i+1}, \cdots, a_{i+3};\,) \text{ of } H_{i+1} = (a_{i+1}, \cdots, a_{i+4}; R_{i+1})$$

in the "natural" way.

The presentation P of G' is $(a_j; R_j, j: 0, \pm 1, \cdots)$, or

$$P = (a_j;\, a_{3+j} a_j a_{2+j}^{-1} a_{1+j}^{-2},\, j: 0, \pm 1, \cdots)\ .$$

In this example one can get rid of all a_j and R_k of P that are not in H_0 by using Tietze transformations. When only H_0 is left, one more Tietze transformation reduces it to H_{01} (or H_{02}). So G' is free of rank 3 and any triple a_i, a_{i+1}, a_{i+2} generates it.

§3. *Summary of results*

Let G be knot-like.

If G' is finitely generated, is G' free? For knot groups the answer is in the affirmative; the rank of G' is then the degree, d, of the Alexander polynomial [7]. The proof is topological, based on the van Kampen theorem, available since there is a knot. If G is only knot-like, it is known only that when G'/G'' is finitely generated then it is free of rank d [8].

More generally, one can ask when is G' free.

For presentations with one defining relation, the answer is complete (Theorem 2): exactly when it is finitely generated. For one relator presentations of these groups M turns out to be an invariant of G and so $d = M$. (Lemma 5.)

On the other hand, if G' is free of rank t, then there exists a presentation with $d = M = t$ since G is extension of a free group F_t by a free cyclic group. In the general case, one would therefore like to have a presentation P^* of G of deficiency 1 for which $M - d$ is as small as possible (this number is non-negative as $d \leq M$). As this runs into the unsolved problem of finding the deficiency of a group [9], the condition $M = d$ had to be used. This condition forces a Freiheitssatz for many-relator presentations by weeding out cases where we cannot tell at the present state of our knowledge of these matters whether certain subgroups are free. If a knot-like group were such that every presentation P^* of deficiency I gave $d < M$ then this group would be weeded out. I do not know whether such a group could have either a free or a finitely generated commutator subgroup. I think not. That is, I suspect that a free G' is finitely generated, as in the one-relator case.

An example of G having a presentation P_1^* with $d < M_1$ and another presentation P_2^* with $d = M_2$ (M_i representing for the moment the value of M for P_i^*) is easily concocted. The following one has $G' = F_2$.

$$P_1^* = (x, a, b, c;\ c^{-1}a^{-1}x^{-1}ax, x^{-1}a^{-1}xb^{-1}x^{-1}bx, x^{-1}b^{-1}cx)$$

gives

$$H_0 = (a_0, b_0, c_0, a_1, b_1, c_1, ;\ c_0^{-1}a_0^{-1}a_1, a_1^{-1}b_0^{-1}b_1, b_1^{-1}c_1)$$

and $M = 3$. Rewriting the relators modulo H_0' gives them the form

$$c^{-1}a^{x-1},\ a^{-x}b^{x-1},\ b^{-x}c^{x}$$

and the determinant of

$$(P_{ij}) = \begin{pmatrix} x-1 & 0 & -1 \\ -x & x-1 & 0 \\ 0 & -x & x \end{pmatrix}$$

is $x(x^2-3x+1)$. Thus $d = 2$. The Alexander polynomial is x^2-3x+1. G is the fundamental group of Listing's knot (4 crossings).

Another presentation

$$P_2^* = (x, a, b;\ b^{-1}a^{-1}x^{-1}ax, x^{-1}a^{-1}xb^{-1}x^{-1}bx)$$

gives

$$H_0 = (a_0, b_0, a_1, b_1;\ b_0^{-1}a_0^{-1}a_1, a_1^{-1}b_0^{-1}b_1)$$

and $M_2 = 2 = d$.

The lemmata in the next section, while they give a little more information, lead to the following theorems.

G will be knot-like, P^* a presentation of deficiency 1, G' the commutator subgroup, and M the ensuing deficiency of the presentation H_0.

THEOREM 1. *Suppose* M *is least possible for all presentations* P^* *of* G. *Then any two of the following three statements imply the third. (1)* G' *is finitely generated. (2)* G' *is free. (3) The degree of the Alexander polynomial of* G *is* M.

THEOREM 2. *For one-relator* P^* *either of the first two statements in Theorem 1 implies the rest.*

THEOREM 3. *Let* $w(a_{ij})$ *and* $w(a_{i,j+1})$ *both be elements of* H_0. *Then* H_0 *is a subgroup of* G' *if and only if for all such pairs, either both are relators in* H_0 *or neither is.*

The last theorem means that from knowledge of H_0 alone, one can determine the factor group H_0/K_0 which is subgroup of G'.

Conversations with colleagues at York University, Downsview, Ontario, especially with Abe Karrass enabled me to put the material of this paper in the present improved form.

§4. *Proofs*

G will be a knot-like group in a fixed presentation, P^*. The proofs will be based on a series of lemmata.

LEMMA 1. *If* G' *is finitely generated then* $H_{0,1}$ *contains a full set of generators of* G'.

Proof. Let H_q^* be the subgroup of G' generated by the elements of H_q. If the lemma holds for $H_{0,1}^*$ then it holds for $H_{0,1}$. I will prove it for the former.

As subgroups of G', H_q^* and H_{q+1}^* are isomorphic under matching each a_{ij} of H_q^* with $a_{i,j+1}$ of H_{q+1}^*. Therefore G' is the free product of the subgroups H_q^* with $H_{q,2}^*$ and $H_{q+1,1}^*$ amalgamated along this isomorphism. Writing $G' = \cup \underset{A}{*} H_q^*$ for this, the segments

$$S_t^* = H_0^* \underset{A}{*} H_1^* \underset{A}{*} \cdots \underset{A}{*} H_t^* \quad \text{and} \quad S_{-t}^* = H_0^* \underset{A}{*} H_{-1}^* \underset{A}{*} \cdots \underset{A}{*} H_{-t}^*$$

are seen to be subgroups of G' for every t.

As G' is finitely generated there is a non-negative number t for which S_t^* contains a full set of generators of G' and so it is isomorphic to G'. Therefore, in $S_{t+1}^* = S_t^* \underset{A}{*} H_{t+1}^*$ there are n relations, one for each i, of the form $a_{i,M_i+t+1} = w_i^*$, and the element w_i^* is in S_t^*.

Now the amalgamation in S_{t+1}^* (as a product of two groups given above) is along an isomorphism identifying the subgroup $H_{t,2}^*$ of S_t^* with $H_{t+1,1}^*$ of H_{t+1}^*. Therefore, if some element w of S_t^* equals an element v of H_{t+1}^* then w must be in $H_{t,2}^*$ and v must be in $H_{t+1,1}^*$. Thus, the n relations $a_{i,M_i+t+1} = w_i^*$ imply that for each i there exist elements v_i^* in $H_{t+1,1}^*$ such that

$$a_{i,M_i+t+1} = v_i^* \text{ is a relation in } H_{t+1}^*$$

and there exist elements u_i^* in $H_{t,2}^*$ such that

$$w_i^* = u_i^* \text{ is a relation in } H_t^* ,$$

and $u_i^* = v_i^*$ under the amalgamation in $S_{t+1}^* = S_t^* \underset{A}{*} H_{t+q}^*$. These give the relations

$$a_{i,M_i+t+1} = v_i^*,\ i : 1,\cdots,n \text{ in } H_{t+1}^* ,$$

expressing a_{i,M_i+t+1} as an element, v_i^*, of $H_{t+1,1}^*$; that is the v_i^* contain no symbols a_{j,M_j+t+1} for any j.

For any integer k, the groups H_0^* and H_k^* are isomorphic under matching $a_{i,r}$ of H_0^* and $a_{i,r+k}$ of H_k^*, $i : 1,\cdots,n$. Then, for any non-negative number p similar results obtain for the subgroups $H_{-p}^* \underset{A}{*} H_{1-p}^* \underset{A}{*} \cdots \underset{A}{*} H_q^*$ whenever $q+p=t$; the latter are conjugate to the subgroup S_t^*. It follows that there exist elements, v_i, in $H_{0,1}^*$ such that

$$a_{i,M_i} = v_i,\ i : 1,\cdots,n \text{ are relations in } H_0^*$$

and the v_i contain none of the a_{j,M_j}.

Similar considerations, starting with S_{-t}^*, yield elements z_i in $H_{0,2}^*$ such that

$$a_{i,0} = z_i,\ i : 1,\cdots,n \text{ are relations in } H_0^*$$

and the z_i contain none of the $a_{j,0}$.

By the same token, like relations hold in each H_k^*.

Then, in G', one can express $a_{i,-1}$ as an element in $H_{0,1}^*$ and so one can express every element of H_{-1}^* in $H_{0,1}^*$. Likewise, the $a_{i,-2}$ are equal to elements of H_{-1}^*, and so H_{-2}^* can be expressed in the generators of $H_{0,1}^*$, etc. Similarly for a_{i,M_i}, a_{i,M_i+1}, and so forth.

This shows: $H_{0,1}^*$, qua subgroup of G' is actually G', so that $H_{0,1}$ contains a full set of generators of G', which proves Lemma 1.

Note that $H_{0,1}^*$ is a factor group of $H_{0,1}$ so that if $H_{0,1}$ itself is a subgroup of G' it is G' in the sense indicated.

LEMMA 2. *$H_{0,1}$ and $H_{0,2}$ are isomorphic under matching the a_{ij} in $H_{0,1}$ with the $a_{i,j+1}$ in $H_{0,2}$ if and only if H_0 is a subgroup of G'.*

Proof. It is clear from the definition of the $H_{i,j}$ that the groups $H_{0,1}$ and $H_{1,1}$ are isomorphic in G' under matching the $a_{i,j}$ of $H_{0,1}$ and the $a_{i,j+1}$ of $H_{1,1}$. If now $H_{0,1}$ is isomorphic to $H_{0,2}$ in the required manner, then matching the $a_{i,j}$ of $H_{0,2}$ with the $a_{i,j}$ of $H_{1,1}$ is an isomorphism. A similar statement holds for all the groups H_q. But then G' is the free product of the H_q with $H_{q,2}$ and $H_{q+1,1}$ amalgamated along this isomorphism, and so H_q is a subgroup of G' for each integer q. As the converse is obvious, this proves Lemma 2.

LEMMA 3. *If* $d = M$ *then* $H_{0,1}$ *and* $H_{0,2}$ *are free of rank* M.

Proof. Let G_0 be the free group generated by the symbols $a_{i,j}$ which belong to $H_{0,1}$, so that G_0 has rank M. Then

$$H_0 = (G_0, a_{i,M_i}, i : 1,\cdots, n; R_0) .$$

R_0 consists of n relators, say $R_0 = w_1,\cdots, w_n$. Let b_{ij} be the exponent sum of a_{i,M_i} in w_j, so that (b_{ij}) is an $n \times n$ matrix; let D_1 be its determinant. It follows from a theorem of Gerstenhaber and Rothaus [2] that if $D_1 \neq 0$ then G_0 is a subgroup of H_0. But in that case $G_0 = H_{0,1}$ and so $H_{0,1}$ is free of rank M.

Replacing $H_{0,1}$ by $H_{0,2}$ in this argument and D_1 by the like determinant, D_2, for the $a_{i,0}$, gives the same result for $H_{0,2}$. If now $d = M$ then by the remarks of Section 1 about the coefficients of the Alexander polynomial, c_d is D_1 and c_0 is D_2, so that $D_1 \neq 0$ and $D_2 \neq 0$. Thus the $H_{0,j}$ are free of rank M, and Lemma 3 is proven.

LEMMA 4. *If* G' *is finitely generated and* H_0 *is subgroup of it, then* G' *is free of rank* $d = M$.

Proof. By Lemma 1, $H_{0,1}$ contains a full set of generators of G' and by assumption it is a subgroup of G'. Therefore $H_{0,1}$ is a presentation of G'.

Since the set of relations $a_{i,M_i} = v_i$, $i: 1,\cdots,n$, with the v_i elements of $H_{0,1}$, found in the proof of Lemma 1, hold in H_0, the words $a_{i,M_i}^{-1}v_i$ are relators in H_0. The determinant of the exponent sums of the a_{i,M_i} in these words is not zero. It follows easily that the like determinant, D_1, for the n-tuple of relators R_0 is also not zero. As in the proof of Lemma 3, one gets that $H_{0,1}$ is free of rank M. Therefore, G' is free of rank M.

As the Abelianized commutator subgroup, G'/G'', is free of rank d [8], this proves: G' is free of rank $M = d$, as Lemma 4 claims.

Concerning the matrix of the exponent sums for the a_{i,M_i} in R_0, respectively of those of the $a_{i,0}$ in R_0, it is true not only that they are non-singular but that their determinants are ± 1. For if the normal closure, N, of n elements, $w_1,\cdots,w_n$, of a free group F contains a subset $s_1,\cdots,s_n$ of some free generating set of F, then it easily follows that the normal closure of $s_1,\cdots,s_n$ is again N. (See for example [4] or [5].)

Lemma 5. *If G' is finitely generated and free then there is a presentation, P^*, of G for which $M = d$.*

Proof. Let G' be free on the generators $a_1,\cdots,a_d$. Then G is defined by the number d and the automorphism which the element x in $(x, a_1,\cdots,a_d; R) = G$ induces in G':

$$x^{-1}a_i x = w_i(a_1,\cdots,a_d),\ i: 1,\cdots,n\ .$$

The relator set R consists of the words $x^{-1}a_i^{-1}xw_i$. Their rewrites as a_{ij}-words have the form $a_{i,1}^{-1}w_i(a_{1,0}, a_{2,0},\cdots,a_{d,0})$, and so $M_i = 1$ and $M = \Sigma M_i = d$, as claimed.

Lemma 6. *If G' is free and there exists a presentation of G for which $M = d$, then G' is finitely generated.*

Proof. Let P^* be the presentation for which $M = d$. By Lemma 3, $H_{0,1}$ and $H_{0,2}$ are free of rank M and isomorphic under matching $a_{i,j}$ of $H_{0,1}$ and $a_{i,j+1}$ of $H_{0,2}$. – Similarly for H_q and H_{-q} for all natural q.

By Lemma 2 then, H_i and every product $S_t = H_0 \underset{A}{*} H_1 \underset{A}{*} \cdots \underset{A}{*} H_t$ with $H_{i,2}$ and $H_{i+1,1}$ amalgamated, is a subgroup of G'. By assumption, G' is free so these groups are free. In particular, $S_0 = H_0$ presents a free group. Since $c_0 \neq 0$, elementary considerations of the *group* H_0/H'_0 show that H_0 has rank $d = M$. Further, since M is the deficiency of the *presentation* H_0 of this free group (there are $M+n$ generators and n defining relators), the defining relator set R_0 must be an independent set (in the strong sense that the normal subgroup N in $H_0 = F/N$ is not the normal closure of any $n-1$ of its elements). The same holds for each n-tuple R_q. The following then is immediate.

As H_0 presents a free group, F_d, of rank d, one can put

$$S_1 = (H_0, a_{i,M_i+1}, i:1,\cdots,n; R_1) = (F_d, a_{i,M_i+1}, i:1,\cdots,n; R_1)$$

and this is a free group. The presentation has deficiency d and the defining relators are independent. Thus the rank of S_1 is again d. Similarly, S_t is free of rank d for each t.

If the sequence $S_t \subseteq S_{t+1} \subseteq \cdots$ of free groups of fixed rank is infinite, then the limit group is not free [6]. Since the latter is contained in G', then G' is not free. Under the assumption that G' is free, the sequence must therefore terminate: there is a non-negative number k such that $S_{k+h} = S_k$ for all natural h.

Now let

$$S_{-1} = (S_k, a_{i,-1}, i:1,\cdots,n; R_{-1}) \quad ,$$

$$S_{-2} = (S_{-1}, a_{i,-2}, i:1,\cdots,n; R_{-2}) \; ,$$

etc. These groups form a sequence

$$S_{-1} \subseteq S_{-2} \subseteq \cdots .$$

Using the fact that $c_0 \neq 0$, the same arguments lead to the conclusion that this sequence breaks off: $S_{-q} = S_{-q-1} = \cdots$, for some q. Thus S_{-q} must be G' and so G' is finitely generated. This proves Lemma 6.

LEMMA 7. *If* G' *is finitely generated and* $M = d$ *then* G' *is free of rank* d.

Proof. By Lemma 3, $H_{0,1}$ and $H_{0,2}$ are free of rank M. Then $H_{0,1}$ is isomorphic to $H_{0,2}$ under matching $a_{i,j}$ of $H_{0,1}$ and $a_{i,j+1}$ of $H_{0,2}$. By Lemma 2, then H_0 is subgroup of G'. Since G' is finitely generated, Lemma 4 applies, and G' is free of rank d, as claimed.

As the last three lemmata cover the statements in Theorem 1, its proof is now complete.

To prove Theorem 2, let $(a_0, a_1, \cdots, a_M; R_0)$ be the one-relator presentation H_0 obtained from the one-relator presentation P^* of G. The word R_0, reduced and cyclically reduced, contains the symbols a_0 and a_M by assumption. Therefore [3] $H_{0,1}$ and $H_{0,2}$ are free subgroups of H_0, isomorphic under matching a_i of $H_{0,1}$ with a_{i+1} of $H_{0,2}$. Similarly for all pairs $H_{q,2}$ and $H_{q+1,1}$. Then Lemma 2 is applicable and so H_0 is a subgroup of G', with G' the free product of the H_q with amalgamation, $\cup \underset{A}{*} H_q$, along the indicated isomorphism.

If G' is finitely generated then Lemma 4 applies and so G' is free of rank $d = M$.

On the other hand, if G' is free, so is the subgroup $H_0 = (a_0, a_1, \cdots, a_M; R_0)$ and its rank is M or $M+1$. Since P^* has deficiency 1, R_0 is not the empty word, and so the rank is M. Similarly for $H_1 = (a_1, \cdots, a_{M+1}; R_1)$. Then $S_1 = (H_0, a_{M+1}; R_1) = H_0 \underset{A}{*} H_1$ is also a subgroup and free, and its rank is M or $M+1$. Were the rank $M+1$, then — H_0 being free of rank M — S_1 would be $(H_0, a_{M+1};) = (a_0, \cdots, a_{M+1}; R_0)$ so that R_1 would be a consequence of R_0 in the free group generated by $a_0, \cdots, a_{M+1}$. But R_0 contains a_0 and R_1 does not, so — again by the Freiheitssatz – this is impossible [3]. It follows that S_1 is free of rank M.

The same holds for all subgroups $S_t = S_{t-1} \underset{A}{*} H_t$, for any natural t. As in Lemma 6, the sequence of the S_t breaks off: $S_k = S_{k+1}$ for some k. Similarly for $S_{-1} = (H_0, a_{-1}; R_{-1})$, etc. Thus G' is finitely generated.

It follows from this that Lemma 4 applies, and so the rank of G' is $M = d$. This proves Theorem 2.

It remains to deduce Theorem 3.

If H_0 is a subgroup of G' then $H_{0,1}$ and $H_{0,2}$ are subgroups of G' and so of G. Then, in G, $H_{0,2} = x^{-1}H_{0,1}x$, with $w(a_{i,j+1}) = x^{-1}w(a_{i,j})x$ for all elements w of G', as claimed.

If both $w(a_{i,j})$ and $w(a_{i,j+1})$ are relators in H_0, or else neither is, then matching of the $a_{i,j}$ of $H_{0,1}$ with the $a_{i,j+1}$ of $H_{0,2}$ produces the isomorphism of Lemma 2 and so H_0 is a subgroup of G'. This completes the proof.

Thus, one can – in theory – read off H_0 what to factor out of it to make it a subgroup of G':

Suppose that $w(a_{i,j})$ contains no symbols a_{k,M_k} for any k and that $v(a_{i,j})$ contains no symbols $a_{k,0}$ for any k, and that both words are relators in H_0. Form the factorgroup $(H_0; w(a_{i,j+1}), v(a_{i,j-1}))$; then apply the same operation to the new group using some further word w^* and/or v^*. Let H_0^* be the largest factorgroup of H_0 that is closed under this operation. Then H_0^* satisfies the assumptions of Theorem 3, and so it is a subgroup of G'.

REMARK. If there are no such w and/or v then H_0 is a subgroup with $H_{0,j}$ free, so G' is the free product of the H_q with amalgamated free subgroups $H_{q,j}$ as in the case $M = d$.

For example, the group

$$(x, a;\ x^{-3}ax^{-2}ax^{5}ax^{-2}a^{-1}x^{-3}a^{-1}x^{4}a^{-2}x)$$

gives

$$H_0 = (a_0, \cdots, a_5;\ a_3 a_5 a_0 a_2^{-1} a_5^{-1} a_1^{-2})$$

with $M = 5$, $d = 3$. Since a_0 and a_5 are present in every relator [3], H_0 is subgroup of G'. This H_0 presents a free group but G' is neither free nor finitely generated.

STATE UNIVERSITY OF NEW YORK, STONY BROOK

BIBLIOGRAPHY

[1] Fox, R. H., Free Differential Calculus I. *Ann. Math.* 57 1953, 547-560.

[2] Gerstenhaber, M., and Rothaus, O. S., The Solution of Sets of Equations in Groups. *Proc. Nat. Acad. Sci. USA* 48 1962, 1531-1533.

[3] Magnus, W., Ueber diskontinuierliche Gruppen mit einer definierenden Relation. (Der Freiheitssatz.) *J. reine u. angew. Math.* 163 1930, 141-165.

[4] ——————, Ueber freie Faktorgruppen und freie Untergruppen gegebener Gruppen. *Monatshefte fuer Math. u. Phys.* 47 1939, 307-313.

[5] Magnus, W., Karrass, A., and Solitar, D., *Combinatorial Group Theory: Presentations of Groups in Terms of Generators and Relations.* Interscience Publishers, New York, 1966.

[6] Neumann, B. H., Some Remarks on Infinite Groups. *J. London Math. Soc.* 12 1937, 120-127.

[7] Neuwirth, L., Interpolating Manifolds for Knots in S^3. *Topology* 2 1964, 359-365.

[8] Rapaport, E. S., On the Commutator Subgroup of a Knot Group. *Ann. Math.* 71 1960, 157-162.

[9] ——————, Finitely Presented Groups. The Deficiency. *J. of Algebra* 24 1973, 531-543.

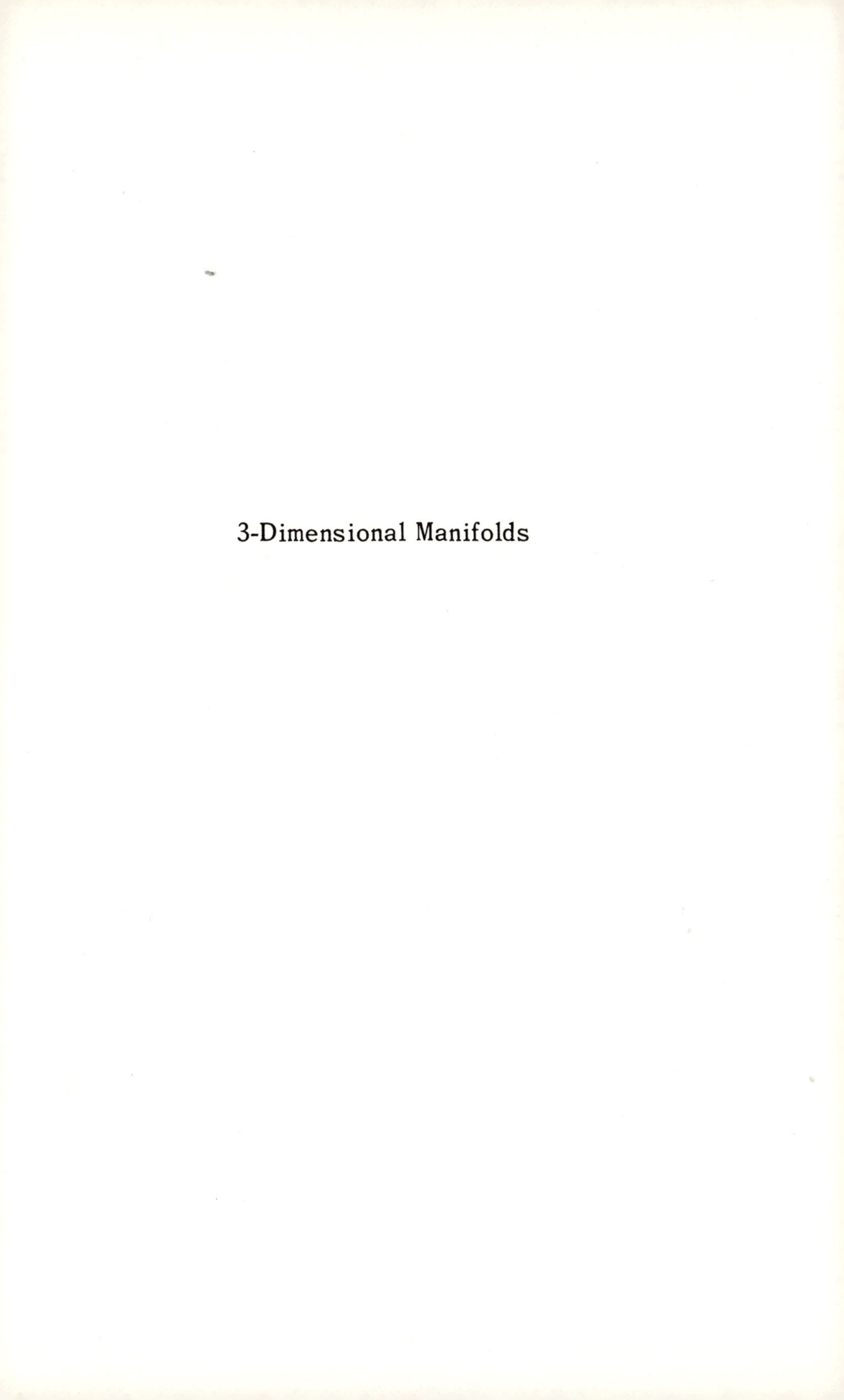

3-Dimensional Manifolds

ON THE EQUIVALENCE OF HEEGAARD SPLITTINGS OF CLOSED, ORIENTABLE 3-MANIFOLDS

Joan S. Birman*

I. Introduction

Let X_g be a oriented handlebody of genus g, and let $X'_g = \tau(X_g)$ be a homeomorphic image of X_g, with an orientation inherited from that on X_g. If ∂X_g and $\partial X'_g$ are identified via an orientation-reversing homeomorphism, then the resulting oriented 3-manifold M is said to be represented by a *Heegaard splitting* of genus g. To make this explicit we will assume that $\delta : \partial X_g \to \partial X_g$ is an arbitrary but henceforth fixed orientation-reversing homeomorphism which extends to an orientation-reversing homeomorphism of $X_g \to X_g$, and that $\phi : \partial X_g \to \partial X_g$ is orientation-preserving. Then we may identify ∂X_g and $\partial X'_g$ by the rule

$$\tau\delta\phi(p) = p, \qquad \forall p \in \partial X_g \tag{1}$$

to obtain a 3-manifold $X_g \cup_\phi X'_g$. If two Heegaard splittings $X_g \cup_\phi X'_g$ and $X_g \cup_\psi X'_g$ define homeomorphic 3-manifolds, we will write $\phi \equiv \psi$. It is immediate that $\phi \equiv \psi$ if the isotopy classes of ϕ and ψ coincide.

The Heegaard splittings $X_g \cup_\phi X'_g$ and $X_g \cup_\psi X'_g$ will be said to be

(i) *strongly equivalent*, denoted $\phi \approxeq \psi$, if there is an orientation-preserving homeomorphism $h : X_g \cup_\phi X'_g \to X_g \cup_\psi X'_g$ such that $h(X_g) = X_g$, $h(X'_g) = X'_g$;

(ii) *equivalent*, denoted $\phi \approx \psi$, if there is an orientation-preserving homeomorphism $h : X_g \cup_\phi X'_g \to X_g \cup_\psi X'_g$ such that either $h(X_g) = X_g$, $h(X'_g) = X'_g$ or $h(X_g) = X'_g$, $h(X'_g) = X_g$.

---

* Supported in part by NSF Grant No. GP 32893.

(iii) *weakly equivalent*, denoted $\phi \sim \psi$, if there is a homeomorphism $h: X_g \cup_\phi X'_g \to X_g \cup_\psi X'_g$ such that either $h(X_g) = X_g$, $h(X'_g) = X'_g$ or $h(X_g) = X'_g$, $h(X'_g) = X_g$.

Note that $\phi \approxeq \psi \Rightarrow \phi \approx \psi \Rightarrow \phi \sim \psi \Rightarrow \phi \equiv \psi$. These definitions place equivalence relations on the class of all Heegaard splittings of any given genus.

It was proved by Waldhausen in [8] that all Heegaard splittings of the same genus of S^3, and also that all Heegaard splittings of the same genus of the n-fold connected sum $\#_n(S^2 \times S^1)$ of n copies of $S^2 \times S^1$, are strongly equivalent. In this paper we study the corresponding question for other closed orientable 3-manifolds. We begin by establishing (Theorem 1) that each strong equivalence class (respectively equivalence class, weak equivalence class) of Heegaard splittings may be identified with a double coset (respectively two double cosets, four double cosets) modulo a certain subgroup $\mathfrak{F}$, in the group $\mathfrak{M}_g$ of isotopy classes of orientation-preserving homeomorphisms of ∂X_g; invariants of these double cosets will then be invariants of Heegaard splittings. We then proceed to study in Section II, computable invariants of double cosets in $\mathfrak{M}_g$ mod $\mathfrak{F}$, (see Theorems 2 and 3). These ideas are applied first in Corollary 2.1 to classify the equivalence classes of Heegaard splittings of genus 1, and again in Corollary 2.2 to prove that Waldhausen's results do not generalize to arbitrary 3-manifolds, by exhibiting a 3-manifold of Heegaard genus 2 which admits two weak equivalence classes of genus 2 Heegaard splittings.[1] Thus, in general, $\phi \equiv \psi$ does not imply $\phi \sim \psi$.

---

[1] After this manuscript was completed, we learned that similar examples of composite manifolds that admit more than one weak equivalence class of Heegaard splittings had been obtained earlier by Engmann [11], using different methods.

In Section III we discuss the use of our methods to determine finer invariants of Heegaard splittings than those given in Theorem 2. We will show that there is a natural generalization of the integer pairs (p,q) which are known to characterize the lens spaces up to homeomorphism to a set of 4 mutually related $g \times g$ matrices (P, Q, R, S) of integers which are invariants of strong equivalence (resp. equivalence, weak equivalence) classes of Heegaard splittings of genus $g > 1$. Theorem 3 treats the problem of distinguishing between classes of Heegaard splittings on the basis of our integer matrices, however the solution is not as neat as the classical solution for the case of the lens spaces, and is not given in closed form.

Methods which are similar to those used here were used earlier by Reidemeister [7] to study topological invariants of closed orientable 3-manifolds. His approach was to utilize the fact that all Heegaard splitting of a 3-manifold are stably-equivalent [7, 10]. The relationship between stable equivalence and equivalence is discussed in Section IV, together with a brief review of Reidemeister's earlier results.

II. Heegaard Splittings and Double Cosets

Let $\mathfrak{H}_g$ denote the group of isotopy classes of self-homeomorphisms of a closed, orientable surface ∂X_g of genus g, and let $\mathfrak{M}_g$ denote the subgroup of those classes which are represented by maps which preserve orientation. Let $\mathfrak{F}_g$ denote the subgroup of $\mathfrak{M}_g$ consisting of those mapping classes which have representatives that extend to homeomorphisms of X_g. Note that $\mathfrak{H}_g$ is naturally isomorphic to the group of outer automorphisms, Out $\pi_1 \partial X_g$, of $\pi_1 \partial X_g$.

If $\phi : \partial X_g \to \partial X_g$, we will use the symbol ϕ_* for the induced automorphism of $\pi_1 \partial X_g$, and Φ for the isotopy class of ϕ.[2] Recall that δ was a fixed orientation reversing homeomorphism of $\partial X_g \to \partial X_g$. We assert:

[2] Similarly, Ψ and Δ denote the isotopy classes of ψ and δ.

THEOREM 1. *Let* $X_g \cup_\phi X'_g$ *and* $X_g \cup_\psi X'_g$ *be Heegaard splittings. Then:*

(i) $\phi \approx \psi$ *if and only if* Φ *and* Ψ *are in the same double coset in* $\mathfrak{M}_g$ (mod $\mathfrak{F}_g$).

(ii) $\phi \approx \psi$ *if and only if* Ψ *is in the same double coset as* Φ *or* $\Delta\Phi^{-1}\Delta^{-1}$.

(iii) $\phi \sim \psi$ *if and only if* Ψ *is in the same double coset as* Φ *or* $\Delta\Phi^{-1}\Delta^{-1}$ *or* Φ^{-1} *or* $\Delta\Phi\Delta^{-1}$.

Proof. Suppose that $\phi \approx \psi$, and that $h: X_g \cup_\phi X'_g \to X_g \cup_\psi X'_g$ is the homeomorphism which defines the equivalence. (Thus h is orientation-preserving.) Let $h_0 = h|X_g$, $h'_0 = h|X'_g$, $h_1 = h_0|\partial X_g$, $h'_1 = h'_0|\partial X'_g$, $\tau_1 = \tau|\partial X_g$. In order for h to be well-defined on $\partial X_g = \partial X'_g$ it is necessary that the diagram

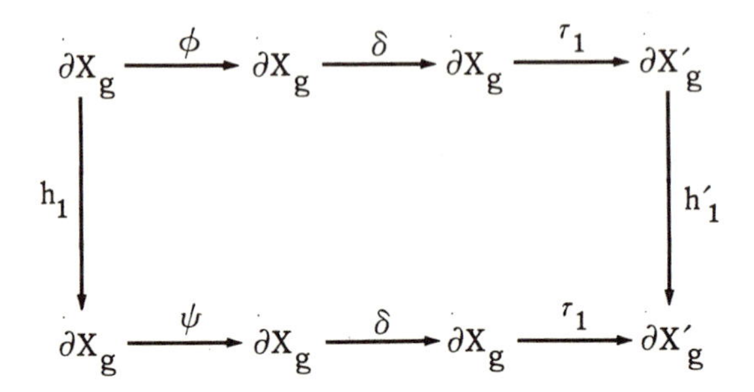

be commutative, that is

$$\psi = \delta^{-1}\tau_1^{-1}h'_1\tau_1\delta\phi h_1^{-1} . \tag{2}$$

Since $\delta^{-1}\tau_1^{-1}h'_1\tau_1\delta$ and h_1^{-1} are each orientation-preserving homeomorphisms of ∂X_g which extend to X_g, this implies

$$\Psi = \xi_1\Phi\xi_2, \qquad \xi_1, \xi_2 \in \mathfrak{F}_g . \tag{3}$$

Conversely, if (3) is satisfied, then (2) is likewise satisfied, and we may use the extensions of $\delta^{-1}\tau_1^{-1}h'_1\tau_1\delta$ and h_1^{-1} to construct a homeomorphism h which defines an equivalence between the Heegaard splittings.

Next, suppose that there is an orientation-preserving homeomorphism $h: X_g \cup_{\phi} X'_g \to X_g \cup_{\psi} X'_g$ such that $h(X_g) = X'_g$ and $h(X'_g) = X_g$. Let $h_0, h_1, h'_0, h'_1, \tau_1$ be defined as before. In order for h to be well-defined on ∂X_g it is now necessary that the diagram

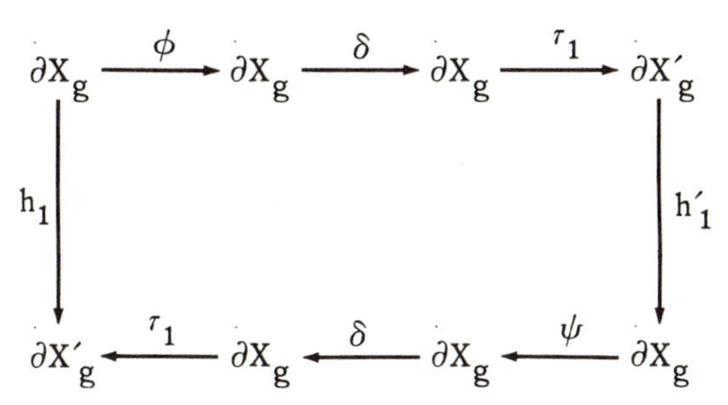

be commutative, that is

$$\psi = \delta^{-1}\tau_1^{-1}h_1\delta^{-1}\delta\phi^{-1}\delta^{-1}\tau_1^{-1}h_1'^{-1} . \tag{4}$$

Since $\delta^{-1}\tau_1^{-1}h_1\delta^{-1}$ and $\tau_1^{-1}h_1'^{-1}$ are each orientation-preserving homeomorphisms of $\partial X_g \to \partial X_g$ which extend to the solid handlebody, we have:

$$\Psi = \xi_1 \Delta \Phi^{-1} \Delta^{-1} \xi_2, \qquad \xi_1, \xi_2 \in \mathfrak{F}_g . \tag{5}$$

Exactly as before, if equation (5) is satisfied, then we may construct a homeomorphism h which defines the required equivalence between the Heegaard splittings, with $h(X_g) = X'_g$ and $h(X'_g) = X_g$.

The remaining cases are similar. If the equivalence is via an orientation-reversing homeomorphism h which maps $X_g \to X_g$, then ψ and ϕ are again related by equation (2), but now $\delta^{-1}\tau_1^{-1}h'_1\tau_1\phi$ and h_1^{-1} are orientation-reversing homeomorphisms of ∂X_g which extend to X_g. However, we may rewrite equation (2) as

$$\psi = (\delta^{-1}\tau_1^{-1}h'_1\tau_1)(\delta\phi\delta^{-1})(\delta h_1^{-1}) . \tag{6}$$

Now, $\delta^{-1}\tau_1^{-1}h'_1\tau_1$ and δh_1^{-1} are each orientation-preserving homeomorphisms of $\partial X_g \to \partial X_g$ which extend, hence

(7) $$\psi = (\xi_1)(\Delta\Phi\Delta^{-1})(\xi_2), \qquad \xi_1, \xi_2 \in \mathfrak{F}_g .$$

In the last case equation (4) applies, but since h is orientation reversing, the products $\delta^{-1}\tau_1^{-1}h_1$ and $\delta^{-1}\tau_1^{-1}h_1'^{-1}$ are each orientation-preserving homeomorphisms of $\partial X_g \to \partial X_g$ which extend to X_g. ||

In order to apply Theorem 1, we will make use of the natural representation of $\mathfrak{H}_g$ in the automorphism group of the first homology group $H_1(\partial X_g)$. Invariants of double cosets (modulo the image of $\mathfrak{F}_g$) in that group will then be invariants of equivalence classes of Heegaard splittings.

To fix conventions, choose generators ω_i and ω_i', $1 \le i \le 2g$, for $\pi_1 \partial X_g$ and $\pi_1 \partial X_g'$, making the choice so that τ maps representatives of ω_i onto representatives of ω_i' for each $1 \le i \le 2g$. Suppose also that the ω_i satisfy the relation $\prod_{j=1}^{g} [\omega_j, \omega_{j+g}] = 1$, and that if $\iota : \partial X_g \to X_g$ is the homomorphism induced by inclusion, that her ι_* is the normal closure N in $\pi_1 \partial X_g$ of $\{\omega_{j+g};\ 1 \le j \le g\}$.

Choose any element $\Phi \in \mathfrak{H}_g$, and let $\phi : \partial X_g \to \partial X_g$ be a homeomorphism which represents Φ. Suppose that the action of ϕ_* on $\pi_1 \partial X_g$ is given by

(8) $$\phi_*(\omega_i) = \omega_1^{\varepsilon_{i1}} \omega_2^{\varepsilon_{i2}} \cdots \omega_{2g}^{\varepsilon_{i,2g}} \mod [\pi_1 \partial X_g, \pi_1 \partial X_g] ,$$

for each $1 \le i \le 2g$. Then we may define an anti-homomorphism $\alpha : \mathfrak{H}_g \to \operatorname{Aut} H_1(\partial X_g)$ by the rule

(9) $$\alpha(\Phi) = \|\varepsilon_{ij}\| , \quad \text{a } 2g \times 2g \text{ matrix of integers.}$$

This definition is independent of the choice of the representative ϕ of Φ because maps which are isotopic to the identity induce the identity automorphism on $H_1(\partial X_g)$.

The symplectic group $Sp(2g, Z)$ is the group of all $2g \times 2g$ matrices of integers M which satisfy the condition

(10) $$M J \overline{M} = \pm J, \text{ where } J = \begin{bmatrix} 0 & I \\ -I & 0 \end{bmatrix}, \quad \overline{M} = \text{transpose of } M .$$

LEMMA 2.1. *The anti-homomorphism* a *maps* $\mathfrak{H}_g$ *onto* Sp(2g,Z). *The subgroup* $a(\mathfrak{M}_g)$ *of* Sp(2g, Z) *is the subgroup of the matrices* M *for which*

$$MJ\bar{M} = +J \ . \tag{11}$$

If $g = 1$, *then* a *is an anti-isomorphism from* $\mathfrak{H}_1$ *onto* GL(2,Z).

Proof. See [5, p. 178]. ||

LEMMA 2.2. *Let*

$$\tilde{\mathfrak{F}}_g = \left\{ \begin{bmatrix} V & W \\ O & U \end{bmatrix} \in Sp^+(2g, Z) \right\} , \tag{12}$$

where O *denotes a* $g \times g$ *block of zeros. Then the group* $\tilde{\mathfrak{F}}_g$ *is the semi-direct product of its normal subgroup* $\tilde{\mathfrak{G}}$ *and subgroup* $\tilde{\mathfrak{U}}$, *where*

$$\tilde{\mathfrak{U}} = \left\{ D(U) = \begin{bmatrix} \bar{U}^{-1} & O \\ O & U \end{bmatrix}, \ \bar{U} = \textit{transpose of } U, \ \det U = \pm 1 \right\} \tag{13}$$

$$\tilde{\mathfrak{G}} = \left\{ F(S) = \begin{bmatrix} I & S \\ O & I \end{bmatrix}, \ S \textit{ symmetric} \right\} . \tag{14}$$

Moreover, a *maps* $\mathfrak{F}_g$ *onto* $\tilde{\mathfrak{F}}_g$.

Proof. The symplectic condition (11) implies that if $M \in \tilde{\mathfrak{F}}_g$, then the matrices U, V in (12) have the property $U\bar{V} = I$, hence $V = \bar{U}^{-1}$. It then follows that we may rewrite $M \in \tilde{\mathfrak{F}}_g$ in the form

$$\begin{bmatrix} V & W \\ O & U \end{bmatrix} = \begin{bmatrix} \bar{U}^{-1} & O \\ O & U \end{bmatrix} \begin{bmatrix} I & \bar{U}W \\ O & I \end{bmatrix} = \begin{bmatrix} I & WU^{-1} \\ O & I \end{bmatrix} \begin{bmatrix} \bar{U}^{-1} & O \\ O & U \end{bmatrix} . \tag{15}$$

Since $M \in Sp^+(2g, Z)$, and also $D(U) \in Sp^+(2g, Z)$, it follows that the matrices $F(\bar{U}W)$ and $F(WU^{-1})$ in (15) are also in $Sp^+(2g, Z)$, and therefore also satisfy equation (11). This implies that $\bar{U}W$ and WU^{-1} are symmetric, hence $\tilde{\mathfrak{F}}_g$ is the semi-direct product of its normal subgroup $\tilde{\mathfrak{G}}$ and subgroup $\tilde{\mathfrak{U}}$.

To see that α maps $\mathfrak{F}_g$ onto $\tilde{\mathfrak{F}}_g$, it suffices to show that $\tilde{\mathfrak{U}}$ and and $\tilde{\mathfrak{G}}$ are generated by the images under α of elements in $\mathfrak{F}_g$. According to [4], an element $\Phi \in \mathfrak{F}_g$ if and only if, for every representative ϕ of Φ, it is true that $\phi_*(N) = N$, where N is the normal closure in $\pi_1 \partial X_g$ of $\{\omega_{j+g};\ 1 \leq j \leq g\}$.

According to [2, p. 85], generators of $\tilde{\mathfrak{U}}$ are the matrices $D(U_i)$, $i = 1,2,3$ whose U_i's are the $g \times g$ matrices:

$$(16)\quad U_1 = \begin{bmatrix} 010\cdots0 \\ 001\cdots0 \\ 000\cdots1 \\ 100\cdots0 \end{bmatrix} \quad U_2 = \begin{bmatrix} -1 & & & \\ & 1 & & \\ & & \ddots & \\ 0 & & & 1 \end{bmatrix} \quad U_3 = \begin{bmatrix} 1 & 0 & & \\ 1 & 1 & & 0 \\ & & \ddots & \\ & 0 & & 1 \end{bmatrix}.$$

These matrices are induced by the following automorphisms of $\pi_1 \partial X_g$:

$$(17)\quad \phi_{1_*} : \omega_i \to \omega_{i+1},\ i = 1,\cdots,g-1,g+1,\cdots,2g-1,\ \omega_g \to \omega_1,\ \omega_{2g} \to \omega_{g+1}$$

$$(18)\quad \begin{aligned} \phi_{2_*} : \omega_1 &\to \omega_1 \omega_{g+1} \omega_1^{-1} \omega_{g+1}^{-1} \omega_1^{-1} \\ \omega_{g+1} &\to \omega_1 \omega_{g+1}^{-1} \omega_1^{-1} \\ \omega_k &\to \omega_k \qquad \text{if } k \neq 1,\ g+1\,. \end{aligned}$$

$$(19)\quad \begin{aligned} \phi_{3_*} : \omega_1 &\to \omega_1 \omega_2 \omega_{g+2} \omega_2^{-1} \omega_1 \omega_2 \omega_{g+2}^{-1} \omega_2^{-1} \omega_1^{-1} \\ \omega_2 &\to \omega_1 \omega_2 \\ \omega_{g+1} &\to \omega_1 \omega_2 \omega_{g+2} \omega_2^{-1} \omega_1^{-1} \omega_2 \omega_{g+2}^{-1} \omega_2^{-1} \omega_{g+1} \omega_1 \omega_2 \omega_{g+2}^{-1} \omega_2^{-1} \omega_1^{-1} \\ \omega_k &\to \omega_k \qquad \text{if } k \neq 1,2,g+1\,. \end{aligned}$$

Since $\phi_{i_*}(N) = N$ for $i = 1,2,3$ the assertion is true for $\tilde{\mathfrak{U}}$.

Generators of $\tilde{\mathfrak{G}}$ are the set of matrices $F(S)$ whose S's are the matrices $S_{rt} = \|s_{ij}\|$, $1 \leq r \leq t \leq g$, where $s_{ij} = 0$ if $i \neq r$ or t, while $s_{rt} = s_{tr} = 1$. The matrix in $\tilde{\mathfrak{G}}$ belonging to $F(S_{rr})$ is induced by the following automorphism of $\pi_1 \partial X_g$:

(20) $$\phi_{rr_*}: \omega_r \to \omega_r\omega_{g+r}, \quad \omega_k \to \omega_k \text{ if } k \neq r$$

while for $1 \leq r < t \leq g$, the matrix $F(S_{rt})$ is induced by:

(21)
$$\begin{aligned}
\phi_{rt_*}: \omega_k &\to \omega_k \text{ if } k \neq r, t, g+r, g+r+1, \cdots, g+t-1,\\
\omega_r &\to (\omega_r)(\omega_{r+1}\cdots\omega_t\omega_{g+t}\omega_t^{-1}\cdots\omega_{r+1}^{-1})\\
\omega_t &\to (\omega_t)(\omega_{g+t}^{-1}\omega_t^{-1}\cdots\omega_{r+1}^{-1}\omega_{g+r}\omega_{r+1}\cdots\omega_t\omega_{g+t})\\
\omega_{g+r} &\to (\omega_{r+1}\cdots\omega_t\omega_{g+t}^{-1}\omega_t^{-1}\cdots\omega_{r+1}^{-1})\omega_{g+r}\\
&\quad(\omega_{r+1}\cdots\omega_t\omega_{g+t}\omega_t^{-1}\cdots\omega_{r+1}^{-1})\\
\omega_{g+k} &\to (\omega_{k+1}\cdots\omega_t\omega_{g+t}^{-1}\omega_t^{-1}\cdots\omega_{k+1}^{-1})\\
&\quad(\omega_k^{-1}\cdots\omega_{r+1}^{-1}\omega_{g+r}\omega_{r+1}\cdots\omega_k)\\
&\quad(\omega_{k+1}\cdots\omega_t\omega_{g+t}\omega_t^{-1}\cdots\omega_{k+1}^{-1})\\
&\quad(\omega_k^{-1}\cdots\omega_{r+1}^{-1}\omega_{g+r}^{-1}\omega_{r+1}\cdots\omega_k)(\omega_{g+k}),\ k = r+1, \cdots, t-1\,.
\end{aligned}$$

Since $\phi_{rt_*}(N) = N$ for each pair (r, t) of interest, the assertion is likewise true for $\tilde{\mathfrak{G}}$. Hence Lemma 2.2 is true. ||

LEMMA 2.3. *Let* $\Phi, \Psi \in \mathfrak{M}_g$, *with* $g \geq 2$. *Then*

(i) $\phi \approxeq \psi$ *only if* $a(\Psi)$ *is in the same double coset in* $Sp(2g, Z)$ modulo $\tilde{\mathfrak{F}}_g$ *as* $a(\Phi)$.

(ii) $\phi \approx \psi$ *only if* $a(\Psi)$ *is in the same double coset as* $a(\Phi)$ *or* $a(\Delta\Phi^{-1}\Delta^{-1})$.

(iii) $\phi \sim \psi$ *only if* $a(\Psi)$ *is in the same double coset as* $a(\Phi)$ *or* $a(\Delta\Phi^{-1}\Delta^{-1})$ *or* $a(\Phi^{-1})$ *or* $a(\Delta\Phi\Delta^{-1})$.

(If $g = 1$, *the conditions above are not only necessary but also sufficient.)*

Proof. This is an immediate consequence of Theorem 1, Lemma 2.1 and Lemma 2.2. ||

LEMMA 2.4. *Let* $\Phi \in \mathfrak{M}_g$, *and suppose that*

$$\alpha(\Phi) = \begin{bmatrix} R & S \\ P & Q \end{bmatrix}. \tag{22}$$

Then

$$\alpha(\Delta\Phi^{-1}\Delta^{-1}) = \begin{bmatrix} \bar{Q} & \bar{S} \\ \bar{P} & \bar{R} \end{bmatrix}, \quad \alpha(\Phi^{-1}) = \begin{bmatrix} \bar{Q} & -\bar{S} \\ -\bar{P} & \bar{R} \end{bmatrix}, \quad \alpha(\Delta\Phi\Delta) = \begin{bmatrix} R & -S \\ -P & Q \end{bmatrix}. \tag{23}$$

Proof. Since $\alpha(\Phi) \in \mathrm{Sp}(2g, Z)$, and $\Phi \in \mathfrak{M}_g$, equation (11) tells us that

$$\alpha(\Phi^{-1}) = (\alpha(\Phi))^{-1} = J^{-1}\alpha(\Phi)\,J = \begin{bmatrix} \bar{Q} & -\bar{S} \\ -\bar{P} & \bar{R} \end{bmatrix}. \tag{24}$$

To see that $\alpha(\Delta\Phi\Delta^{-1})$ and $\alpha(\Delta\Phi^{-1}\Delta^{-1})$ have the stated forms, recall that δ can be any orientation-reversing homeomorphism of ∂X_g which extends to an orientation-reversing map of X_g. By Lemma 2.1, every symplectic matrix M which satisfies the condition $MJ\bar{M} = -J$ lifts to an orientation-reversing homeomorphism of ∂X_g. It then follows that the matrix

$$\alpha(\Delta) = \begin{bmatrix} -I & O \\ O & I \end{bmatrix} \tag{25}$$

lifts to an orientation-reversing homeomorphism of ∂X_g which extends to X_g. The remaining assertions of Lemma 2.4 follow immediately. ||

We are now ready to determine invariants of double cosets modulo $\tilde{\mathfrak{F}}_g$ in $\mathrm{Sp}(2g, Z)$. Note first that the negative identity matrix belongs to $\tilde{\mathfrak{F}}_g$; and second, that, for any $g \times g$ matrix P, one may always find unimodular matrices U_0, V_0 such that

$$U_0PV_0 = \begin{pmatrix} p_1 & & & & \\ & p_2 & & & \\ & & \cdot & & \\ & & & \cdot & \\ & & & & p_g \end{pmatrix}; \text{ where } p_i \mid p_{i+1}, 1 \le i \le r-1 \le g,\ p_{r-1} \ne 0, \tag{26}$$

and $p_r = \cdots = p_g = 0$.

(See, for example, [6, p. 26].) Hence, by multiplying the four matrices given in Lemma 2.4 on the left by $D(U_0)$ and on the right by $D(V_0)$, and possibly also on the right by the negative identity matrix, we may choose representatives of the double cosets of $\alpha(\Phi)$, $\alpha(\Delta\Phi^{-1}\Delta^{-1})$, $\alpha(\Phi^{-1})$, and $\alpha(\Delta\Phi\Delta)$ of the form:

$$(27)\qquad \begin{pmatrix} R_1 & S_1 \\ P_1 & Q_1 \end{pmatrix}, \begin{pmatrix} \bar{Q}_1 & \bar{S}_1 \\ P_1 & \bar{R}_1 \end{pmatrix}, \begin{pmatrix} -\bar{Q}_1 & \bar{S}_1 \\ P_1 & -\bar{R}_1 \end{pmatrix}, \begin{pmatrix} -R_1 & S_1 \\ P_1 & -Q_1 \end{pmatrix},$$

where $P_1 = U_0 P \bar{V}_0^{-1}$ has the form given in (26), and where $Q_1 = U_0 P V_0$, $R_1 = \bar{U}_0^{-1} R \bar{V}_0^{-1}$, $S_1 = \bar{U}_0^{-1} S V_0$.

LEMMA 2.5. *The diagonal entries in* P_1 *are invariants of the double cosets modulo* $\tilde{\mathfrak{F}}_g$ *of the four matrices in (27).*

Proof. By Lemma 2.2, the group $\tilde{\mathfrak{F}}$ is the semi-direct product of its normal subgroup $\tilde{\mathfrak{U}}$ and subgroup $\tilde{\mathfrak{G}}$, hence every element $\xi \in \tilde{\mathfrak{F}}$ can be written in the form

$$(28)\qquad \xi = F(S)D(U) = D(U)F(\bar{U}^{-1}SU^{-1})$$

for some unimodular matrix U and some symmetric matrix S. Now, if $M \in Sp(2g, Z)$ is any of the matrices in (27), the effect of multiplying M on the left by $F(S_1)D(U_1)$ and on the right by $D(U_2)F(S_2)$ will be to replace P_1 by $P_2 = U_1 P_1 \bar{U}_2^{-1}$. Since P_1 and P_2 have the same elementary invariants, Lemma 2.5 is true. ||

(REMARK. It is easy to see that the matrix P_1 in each lower left box in (27) is a matrix of integral one-dimensional homology for $X_g \cup_\phi X'_g$. Hence if $p_1 = \cdots = p_k = 1$, $1 < p_{k+1}, \cdots, p_{r-1}$, and $p_r = \cdots = p_g = 0$, then $g - r + 1$ is the Betti number of $X_g \cup_\phi X'_g$ and $p_{k+1}, \cdots, p_{r-1}$ are the coefficients of torsion.)

THEOREM 2. *Let* $\Phi, \Phi' \in \mathfrak{M}_g$, *and suppose that*

$$(29) \qquad \alpha(\Phi) = \begin{bmatrix} R & S \\ P & Q \end{bmatrix} \qquad \alpha(\Phi') = \begin{bmatrix} R' & S' \\ P' & Q' \end{bmatrix}.$$

A necessary condition for $\Phi \equiv \Phi'$ *is that the elementary invariants of the submatrices* P *and* P′ *coincide. If they coincide, and if the elementary invariants of* P *are not all* 0, *and if* $p_1 = p$, *then* $\phi \approxeq \phi'$ *only if*

$$(30) \qquad \det Q' \equiv \det Q (\mathrm{mod}\ p) \quad \textit{and} \quad \det R' \equiv \det R (\mathrm{mod}\ p);$$

$\phi \approx \phi'$ *only if either (30) is satisfied, or*

$$(31) \qquad \det Q' \equiv \det R (\mathrm{mod}\ p) \quad \textit{and} \quad \det R' \equiv \det Q (\mathrm{mod}\ p);$$

$\phi \sim \phi'$ *only if either (30) or (31) are satisfied, or*

$$(32) \qquad \det Q' \equiv - \det Q (\mathrm{mod}\ p) \ \textit{and} \ \det R' \equiv - \det R (\mathrm{mod}\ p),$$

or

$$(33) \qquad \det Q' \equiv - \det R (\mathrm{mod}\ p) \ \textit{and} \ \det R' \equiv - \det Q (\mathrm{mod}\ p).$$

Proof. Let U_1, U_2 be any pair of unimodular matrices which satisfy the condition

$$(34) \qquad U_1 P_1 \overline{U}_2^{-1} = P_1 ,$$

and let S_1 and S_2 be arbitrary symmetric matrices. Then, left multiplication of the matrices in (27) by $F(S_1)D(U_1)$ and right multiplication by $D(U_2)F(S_2)$ will not disturb the normal form of P_1 in any of the matrices in (27). Consider, first, the effect of the latter operation on $\alpha(\Phi)$ in (27). It will be replaced by

$$(35) \qquad \begin{bmatrix} R_2 & * \\ P_1 & Q_2 \end{bmatrix}, \qquad \begin{aligned} R_2 &= U_1 R_1 U_2 + S_1 P_1 , \\ Q_2 &= \overline{U}_1^{-1} Q_1 \overline{U}_2^{-1} + P_1 S_2 . \end{aligned}$$

Since $p | p_i$ for each diagonal entry p_i in P_1, it then follows that:

(36) $$\det Q_2 \equiv \det Q_1 (\text{mod } p) \equiv \det Q (\text{mod } p)$$

(37) $$\det R_2 \equiv \det R_1 (\text{mod } p) \equiv \det R (\text{mod } p) .$$

A similar argument applies to the remaining cases. ||

We wish to use Theorem 2 to establish that there are 3-manifolds which admit more than one weak equivalence class of Heegaard splittings. The example we construct will be a connected sum of lens spaces, hence as a first step we use Theorem 2 to examine equivalence classes of Heegaard splittings of lens spaces.

Lens spaces are 3-manifolds of Heegaard genus 1 which have finite cyclic fundamental groups. We consider, then, two 3-manifolds $X_1 \cup_\phi X'_1$ and $X_1 \cup_{\phi'} X'_1$, where $\phi, \phi' \in \mathfrak{M}_1$, and

(38) $$\phi_* : \omega_1 \to \omega_1^r \omega_2^s \qquad \phi'_* : \omega_1 \to \omega_1^{r'} \omega_2^{s'}$$
$$\omega_2 \to \omega_1^p \omega_2^q \qquad \omega_2 \to \omega_1^p \omega_2^{q'}$$

with $p > 1$. Thus, $X_1 \cup_\phi X'_1$ is the lens space $L(p,q)$ and $X_1 \cup_{\phi'} X'_1$ is the lens space $L(p,q')$. We assert:

COROLLARY 2.1.

(i) $\phi \approxeq \phi'$ iff $q' \equiv q(\text{mod } p)$,

(ii) $\phi \approx \phi'$ iff $q' \equiv q(\text{mod } p)$ *or* $q'q \equiv 1(\text{mod } p)$,

(iii) $\phi \sim \phi'$ iff $q' \equiv \pm q(\text{mod } p)$ *or* $q'q \equiv \pm 1(\text{mod } p)$.

Moreover, $\phi \sim \phi'$ iff $\phi \equiv \phi'$.

Proof. By results in [1], the lens spaces $L(p,q)$ and $L(p',q')$ are homeomorphic iff $q' \equiv \pm q(\text{mod } p)$ or $q'q \equiv \pm 1(\text{mod } p)$, and in view of the fact that $rq - ps = r'q' - p's' = 1$, these are precisely the conditions given in Theorem 2 for weak equivalence of ϕ and ϕ'. To see that the conditions of Theorem 2 for $\phi \approxeq \phi'$, $\phi \approx \phi'$ and $\phi \sim \phi'$ are not only necessary but also sufficient if $g = 1$, one need only produce the homeomorphisms which define the equivalence, and this is easily done by examining the proof of Theorem 1, which is constructive, and using the fact that α is an anti-isomorphism if $g = 1$. ||

As a second application of Theorem 2, we prove that Heegaard splittings are not unique, by exhibiting a 3-manifold of Heegaard genus 2 which admits two weak equivalence classes of Heegaard splittings.

COROLLARY 2.2. *Let $\Phi, \Phi' \in \mathfrak{M}_2$ be represented by homeomorphisms ϕ, ϕ' which induce the automorphisms:*

$$\begin{aligned} \phi_* : \omega_1 &\to \omega_1\omega_3\omega_1^3 \\ \omega_2 &\to \omega_2\omega_4\omega_2^3 \\ \omega_3 &\to \omega_1^{-1}(\omega_1\omega_3\omega_1^3)^2 \\ \omega_4 &\to \omega_2^{-1}(\omega_1\omega_3\omega_1^3)^2 \end{aligned} \tag{41}$$

$$\begin{aligned} \phi'_* : \omega_1 &\to \omega_1\omega_3\omega_1^3 \\ \omega_2 &\to \omega_2\omega_4\omega_2 \\ \omega_3 &\to \omega_1^{-1}(\omega_1\omega_3\omega_1^3)^2 \\ \omega_4 &\to (\omega_4\omega_2^2)^4\omega_2^{-1} \; . \end{aligned} \tag{42}$$

Then $X_2 \cup_\phi X'_2$ is homeomorphic to $X_2 \cup_{\phi'} X'_2$, but the Heegaard splittings ϕ and ϕ' are not weakly equivalent.

Proof. Consider the lens space $L(7,2)$, which admits equivalent Heegaard splittings $X_1 \cup_\beta X'_1$ and $X_1 \cup_\delta X'_1$, defined by the automorphisms

$$\beta_* : \omega_1 \to \omega_1^4\omega_2 \qquad \delta_* : \omega_1 \to \omega_1^2\omega_2$$
$$\omega_2 \to \omega_1^7\omega_2^2 \qquad \omega_2 \to \omega_1^7\omega_2^4 \; . \tag{43}$$

By Corollary 2.1, β and δ are equivalent, but not strongly equivalent. Hence there is an orientation-preserving homeomorphism $h: X_1 \cup_\beta X'_1 \to X_1 \cup_\delta X'_1$ such that $h(X_1) = X'_1$, $h(X'_1) = X_1$.

The connected sum $M \# M'$ of two closed oriented 3-manifolds M, M' is defined in the following manner: remove a 3-cell D from M, and a 3-cell D' from M', and identify ∂D with $\partial D'$ by an orientation-reversing homeomorphism. If M and M' are defined by Heegaard splittings, one may always choose the 3-cells D and D' so that they intersect the Heegaard surfaces in discs on ∂D and $\partial D'$ respectively. Then $M \# M'$ will also have a natural representation as a Heegaard splitting, induced by the Heegaard splittings of M and M'. We will carry this out explicitly in the case where M and M' are each copies of $L(7,2)$.

Let $\pi_1(\partial X_1\text{-disc})$ be the free group freely generated by $\hat{\omega}_1, \hat{\omega}_2$, where the boundary of the deleted disc represents the homotopy class of $\hat{\omega}_1 \hat{\omega}_2 \hat{\omega}_1^{-1} \hat{\omega}_2^{-1}$. Then β_*, δ_* lift to automorphisms $\hat{\beta}_*, \hat{\delta}_*$ of $\pi_1(\partial X_1\text{-disc})$ defined by:

$$\hat{\beta}_* : \hat{\omega}_1 \to \hat{\omega}_1 \hat{\omega}_2 \hat{\omega}_1^3 \qquad \hat{\delta}_* : \hat{\omega}_1 \to \hat{\omega}_1 \hat{\omega}_2 \hat{\omega}_1$$

$$\hat{\omega}_2 \to \hat{\omega}_1^{-1} (\hat{\omega}_1 \hat{\omega}_2 \hat{\omega}_1^3)^2 \qquad \hat{\omega}_2 \to (\hat{\omega}_1 \hat{\omega}_2^2)^4 \hat{\omega}_1^{-1} .$$

We may now define two isomorphisms from $\pi_1(\partial X_1\text{-disc})$ into $\pi_1(\partial X_2)$ by the rules

$$j_* : \pi_1(\partial X_1\text{-disc}) \to \pi_1(\partial X_2) \qquad k_* : \pi_1(\partial X_1\text{-disc}) \to \pi_1(\partial X_2)$$

$$\hat{\omega}_1 \to \omega_1 \qquad \hat{\omega}_1 \to \omega_2$$

$$\hat{\omega}_2 \to \omega_3 \qquad \hat{\omega}_2 \to \omega_4 .$$

Now define automorphisms ϕ_* and ϕ'_* of $\pi_1(\partial X_2)$ by

$$\begin{aligned} \phi_*(\omega_i) &= j_* \hat{\beta}_* j_*^{-1} (\omega_i), \ i = 1,3 \\ &\quad k_* \hat{\beta}_* k_*^{-1} (\omega_i), \ i = 2,4 \\ \phi'_*(\omega_i) &= j_* \hat{\beta}_* j_*^{-1} (\omega_i), \ i = 1,3 \\ &\quad k_* \hat{\delta}_* k_*^{-1} (\omega_i), \ i = 2,4 . \end{aligned}$$

Then ϕ_* and ϕ'_* each define Heegaard splittings of genus 2 of the 3-manifold $L(7,2) \# L(7,2)$.[3] To see that these are not weakly equivalent Heegaard splittings, we apply Theorem 2. Observe that

$$\alpha(\Phi) = \begin{pmatrix} 4 & 0 & 1 & 0 \\ 0 & 4 & 0 & 1 \\ 7 & 0 & 2 & 0 \\ 0 & 7 & 0 & 2 \end{pmatrix}, \qquad \alpha(\Psi) = \begin{pmatrix} 4 & 0 & 1 & 0 \\ 0 & 2 & 0 & 1 \\ 7 & 0 & 2 & 0 \\ 0 & 7 & 0 & 4 \end{pmatrix}. \tag{44}$$

Then, $p = 7$, $\det Q = 4$, $\det R = 16$, $\det Q' = 8$, $\det R' = 8$. Since none of the congruences (30)-(33) is satisfied, it follows that $\Phi \not\sim \Psi$. This proves Corollary 2.2. ||

REMARK. After this manuscript was completed, we learned that similar examples of composite 3-manifolds which admit more than one weak equivalence class of Heegaard splittings had been obtained earlier by Engmann [11]. Her methods are different than those used here.

III. Finer Invariants of Equivalence Classes of Heegaard Splittings

In this section we study the question of determining finer invariants of the double coset modulo $\tilde{F}_g$ of an element $A = \alpha(\Phi)$ in the group $Sp(2g, Z)$. In view of the results in Section II, any such invariants will also be invariants of the equivalence class of Heegaard splittings which are strongly equivalent to ϕ.[4] If

---

[3] Note that care is needed in this definition, because if M' denotes the 3-manifold which is homeomorphic to M but oppositely oriented, then $M \# M'$ and $M \# M'$ are not homeomorphic. In our definition, the first and second copies of L(7,2) in the two cases are coherently oriented, so that our connected sums define homeomorphic 3-manifolds. This would not be the case if β and δ were weakly equivalent, but not equivalent splittings.

[4] We restrict our attention here to strong equivalence, however the results are easily modified to include the four cases considered earlier.

(44) $$\alpha(\Phi) = A = \begin{bmatrix} R & S \\ P & Q \end{bmatrix}$$

then one such set of invariants were shown in Lemma 2.5 to be the elementary divisors of P. Therefore we may without loss of generality assume that P is in Smith normal form, and restrict our attention to multiplication of A on the right and the left by elements in $\tilde{\mathfrak{F}}_g$ which leave P invariant. This is equivalent to the restriction that in considering multiplication by elements of the subgroup $\tilde{\mathfrak{U}}_g$ of $\tilde{\mathfrak{F}}_g$ we restrict our attention to left multiplication by elements $D(U)$ and right multiplication by elements $D(V)$, where U and V satisfy the condition

(45) $$UP = P\bar{V} .$$

We begin by considering the case where P is singular. The diagonal entries of P will be denoted $p_1, \cdots, p_g$.

LEMMA 3.1. *Suppose that* $p_r = p_{r+1} = \cdots = p_g = 0$, but $p_{r-1} \neq 0$. *Let* A_1 *be matrix obtained from* A *by deleting rows* r *through* g *and* $g+r$ *through* $2g$ *from* A. *Then* A_1 *is in* $Sp(2(r-1), Z)$ *and invariants of* A_1 mod $\tilde{\mathfrak{F}}_{r-1}$ *are invariants of* A mod $\tilde{\mathfrak{F}}_g$.

Proof. Since A is in $Sp(2g, Z)$, it satisfies equation (11). Equivalently, the $g \times g$ blocks R, S, P, Q satisfy:

(46)
$$\begin{aligned} &46.1\ P\bar{Q} = Q\bar{P} && 46.4\ \bar{R}Q - \bar{P}S = I \\ &46.2\ \bar{R}P = \bar{P}R && 46.5\ R\bar{S} = S\bar{R} \\ &46.3\ R\bar{Q} - S\bar{P} = I && 46.6\ \bar{Q}S = \bar{S}Q \ . \end{aligned}$$

Since $p_r = p_{r+1} = \cdots = p_g = 0$, equations 46.1 and 46.2 imply that A decomposes as:

$$A = \begin{bmatrix} R_1 & O_2 & S_1 & S_2 \\ R_3 & R_4 & S_3 & S_4 \\ P_1 & O_2 & Q_1 & Q_2 \\ O_3 & O_4 & O_3 & Q_4 \end{bmatrix} \quad \begin{array}{ll} R_1, S_1, P_1, Q_1 & \text{are } (r-1)x(r-1) \\ O_2, S_2, Q_2 & \text{are } (g-r+1)x(r-1) \\ R_3, S_3, O_3 & \text{are } (r-1)x(g-r+1) \\ R_4, S_4, O_4, Q_4 & \text{are } (g-r+1)x(g-r+1) \end{array}$$

where the blocks labeled O denote blocks of zeros. Equation (46.3) now implies that $R_4\bar{Q}_4 = I_4$, the $(g-r+1)x(g-r+1)$ identity matrix. Hence $\det R_4 = \pm 1$ and $R_4 = \bar{Q}_4^{-1}$. Define the unimodular matrix $U^{\#}$ by

$$U^{\#} = \begin{bmatrix} I_1 & O_2 \\ O_3 & \bar{R}_4 \end{bmatrix}.$$

The $D(U^{\#})A$ is equivalent to A and has the simpler block decomposition

$$A^{\#} = D(U^{\#})A = \begin{bmatrix} R_1 & O_2 & S_1 & S_2 \\ R_3^{\#} & I_4 & S_3^{\#} & S_4^{\#} \\ P_1 & O_2 & Q_1 & Q_2 \\ O_3 & O_4 & O_3 & I_4 \end{bmatrix}. \tag{47}$$

In any further modifications of $A^{\#}$ mod $\tilde{\mathfrak{F}}_g$ we may now restrict ourselves to modifications which not only preserve the normal form of P, but also preserve the blocks O_2 and I_4 in R, and the blocks O_3 and I_4 in Q.

Let A_1 be the $2(r-1)x2(r-1)$ matrix

$$A_1 = \begin{bmatrix} R_1 & S_1 \\ P_1 & Q_1 \end{bmatrix}.$$

The fact that $A^{\#}$ satisfies equations (46) is now seen to imply that A_1 satisfies the corresponding conditions, hence A_1 is an element of $Sp(2(r-1), Z)$.

We consider now the effect on A_1 of left and right multiplication of $A^{\#}$ by elements $D(U), D(V)$ which preserve the normal form of P and the partial normal forms of R and Q. The condition that left multiplication of $A^{\#}$ by $D(U)$ and right multiplication by $D(V)$ not alter P is given by equation (45). This implies that U and V have the block decompositions

$$U = \begin{bmatrix} U_1 & U_2 \\ O_3 & U_4 \end{bmatrix} \qquad V = \begin{bmatrix} V_1 & V_2 \\ O_3 & V_4 \end{bmatrix}$$

where $U_1P_1 = P_1\bar{V}_1$. The condition that the normal forms of Q and R be preserved then implies that $U_4V_4 = I_4$, hence $\det U_4 = \det V_4 = \pm 1$ and $V_4 = U_4^{-1}$. This, in turn, implies that $\det U_1 = \det V_1 = \pm 1$. It then follows that left (respectively right) multiplication of $A^\#$ by $D(U)$ (respectively $D(V)$) has the same effect on A_1 as left (right) multiplication of A_1 by $D(U_1)$ $(D(V_1))$, where $D(U_1)$ and $D(V_1)$ denote elements of $Sp(2(r-1), Z)$.

Left (respectively right) multiplication of $A^\#$ by elements $F(L)$ (or $F(K)$) will not change the normal form of P or the partial normal forms of Q and R, no matter how one chooses the symmetric matrices L and K. Moreover, if L and K are partitioned as before into blocks, then the effect on A_1 of replacing $A^\#$ by $F(L)\,A^\#\,F(K)$ is the same as the effect of replacing A_1 by $F(L_1)\,A_1\,F(K_1)$. This proves Lemma 3.1. ||

REMARK. Modifications in A_1 mod $\tilde{\mathfrak{F}}_{r-1}$ do not, in general, preserve the subblocks $Q_2, R_3^\#, S_1, S_2, S_3^\#$ or $S_4^\#$ of $A^\#$. However, each modification of A_1 lifts to a modification of $A^\#$, defined by setting $U_2 = V_2 = 0_2$ and $U_4 = V_4 = I_4$. Moreover, subsequent modifications of $A^\#$ with $U_1 = V_1 = I_1$ and $L_1 = K_1 = 0_1$ will then leave any normal form which we find for A_1 invariant, hence it is possible to treat the subblock A_1 separately from the rest of $A^\#$.

LEMMA 3.2. *The matrix* S_1 *in the upper right corner of any matrix which is* $\tilde{\mathfrak{F}}_{r-1}$*-equivalent to* A_1 *and has its* P_1 *in Smith normal form is determined by the remaining entries.*

Proof. This is an immediate consequence of the fact that every symplectic matrix must satisfy the six equations (46.1)-(46.6), and since P_1 is diagonal and non-singular, equations (46.3) and (46.4) determine S_1 uniquely.

Since P_1 is in normal form, and since by Lemma 3.2 the matrix S_1 need not be considered further, we have reduced the problem of finding invariants of A_1 mod $\tilde{\mathfrak{F}}_{r-1}$ to that of studying the effect of equivalence mod $\tilde{\mathfrak{F}}_{r-1}$ on Q_1 and R_1. The most general type of modification which we must consider is one which replaces Q_1 and R_1 with

$$
\begin{aligned}
(48) \qquad Q^* &= U_1 Q_1 V_1 \quad + P_1 K_1 \\
R^* &= \bar{U}_1^{-1} R_1 \bar{V}_1^{-1} + L_1 P_1
\end{aligned}
$$

where U_1 and V_1 satisfy (45), i.e. $U_1 P_1 = P_1 \bar{V}_1$, also $\det U_1 = \det V_1 = 1$, and where K_1 and L_1 are arbitrary $(r-1)x(r-1)$ symmetric matrices.

It will now be helpful to note the effect of (45) and (46) on the individual entries of admissible U_1, V_1, Q_1, R_1. Let $m_1, \cdots, m_{r-2}$ denote the ratios of the diagonal entries of P_1, i.e.

$$
(49) \qquad p_{t+1} = p_t m_t, \qquad t = 1, 2, \cdots, r-2 .
$$

Let $U_1 = \|u_{ij}\|$, $V_1 = \|v_{ij}\|$, $Q_1 = \|q_{ij}\|$, $R_1 = \|r_{ij}\|$. Then (45) is equivalent to the conditions

$$
\begin{aligned}
(50) \qquad v_{ji} &= u_{ij} m_i m_{i+1} \cdots m_{j-1} \quad \text{if} \quad i < j \\
v_{ii} &= u_{ii} \\
u_{ij} &= v_{ji} m_j m_{j+1} \cdots m_{i-1} \quad \text{if} \quad i > j .
\end{aligned}
$$

Similarly, the symmetries imposed on Q_1 and R_1 by virtue of equations (46.1) and (46.2) imply that

$$
\begin{aligned}
q_{ji} &= q_{ij} m_i m_{i+1} \cdots m_{j-1} \quad \text{if} \quad i < j \\
r_{ij} &= r_{ji} m_j m_{j+1} \cdots m_{i-1} \quad \text{if} \quad i > j .
\end{aligned}
$$

LEMMA 3.3. *Let* A_1 A'_1 *be matrices in* $Sp(2(r-1), Z)$ *which have the block form*

$$
A_1 = \begin{bmatrix} R_1 & S_1 \\ P_1 & Q_1 \end{bmatrix} \qquad A'_1 = \begin{bmatrix} R'_1 & S'_1 \\ P_1 & Q'_1 \end{bmatrix} .
$$

Suppose also that P_1 *is non-singular, and is in Smith normal form. Con-*

sider the sets C *of all ordered pairs of matrices* (Q^*, R^*), *as defined in (48) and also the corresponding set* C′ *of all ordered pairs* (Q'^*, R'^*). *Let* E, E′ *denote the congruence class of* C, *respectively* C′ mod p_{r-1}, *where congruence means congruence elementwise of the individual matrices in* C, C′. *Then* A_1, A'_1 *are in the same double coset* mod $\tilde{\mathfrak{F}}_{r-1}$ *if and only if the sets* E *and* E′ *coincide.*

Proof. It is a consequence of the remarks following the proof of Lemma 3.2 that A_1 and A'_1 are in the same double coset mod $\tilde{\mathfrak{F}}_{r-1}$ if and only if C coincides with C′. If C and C′ coincide, then it is clearly necessary that E and E′ coincide. To see that the converse is also true, suppose that E and E′ coincide. Then, for some admissible U_1, V_1, K_1, L_1 as above it must be true that

$$Q'_1 \equiv U_1 Q_1 V_1 + P_1 K_1 (\text{mod } p_{r-1}) \tag{52}$$

$$R'_1 \equiv \bar{U}_i^{-1} R_1 \bar{V}_1^{-1} + L_1 P_1 (\text{mod } p_{r-1}) \,. \tag{53}$$

Since $p_i | p_{r-1}$ for each $i = 1, \cdots, r-2$, and since K_1 and L_1 range over all possible symmetric integer matrices, we now assert that by possibly choosing a new pair of symmetric matrices K_2, L_2 the congruences of (52) can be made equalities:

$$Q'_1 = U_1 Q_1 V_1 + P_1 K_1 + P_1 K_2 \tag{54}$$

$$R'_1 = \bar{U}_1^{-1} R_1 \bar{V}_1^{-1} + L_1 P_1 + L_2 P_1 \,. \tag{55}$$

This is immediate for entries which are on and above the main diagonal in (52) and those which are on and below the main diagonal in (53). It is true for every entry because of the symmetries imposed by equations (51). Thus C and C′ have an element in common, which implies that the entire sets C and C′ coincide, since the entire set can be computed from any one entry. ||

In the computation of the sets E and E′, it was necessary to first calculate the matrices Q^* and R^* in (48) over the integers, and then reduce elementwise mod p_{r-1}. This means that U_1 and V_1 are required to range over the infinite set of pairs U_1, V_1 in $SL(r-1, Z)$ which satisfy (50). However, it is immediate that we can restrict our attention to admissible pairs in the *finite* group $SL(r-1, Z_{p_{r-1}})$, where U_1 and V_1 are now restricted to matrices which satisfy equations (50) mod p_{r-1}. Thus the sets E and E′ of Lemma 3.3 may be computed by a finite procedure.

The finite set E may now be replaced by a particular member of E, which will be regarded as a representative of the class. For example, one might select such a representative by ordering the matrices Q^* and choosing a "smallest" one; such an ordering may be based on an ordering of the individual entries q_{ij} of the array of matrices in E. This representative then defines a unique matrix

$$A_{10} = \begin{bmatrix} R_{10} & S_{10} \\ P_1 & Q_{10} \end{bmatrix} \tag{56}$$

which we will define to be the normal form for A_1.

THEOREM 3. *The matrix* A_{10} *in (56) is an invariant of the class of Heegaard splittings which are strongly equivalent to* ϕ.

Proof. This is an immediate consequence of Lemmas 3.1, 3.2 and 3.3 and the discussion following the proof of Lemma 3.3. ||

REMARK. Having chosen the matrix $A_{10} = F(L_1)D(U_1)\, A_1\, D(V_1)F(K_1)$ one may now enlarge A_{10} to a suitable modification of the matrix $A^{\#}$ of (47), by replacing the deleted blocks (appropriately modified) to obtain a new representative A_0 of the equivalence class of A mod $\tilde{\mathfrak{F}}_g$ in $Sp(2g, Z)$. We are then free to make further modifications in A_0, but subject to the new restriction that the subblocks corresponding to A_{10}

remain unaltered. It is possible to solve this problem, in order to select a unique double coset representative for A mod $\tilde{\mathfrak{F}}_g$, however we omit that derivation because it is complicated, and not of sufficient interest since we do not know when the procedure will yield *topological* invariants of $X_g \cup_\phi X'_g$. That question cannot be settled until one settles the difficult question of uniqueness of Heegaard splittings which is raised by the example of Corollary 2.2.

A weaker set of invariants, which may also be computed from our symplectic matrices, and which are true topological invariants, will be discussed in the next section.

IV. Stable Equivalence

If $X_g \cup_\phi X'_g$ is a Heegaard splitting of a 3-manifold M, then it is always possible to increase the genus of the Heegaard splitting by forming the connected sum $M \# S^3$, where S^3 denotes the 3-sphere, which is assumed to be represented by a Heegaard splitting $X_1 \cup_\beta X'_1$, and where the 3-balls B and B′ which are removed from M and S^3 in order to define $M \# S^3$ are chosen in such a way that $B \cap \partial X_g$ and $B' \cap \partial X_1$ are each discs. Iterating this process, we may form a splitting we denote $\phi \# \beta_1 \# \cdots \# \beta_n$ of M of any genus $g+n$. Two Heegaard splittings ϕ, ϕ' are of M of genus g and h are said to be *stably* equivalent if there exist integers n, m, with $g+n = h+m$, and splittings $\beta_1, \cdots, \beta_n$ and $\beta'_1, \cdots, \beta'_n$ of S^3 such that $\phi \# \beta_1 \# \cdots \# \beta_n \approx \phi' \# \beta'_1 \# \cdots \# \beta'_m$. This concept is of some interest because it was proved by Singer [10] that any two Heegaard splittings whatsoever of a 3-manifold are stably equivalent. Thus stable equivalence implies topological equivalence.

We now wish to determine how the additional freedom which is allowed under stable equivalence alters the admissible operations which preserve the equivalence class of the symplectic matrix $A = \alpha(\Phi)$ associated with a Heegaard splitting ϕ. Recall that if ϕ, ϕ' are Heegaard splittings of genus g, with $A = \alpha(\Phi)$, $A' = \alpha(\Phi')$, then $\phi \approx \phi'$ only if

A and A' have identical P-blocks, and also only if there exist $g \times g$ symmetric matrices L and K, and $g \times g$ unimodular matrices U, V satisfying equation (45), such that $A = F(L)D(U)A'D(V)F(K)$. (See Theorem 1 and Lemma 2.2.) Our next result says, essentially, that the identical condition is necessary for $\phi \equiv \phi'$, however the requirement that U and V be unimodular is replaced by the less restrictive condition that there exist $n \times n$ unimodular matrices, for some integer $n \geq g$, say U_0, V_0, such that matrices which play the roles of D(U) and D(V) can be obtained from $D(U_0)$ and $D(V_0)$ by striking out rows 1 through $n-g$ and $n+1$ through $2n-g+1$, and the corresponding columns.

We illustrate this with the example of Corollary 2.2. Recall that the Heegaard splittings ϕ and ϕ' of Corollary 2.2 both define the manifold L(7,2) # L(7,2). The desired equivalence between the matrices $\alpha(\Phi)$ and $\alpha(\Phi')$ may be obtained by choosing $n = 3$ and

$$U_0 = V_0 = \begin{bmatrix} -3 & 0 & -2 \\ 0 & 1 & 0 \\ 14 & 0 & 9 \end{bmatrix} \qquad L = \begin{bmatrix} 0 & 0 \\ 0 & -2 \end{bmatrix} \qquad K = \begin{bmatrix} 0 & 0 \\ 0 & -46 \end{bmatrix}.$$

The precise statement of the conditions for stable equivalence is given below. It includes the additional freedom that ϕ and ϕ' may be splittings of distinct genus.

Let $X_k \cup_\phi X'_k$ and $X_h \cup_{\phi'} X'_h$ be Heegaard splittings, with

$$A_0 = \alpha(\Phi) = \begin{bmatrix} R_0 & S_0 \\ P_0 & Q_0 \end{bmatrix}, \qquad A'_0 = \alpha(\Phi') = \begin{bmatrix} R'_0 & S'_0 \\ P'_0 & Q'_0 \end{bmatrix}.$$

It will be assumed that P_0, P'_0 are in diagonal form, with diagonal entries $p_1, \cdots, p_k$ and $p'_1, \cdots, p'_h$. If $p_1 = \cdots = p_t = 1$, but $p_{t+1} \neq 1$, delete rows 1 through t and $k+1$ through $k+t+1$ and the corresponding columns from A_0 to obtain a new matrix A. Similarly, for A'. Suppose then that

$$A = \begin{bmatrix} R & S \\ P & Q \end{bmatrix} \qquad A' = \begin{bmatrix} R' & S' \\ P' & Q' \end{bmatrix}.$$

Let $g = k-t$ and let $g' = k'-t'$. A necessary condition for $\phi \equiv \phi'$ is that $P = P'$. This implies that $g' = g$.

THEOREM 4. *A necessary condition for* $\phi \equiv \phi'$ *is that there exist* $g \times g$ *symmetric matrices* L, K *and* $n \times n$ *unimodular matrices* U_0, V_0, *for some* $n \geq g$, *such that if* $\hat{D}(U)$ and $\hat{D}(V)$ *are the* $2g \times 2g$ *matrices obtained by striking out rows 1 through* $n-g$ *and* $n+1$ *through* $2n-g+1$ *and the corresponding columns from* $D(U_0)$ *and* $D(V_0)$, *then*

$$A = F(L)\,\hat{D}(U)\,A'\,\hat{D}(V)\,F(K) .$$

In the above, the submatrices U, V *obtained from* U_0, V_0 *by striking out the first* $n-g$ *rows and columns are required to satisfy equations (50).*

Proof. Note that if $\phi = \phi'$, then $H_1(X_k U_\phi X'_k)$ $H_1(X_h U_\phi, X'_h)$, hence the diagonal matrices P_0, P'_0 can differ only in having a different number of unit entries on the main diagonal. Considering A_0 first, suppose $Q_0 = \|q_{ij}\|$ and $R_0 = \|r_{ij}\|$.

Define symmetric matrices $K_0 = \|k_{ij}\|$ and $L_0 = \|\ell_{ij}\|$ by the rules $k_{ij} = -q_{ij}$ if $i \leq j$, $i = 1, \cdots, t$ and $k_{ij} = 0$ if $i \leq j$ and $i = t+1, \cdots, g$; also $\ell_{ij} = -r_{ij}$ if $i \geq j$ and $j = 1, \cdots, t$ and $\ell_{ij} = 0$ if $i \geq j$ and $j = t+1, \cdots, g$. Let $A_0^* = F(L_0)\,A_0\,F(K_0) = \begin{bmatrix} R_0^* & S_0^* \\ P_0^* & Q_0^* \end{bmatrix}$. Then A_0^* is the symplectic matrix associated with a Heegaard splitting of genus k which is strongly equivalent to ϕ. By our choice of L_0 and K_0 the matrices R_0^* and Q_0^* will be bordered top and left by t rows and t columns of zeros. It then follows from (46.3) that S_0^* will be bordered top and left by t rows and columns of zeros, except for -1's on the main diagonal. Since any matrix which arises from a Heegaard splitting of this same 3-manifold, e.g. A'_0, may be brought to a similar partial normal form, except possibly with additional borders of zeros and 1's, we may without loss of generality assume that any further modifications do not alter the blocks of zeros and 1's. We may therefore concentrate our attention on the submatrices A and A' defined in the statement of the theorem (which were left unaltered in the multiplication by $F(L_0)$ and $F(K_0)$). We must however keep in mind that the freedom to change the size of A and A' may introduce a new freedom in the choice of L, K, U and V. The choice of L and K will not present any problem, since every symmetric matrix restricts to a symmetric matrix when one deletes the first row and column. However the matrix obtained from a unimodular matrix by a similar deletion may no longer be unimodular, hence we require the condition given in Theorem 4 for the choice of U and V. ||

THEOREM 5 (a generalization of a result due to Reidemeister, [7]). *Let ϕ, ϕ' be Heegaard splittings of genus k and h, and let A and A' be the deleted matrices defined before the statement of Theorem 4, with $P = P'$. If P is singular, perform the additional deletions described in Lemma 3.2, to obtain submatrices of A and A' which we will denote by the symbols A_1 and A'_1. Denote the diagonal entries of P_1 by $p_1, \cdots, p_g$, with $p_i = m_{i-1}p_{i-1}$, $i = 2, \cdots, g$, and let $e_i = \gcd(m_i, m_{i-1})$, $i = 2, \cdots, g$, with $e_1 = p_1$. Let Π_{ix}, $x = 1, \cdots, x_i$ be the ordered array of distinct prime factors of e_i. Let $Q_1 = \|q_{ij}\|$ and $R_1 = \|r_{ij}\|$. Define two arrays of quadratic characters y_{ix} and z_{ix} by the rules:*

1. *If* $\Pi_{ix} | q_{ii}$, *then* $y_{ix} = 0$
   If $\Pi_{ix} | r_{ii}$, *then* $z_{ix} = 0$
2. *If* $\Pi_{ix} \nmid q_{ii}$, *then* $y_{ix} = (q_{ii} | \Pi_{ix})$
   If $\Pi_{ix} \nmid r_{ii}$, *then* $z_{ix} = (r_{ii} | \Pi_{ix})$,

where the symbol $(a|b)$ is the Legendre symbol.[5] *Then, $\phi \equiv \phi'$ only if the ordered arrays y_{ix} and z_{ix} coincide.*

Proof. We examine the manner in which q_{ii} and r_{ii} are altered by the admissible operations in Theorem 4. Note that, by Lemma 3.1, we may

---

5 Let a, b be coprime integers. Then $(a|b) = 1$ if there exists an integer x such that $x^2 \equiv a (\text{mod } b)$. If no such integer x exists, then $(a|b) = -1$.

restrict our attention to the submatrix which is obtained by deleting the rows and columns which are associated with zero diagonal entries in P. (Thus our invariants are associated with the torsion subgroup of the homology group of the manifold.)

Replacing A by $F(L)D(U)\,A\,D(V)F(K)$, we find that Q_1, R_1 go over to $UQ_1V + PK$, $\bar{U}^{-1}R_1\bar{V}^{-1} + LP$. Since $p_ik_{ii} = \ell_{ii}p_i \equiv 0 \pmod{\Pi_{ix}}$, we may restrict our attention to the matrices UQ_1V and $\bar{U}^{-1}R_1\bar{V}^{-1}$.

The entry q_{ii} will be replaced by

$$q_{ii}^* = \sum_{k=1}^{g} \sum_{t=1}^{g} u_{ik}q_{kt}v_{ti} \,.$$

This sum decomposes as:

$$q_{ii}^* = u_{ii}q_{ii}v_{ii} + \sum_{k<i} \sum_{t=1}^{g} u_{ik}q_{kt}v_{ti} + \sum_{k>i} \sum_{t>i} u_{ik}q_{kt}v_{ti}$$

$$+ \sum_{k>i} \sum_{t<i} u_{ik}q_{kt}v_{ti} + \sum_{k>i} u_{ik}q_{ki}v_{ii}$$

$$+ \sum_{t>i} u_{ii}q_{it}v_{ti} + \sum_{t<i} u_{ii}q_{it}v_{ti} \,.$$

From equations (50) and (51), we now note that

$$\begin{aligned} u_{ii} = v_{ii} \quad & \text{if } k < i, \text{ then } m_{i-1} | u_{ik} \\ & \text{if } t < i, \text{ then } m_{i-1} | q_{it} \\ & \text{if } t > i, \text{ then } m_i | v_{ti} \\ & \text{if } k > i, \text{ then } m_i | q_{ki} \,. \end{aligned}$$

Thus $q_{ii}^* \equiv q_{ii}u_{ii}^2 \pmod{\Pi_{ix}}$, and our assertion follows.

If $R_1^* = \bar{U}^{-1}R_1\bar{V}^{-1}$, then $\bar{R}_1 = VR_1^*U$. Thus, exactly as above, we find that $r_{ii} \equiv r_{ii}^*u_{ii}^2 \pmod{\Pi_{ix}}$, and our theorem is proved. ||

The quadratic characters y_{ix} discovered by Reidemeister, and described in Theorem 5, were interpreted by Seifert [9] to be linking invariants of the torsion subgroup of $H_1(X_k U_\phi X'_k)$. One expects that other integer invariants can be associated with those blocks of the matrix $A^\#$ in (47) which are related to the infinite part of $H_1(X_k U_\phi X'_k)$, i.e. the subblocks $R_3^\#$, Q_2, and S_2, $S_3^\#$, $S_4^\#$. However, lacking a normal form for the submatrix A_1 of $A^\#$ under the general operations allowed in Theorem 4, it appears to be a difficult problem to determine such invariants.[6]

COLUMBIA UNIVERSITY and BARNARD COLLEGE
NEW YORK, NEW YORK 10027

REFERENCES

[1] Brody, "The topological classification of the lens spaces," *Annals of Math.*, 71, *No. 1*, Jan. 1960, 163-184.

[2] Coxeter and Moser, *Generators and Relations for Discrete Groups*. Springer-Verlag, 1965.

[3] Hua, L. K., and Reiner, I., "On the generators of the symplectic modular group," *Trans. Amer. Math. Soc.*, 69 (1949), 415-426.

[4] MacMillan, D. R., "Homeomorphisms on a solid torus." *Proc. AMS*, 14 (1963), 386-390.

[5] Magnus, Karass and Solitar, *Combinatorial Group Theory*. Interscience, John Wiley, 1966.

[6] Newman, Morris, *Integral Matrices*. Academic Press, 1972.

[7] Reidemeister, Kurt, "Heegaarddiagramme und Invarianten von Mennigfaltigkeiten." Hamburg Abh. 10 (1933), 109-118.

[8] Waldhausen, Friedhelm, "Heegaard-Zerlegungen der 3-sphäre." *Topology*, Vol. 7 (1968), 195-203.

[9] Seifert, Herbert, "Verschlingungsinvarianten." Sitzumgsberichte der Preuss. Akad. der Wiss., Berlin, 1933, 811-828.

[10] Singer, J., "Three dimensional manifolds and their Heegaard diagrams." Trans. AMS, Vol. 35 (1933), 88-111.

[11] Engmann, Renate, "Nicht-homöomorphe Heegaard-Zerlegungen vom Geschlecht 2 der zusammenhängen Summe zweier Linsenräume." Abh. Math. Sem. Univ. Hamburg 35 (1970), 33-38.

---

[6] Recent results indicate that the signature and type of the symmetric matrices QP and PR, where P is diagonal, are also topological invariants.

BRANCHED CYCLIC COVERINGS

Sylvain E. Cappell* and Julius L. Shaneson*

The present paper outlines a solution to the following problem of Fox [4]: *When is the* p*-fold branched cyclic covering space of a manifold, with a manifold branching set, again a manifold?* The solution of this problem has many consequences for the study of cyclic group actions on manifolds. A few examples of applications are described below in Section 1. In Section 2 we study branched cyclic covers of S^3 and relate a result on these to the classical P. A. Smith conjecture and the above problem of Fox.

§1. *Solution of Fox's problem*

Let M^n and W^{n+2} be P.L. manifolds with M compact and $f: M \to W$ a P.L. embedding which is proper, i.e. $f(\partial M) = \partial W \cap f(M)$. A branched cyclic covering space of W along M is a simplicial complex Y equipped with a simplicial map $\pi: Y \to W$ so that Y is a branched cover of W along M [4] with $\pi^{-1}(M) \cong M$ a P.L. homeomorphism and $Y - M \to W - M$ a regular covering space with a finite cyclic group of covering translations. Note that we do not assume that f(M) is a locally-flat submanifold of W. It is easy to see that in general W has a p-fold cyclic cover branched along M if and only if there is a class of order p in $H^1(W-M; Z_p)$ which under the composition of the natural maps $H^1(W-M; Z_p) \to H^1(\bar{\partial} E; Z_p) \to H^2(E, \bar{\partial} E; Z_p)$ goes to a mod p Thom class [3] of the regular neighborhood E of M in W, with $\bar{\partial} E = \partial E -$ interior $(\partial W \cap E)$. If $W = E$ is a regular neighborhood of M, this condition just means that the integral Thom class of E, defined by analogy with the Thom class of a bundle, is divisible by p.

---

* This research was supported by NSF grants.

Fox observed that if M^n is a locally-flat submanifold of W^{n+2}, Y is certainly a manifold [4]. A precise but entirely local answer to Fox's problem for P.L. but not necessarily locally-flat submanifolds can be given as follows. Regard M as a subcomplex of a triangulation of W. For any point in the interior of an i-simplex Δ_α^i of M, the link pair in (M, W) is the i-th suspension of a P.L. locally flat knot pair (S^{n-1-i}, S^{n+1-i}) [11]. Let X_α be the manifold which is the p-fold cyclic cover of S^{n+1-i} along this locally-flat S^{n-1-i}. It is easy to see that Y is a P.L. manifold if and only if each such X_α is a sphere. While this reduces Fox's problem to questions about locally-flat P.L. submanifolds, it is too local to be very useful in applications.

We are thus led to a reformulation of Fox's problem. First note that outside of a regular neighborhood of the branching set M, Y is certainly a manifold. Thus, Fox's problem is solved by determining which branched cyclic covers of a manifold regular neighborhood E^{n+2} of M^n are again P.L. manifolds. Two manifold oriented regular neighborhoods E_0^{n+2} and E_1^{n+2} of M^n are said to be concordant if there is an oriented regular neighborhood V of $M \times 1$ which restricts to regular neighborhoods E_1 of $M \times 1$ and $-E_0$ of $M \times 0$. Recall the classifying space for oriented regular neighborhoods $BSRN_2$ constructed in [3] using results of [11] and analyzed using methods of [2]. See also [6], [1]. Concordance classes of manifold oriented codimension two regular neighborhoods of M are in 1 to 1 correspondence with elements of $[M, BSRN_2]$. Theorem 1 provides a global answer to the following formulation of Fox's problem. *Which* P.L. *oriented manifold regular neighborhoods of* M *are concordant to regular neighborhoods which have manifold* p-*fold cyclic covers branched along* M? In applying Theorem 1 it is useful to recall that if $f_0 : M \to W^{n+2}$ is a P.L. embedding, $n \geq 4$, with E_0 the regular neighborhood of $f_0(M)$ in W, and E_1 is a manifold regular neighborhood of M which is concordant to E_0, then there is an ambient concordance of f_0 in W to a P.L. embedding f_1 with E_1 P.L. homeomorphic to a regular neighborhood of $f_1(M)$ [3].

THEOREM 1. *There exists a classifying space* $BSRN_2^{Z_p}$ *equipped with a natural map* $\pi: BSRN_2^{Z_p} \to BSRN_2$ *such that an oriented manifold regular neighborhood* E_0^{n+2} *of the* P.L. *manifold* M^n *is concordant to an oriented manifold regular neighborhood* E_1 *of* M *with a manifold* p-*fold branched cyclic cover if and only if the map* $g: M \to BSRN_2$ *which classifies* E_0 *lifts to a map* $\bar{g}: M \to BSRN_2^{Z_p}$ *so that* $\pi\bar{g}$ *is homotopic to* g.

This classifying space for branched cyclic covers $BSRN_2^{Z_p}$ has non-finitely generated homotopy groups in even dimensions greater than 2. To see this, note that an element of $\pi_i(BSRN_2^{Z_p})$ is represented by a regular neighborhood E^{i+2} of S^i, which after being modified within its concordance class may be assumed to be locally-flat except possibly at one point P of S^i. The p-fold cyclic branched cover of the link pair of P in (S^i, E^{i+2}) is then, by the local criteria for branched covering spaces to be manifolds discussed above, a sphere equipped with a semi-free Z_p action with a knot as fixed points. This construction defines a map which is an isomorphism (except for $i = 2$, when it has kernel Z) of $\pi_i(BSRN_2^{Z_p})$ to the groups of concordance classes of "(i+2)-dimensional counterexamples to the P. A. Smith conjecture" defined and algebraically analyzed in [2, §11]. In particular, $\pi_{2i}(BSRN_2^{Z_p})$ is not finitely generated for $i > 1$, $\pi_{2i+1}(BSRN_2^{Z_p}) = 0$ for p odd, and $\pi_2(BSRN_2^{Z_p}) = Z$ if the classical P.A. Smith conjecture is true for Z_p actions on S^3. Thus, as a consequence of [13], $\pi_2(BSRN_2^{Z_2}) = Z$.

The detailed homotopy type of $BSRN_2^{Z_p}$ can be studied by combining the homology surgery method of studying codimension two embedding problems of [2], the global approach to non-locally flat embeddings developed in [3] and generalizations of the characteristic variety theorem developed by Sullivan [12] to study G/PL. That the characteristic variety theorem could be generalized to spaces other than G/PL was observed by J. Morgan and by L. Jones.

As an application of Theorem 1 we will consider the following problem: Which oriented closed manifolds M^n are the codimension two fixed points of semi-free actions on S^{n+2}? Theorem 3 which answers this problem will combine the following criteria for M to have a P.L. embedding in S^{n+2} with the condition that M be a Z_p-homology sphere which is imposed by P.A. Smith theory.

THEOREM 2 [3]. *Let* M^n *be a closed* P.L. *manifold with* $\pi_{n+1}(\Sigma M) \to H_{n+1}(\Sigma M)$ *onto. Then there is a* P.L. *embedding* $M \subset S^{n+2}$.

The relationship between the dimension k of the non-locally flat points of the embedding of M^n in S^{n+2} and the characteristic classes of M, developed in [3] shows that in many cases k must be at least $n-4$. Note that if M^n does have a P.L. embedding in S^{n+2}, then by the Thom-Pontrjagen construction, $\pi_{n+2}(\Sigma^2 M) \to H_{n+2}(\Sigma^2 M)$ is onto.

The following result is a kind of converse to P.A. Smith theory. Related results were obtained by L. Jones in high codimensions [7].

THEOREM 3. *Let* M^n *be a* Z_p *homology sphere with* $\pi_{n+1}(\Sigma M) \to H_{n+1}(\Sigma M)$ *surjective. Assume that* $H^2(M; Z_2) = 0$. *Then there exists a semi-free* P.L. *action of* Z_p *on* S^{n+2} *with* M *as fixed points.*

Many M which satisfy the hypothesis of this theorem do not have locally-flat embeddings in S^{n+2}. If the condition in Theorem 3 on the surjectivity of the Hurewicz map is dropped, we can still show, for n odd, that there is a Z_p homology sphere V^{n+2} with a semi-free Z_p action and with M as fixed points. The condition on $H^2(M; Z_2)$ arises from the 3-dimensional P.A. Smith conjecture in a manner which will be described below.

Another result which follows from Theorem 1, an analysis of $BSRN_2^{Z_p}$ and methods of [2], [3] is the following:

THEOREM 4. *Let* W^{n+2} *be an oriented compact* P.L. *manifold equipped with a semi-free* Z_p *action,* p *odd, with fixed points* $M^n \subset$ *interior* (W),

M *an oriented closed* P.L. *manifold with* $H^2(M; Z_2) = 0$. *Then if* n *is odd or if* $\pi_1(W) = 0$, *for every closed* P.L. *manifold* M' *homotopy equivalent to* M, *there exists a compact* P.L. *manifold* W', *equipped with a semi-free* P.L. Z_p *action with* M' *as fixed points, with* $(W', \partial W')$ *equivariantly homotopy equivalent to* $(W, \partial W)$.

The conditions on $H^2(M; Z_2)$ in the above results arise in the following way. In proving Theorems 3 and 4, we study a natural map of $BSRN_2^{Z_p}$ to G/PL and attempt to find a splitting of it. In particular, on the level of the second homotopy groups, we are trying to find a splitting of the map which assigns to a knot which is a counterexample to the classical P.A. Smith conjecture its Arf invariant. We thus propose the following weak form of the P.A. Smith conjecture, whose truth would imply the necessity of the conditions on $H^2(M; Z_2)$.

WEAK P.A. SMITH CONJECTURE. *Let* $K \subset S^3$, $K \cong S^1$ *be the fixed points of a* P.L. Z_p *action on* S^3, p *odd. Then is* $\Delta_K(-1) \equiv \pm 1$ *(modulo 8), where* $\Delta_K(t)$ *is the Alexander polynomial of the knot* $K \subset S^3$?

Fox [5] studied restrictions on $\Delta_K(t)$. However as his methods, which involve expressing homology in terms of $\Delta_K(t)$, apply in high dimensions, where for p odd the weak P.A. Smith conjecture is false [2], they alone will not suffice.

A result on $\pi_2(BSRN_2^{Z_p})$ is indicated at the end of Section 2 below.

§2. *Cyclic branched covering of* S^3

Let $\beta \subset S^3$ be a knot. Let V be a Seifert surface of β, with linking form L_V. Let L be a matrix for L_V with respect to some basis. Let L' denote the transpose of L. If ξ is complex number of norm 1, let

$$K_\xi = L + L' - \xi L - \xi^{-1} L' .$$

Then K_ξ is a Hermitian form over the complex numbers; let $\sigma_\xi(\beta)$ denote its signature. Let $\Sigma(\beta, p)$ be the p-fold cyclic branched cover of S^3 along β, with the induced orientation.

THEOREM 5. *The* p*-fold cyclic cover* $\Sigma(\beta, p)$ *bounds a parallelizable manifold with signature*

$$\sum_{i=1}^{p-1} \sigma_{\xi^i}(\beta) \ ,$$

ξ *a primitive* p^{th} *root of unity.*

NOTES:

1. $\sigma_\xi(\beta) = \sigma_{\xi^{-1}}(\beta)$.
2. The function $\sigma_\xi(\beta)$ is actually a cobordism invariant of β.
3. $\sigma_\xi(\beta)$ is continuous in ξ, except possibly at the negatives of the roots of the Alexander polynomial of β.
4. The manifold constructed to bound $\Sigma = \Sigma(\beta, p)$ is simply connected and has even middle betti number.
5. Analogous results are true in high dimensions.

Theorem 5 has been obtained independently by L. Kauffman [14].

Proof of Theorem 5. Consider

$$P = \beta \times I \bigcup_{\beta \times 0} V \times 0 \subset S^3 \times I \bigcup_{S^3 \times 0} D^4 \cong D^4 \ .$$

Then the p-fold cyclic branched cover Q of D^4 along P is a 4-manifold with boundary $\Sigma(\beta, p)$. Clearly

$$Q = (\hat{\Sigma} \times I) \cup p(D^4) \bigcup_{P \times S^1} P \times D^2$$

where $\hat{\Sigma}$ is the part of Σ lying over the closure of the complement of a tubular neighborhood of β, and $p(D^4)$ is attached to $\hat{\Sigma} \times 1$ along the subset of its boundary $pS^3 = S^3 \cup \cdots \cup S^3$ consisting of p copies of the closure of the complement of a tubular neighborhood of V in S^3. (If $\pi : \to S^3$ is the projection, $\pi|\pi^{-1}(S^3 - V)$ is the trivial p-fold covering space.)

Let $\hat{Q} = \hat{\Sigma} \times I \cup pD^4 \subset Q$. By excision $H_2(\hat{Q}) \cong H_2(\hat{Q}, pD^4) \cong H_2(p(V \times I, V \times \partial I)) \cong H_1(pV)$. Moreover, the mapping $H_2(\hat{Q}) \to H_2(Q)$ is surjective, as the composite

$$H_2(Q, \hat{Q}) \xrightarrow{\cong} H_2(P \times D^2, P \times S^1) \longrightarrow H_1(P \times S^1) \longrightarrow H_1(\hat{Q}) \xrightarrow{\ell} Z$$

ℓ = linking number with P, is a monomorphism. Hence $\hat{Q}$ and Q have the same index. Since $\hat{Q}$ is an unbranched cover of a subset of S^3, it is parallelizable, i.e. for $x \in H_2(\hat{Q})$, $x \cdot x \equiv 0(2)$. Hence Q is also parallelizable.

A basis of $H_2(Q)$ is obtained by pushing circles representing a basis of $H_1(V)$ in each component of $\pi^{-1}V$ in $\Sigma \times 0$ to each of the boundary components of a neighborhood of $\pi^{-1}V$ and making the results bound in the corresponding copy of D^4. With respect to the basis thus obtained from the basis of $H_1 V$ used to obtain L from L_V, it is easy to see that the intersection form on $H_2(\hat{Q})$ has the matrix

$$K = \begin{pmatrix} L+L' & -L & 0 & \cdots & & & 0 & -L' \\ -L' & L+L' & 0 & \cdots & & & 0 & 0 \\ 0 & -L' & L+L' & -L & & \cdots & 0 \cdots & 0 \\ \vdots & & & & & & & 0 \\ \vdots & & & & & & & -L \\ -L & 0 & 0 & \cdots & & & -L' & L+L' \end{pmatrix}.$$

Let H be the matrix

$$\begin{pmatrix} I & I & \xi^2 I & \cdots & \xi^{p-1} I \\ & \xi^2 I & \xi^4 I & \cdots & \xi^{2p-1} I \\ \vdots & & & & \\ I & \xi^{p-1} I & & \cdots & \xi^{(p-1)^2} I \end{pmatrix}$$

where ξ is a primitive p^{th} root of 1 and I is the identity matrix of the same size as L. Then $\overline{H}'KH$ is the matrix

$$p\begin{pmatrix} K_{\xi^0} & & & 0 \\ & K_{\xi^1} & & \\ & & \ddots & \\ 0 & & & K_{\xi^{p-1}} \end{pmatrix}$$

and $\overline{H}'H = pI$. The theorem follows.

NOTE. One can easily show that the intersection form on $H_2(Q)$ has the matrix

$$\begin{pmatrix} K_{\xi} & & & 0 \\ & K_{\xi^1} & & \\ & & \ddots & \\ 0 & & & K_{\xi^{p-1}} \end{pmatrix}.$$

From this and Poincaré Duality, we may recover all known results on $H_1(\Sigma)$.

EXAMPLE: β = trefoil knot, $p = 5$. Then

$$L = \begin{pmatrix} -1 & 1 \\ 0 & -1 \end{pmatrix},$$

so that $\sigma_{\xi}(\beta) = -2 = \sigma_{-1}(\beta)$ for

$$\xi = e^{2\pi it}, \quad 1/6 < t < 5/6 ,$$

and

$$\sigma_{\xi}(\beta) = 0 \quad \text{if} \quad -1/6 < t < 1/6 .$$

Thus $\sum_{i=1}^{4} \sigma_{\xi}(\beta) = -8.$

In fact, it is well known [10] that the 5-fold branched cyclic cover of 3_1 is binary dodecahedral space ("Poincaré space").

As a consequence of Theorem 5 and Rohlin's Theorem [9], and results of [2] the natural periodicity map $\pi_2(BSRN_2^{Z}p) \to \pi_6(BSRN_2^{Z}p)$ is seen to be not surjective.

BIBLIOGRAPHY

[1] Cappell, S. E., and Shaneson, J. L., "Nonlocally flat embeddings, smoothings and group actions." Bull. Amer. Math. Soc. 79 (1973), 577-582.

[2] ———————————, "The codimension two placement problem and homology equivalent manifolds." Ann. of Math. 99 (1974), 277-348.

[3] ———————————, "P.L. embeddings and their singularities." To appear.

[4] Fox, R. H., "Covering spaces with singularities," in "Algebraic Geometry and Topology, A Symposium in honor of S. Lefschetz." Ed. R. H. Fox, et al., Princeton, 1957, 243-257.

[5] ————, "A quick trip through knot theory." Topology of 3-manifolds and related topics. Ed. M. K. Fort, Jr., Prentice Hall 1962, 95-99.

[6] Jones, L., "Three characteristic classes measuring the obstruction to P.L. local unknottedness." Bull.Amer. Math. Soc. 78 (1972), 979-980.

[7] ————, A converse to the fixed point theory of P. A. Smith. Ann. of Math. 94 (1971), 52-68.

[8] Neuwirth, L. P., "Knot groups." Ann. of Math. Studies 56, Princeton 1965.

[9] Rohlin, V. A., "A new result in the theory of 4-dimensional manifolds." Doklady 8 (1952), 221-224.

[10] Seifert, H., and Threlfall, W., "Lehrbach der Topologie." Akad. Verlagsgesesellschaft, B. G. Teubner, Leipzig, 1934.

[11] Stone, D., "Stratified Polyhedra." Springer Lecture Notes in Math. 252, New York and Berlin 1972.

[12] Sullivan, D., "Geometric Topology Seminar Notes." Princeton University 1967 (mimeographed notes).

[13] Waldhausen, F., "Uber Involutionen der 3-Sphare." Topology 8 (1969), 81-91.

[14] Kauffman, L., to appear.

ON THE 3-DIMENSIONAL BRIESKORN MANIFOLDS $M(p, q, r)$

John Milnor

§1. Introduction

Let $M = M(p, q, r)$ be the smooth, compact 3-manifold obtained by intersecting the complex algebraic surface

$$z_1{}^p + z_2{}^q + z_3{}^r = 0$$

of Pham and Brieskorn with the unit sphere $|z_1|^2 + |z_2|^2 + |z_3|^2 = 1$. Here p, q, r should be integers ≥ 2. In strictly topological terms, M can be described as the r-fold cyclic branched covering of the 3-sphere, branched along a torus knot or link of type (p, q). See 1.1 below.

The main result of this paper is that M is diffeomorphic to a coset space of the form $\Pi\backslash G$ where G is a simply-connected 3-dimensional Lie group and Π is a discrete subgroup. In particular the fundamental group $\pi_1(M)$ is isomorphic to this discrete subgroup $\Pi \subset G$. There are three possibilities for G, according as the rational number $p^{-1} + q^{-1} + r^{-1} - 1$ is positive, negative, or zero. In the positive case discussed in Section 4, G is the unit 3-sphere group $SU(2)$, and Π is a finite subgroup of order $4(pqr)^{-1}(p^{-1} + q^{-1} + r^{-1} - 1)^{-2}$. (See Section 3.2.) In the negative case discussed in Section 6, G is the universal covering group of $SL(2, \mathbb{R})$. The proof in this case is based on a study of automorphic forms of fractional degree. In both of these cases the discrete subgroup $\Pi \cong \pi_1(M)$ can be characterized as the commutator subgroup $[\Gamma, \Gamma]$ of a certain "centrally extended triangle group" $\Gamma \subset G$. [See Section 3. This result has also been obtained by C. Giffen (unpublished).] The centrally extended triangle group Γ has a presentation with generators $\gamma_1, \gamma_2, \gamma_3$ and relations

$$\gamma_1{}^p = \gamma_2{}^q = \gamma_3{}^r = \gamma_1 \gamma_2 \gamma_3 \ .$$

(Compare [Coxeter].) It follows that M is diffeomorphic to the maximal abelian covering space of the 3-manifold $\Gamma \backslash G$.

These statements break down when $p^{-1} + q^{-1} + r^{-1} = 1$. However, it is shown in Section 8 that M can still be described as a coset space $\Pi \backslash G$ where G is now a nilpotent Lie group, and Π is a (necessarily nilpotent) discrete subgroup. The proof is based on a more general fibration criterion. (Section 7.)

The author is indebted to conversations with J. -P. Serre, F. Raymond, and J. Joel.

HISTORICAL REMARKS. The triangle groups were introduced by H. A. Schwarz in the last century. [Three-dimensional analogues have recently been constructed by W. Thurston (unpublished).] The study in Section 5 of automorphic forms clearly is based on the work of Klein, Fricke, Poincaré and others. The manifolds $M = M(p, q, r)$ and their (2n–1)-dimensional analogues were introduced by [Brieskorn, 1966]. He computed the order of the homology group $H_1(M; \mathbf{Z})$, showing that M has the homology of a 3-sphere if and only if the numbers p, q, r are pairwise relatively prime. From the point of view of branched covering manifolds, this same result had been obtained much earlier by [Seifert, p. 222]. Those Brieskorn manifolds with $p^{-1} + q^{-1} + r^{-1} > 1$ have long been studied by algebraic geometers: Compare the discussion in [Milnor, 1968, §9.8] as well as [Milnor, 1974]. Those singular points of algebraic surfaces with finite local fundamental group have been elegantly characterized by [Prill] and [Brieskorn, 1967/68]. Those with infinite nilpotent local fundamental group have been elegantly classified by [Wagreich]. For other recent work on such singularities see [Arnol'd], [Conner and Raymond], [Orlik], [Saito], and [Siersma]. The work of [Dolgačev] and [Raymond and Vasquez] is particularly close to the present manuscript.

To conclude this introduction, here is an alternative description of $M(p, q, r)$. Recall that the *torus link* $L(p, q)$ of type (p, q) can be defined as the set of points (z_1, z_2) on the unit 3-sphere which satisfy the equation

$$z_1{}^p + z_2{}^q = 0 .$$

This link has d components, where d is the greatest common divisor of p and q. The n-th component, $1 \le n \le d$, can be parametrized by setting

$$z_1 = e(t/p), \quad z_2 = e((t+n+\tfrac{1}{2})/q)$$

for $0 \le t \le pq/d$, where $e(a)$ stands for the exponential function $e^{2\pi i a}$. Note that this link $L(p,q)$ has a canonical orientation.

LEMMA 1.1. *The Brieskorn manifold* $M(p,q,r)$ *is homeomorphic to the* r*-fold cyclic branched covering of* S^3, *branched along a torus link of type* (p,q).

Proof. Let $V \subset \mathbf{C}^3$ be the Pham-Brieskorn variety $z_1{}^p + z_2{}^q + z_3{}^r = 0$, non-singular except at the origin. Consider the projection map

$$(z_1, z_2, z_3) \mapsto (z_1, z_2)$$

from $V - \mathbf{0}$ to $\mathbf{C}^2 - \mathbf{0}$. If we stay away from the branch locus $z_1{}^p + z_2{}^q = 0$, then clearly each point of $\mathbf{C}^2 - \mathbf{0}$ has just r pre-images in V. In fact these r pre-images are permuted cyclically by the group Ω of r-th roots of unity, acting on $V - \mathbf{0}$ by the rule

$$\omega : (z_1, z_2, z_3) \mapsto (z_1, z_2, \omega z_3)$$

for $\omega^r = 1$. Thus the quotient space $\Omega \backslash (V - \mathbf{0})$ maps homeomorphically onto $\mathbf{C}^2 - \mathbf{0}$. It follows easily that $V - \mathbf{0}$ is an r-fold branched cyclic covering of $\mathbf{C}^2 - \mathbf{0}$, branched along the algebraic curve $z_1{}^p + z_2{}^q = 0$.

Now let the group $\mathbf{R}^+$ of positive real numbers operate freely on $V - \mathbf{0}$ by the rule

$$t : (z_1, z_2, z_3) \mapsto (t^{1/p} z_1, t^{1/q} z_2, t^{1/r} z_3)$$

for $t > 0$. Since every R^+-orbit intersects the unit sphere transversally and precisely once, it follows that $V - 0$ is canonically diffeomorphic to $R^+ \times M(p,q,r)$. Note that this action of R^+ on $V - 0$ commutes with the action of Ω.

Similarly, letting R^+ act freely on $C^2 - 0$ by the rule $t:(z_1, z_2) \mapsto (t^{1/p}z_1, t^{1/q}z_2)$, it follows that $C^2 - 0$ is canonically diffeomorphic to $R^+ \times S^3$. The projection map $V - 0 \to C^2 - 0$ is R^+-equivariant. Therefore, forming quotient spaces under the action of R^+, it follows easily that $M(p,q,r)$ is an r-fold cyclic branched covering of S^3 with branch locus $L(p,q)$. ∎ (Compare [Durfee and Kauffman], [Neumann].)

§2. The Schwarz triangle groups $\Sigma^* \supset \Sigma$

This section will be an exposition of classical material due to H. A. Schwarz and W. Dyck. (For other presentations see [Caratheodory], [Siegel], [Magnus].) We will work with any one of the three classical simply connected 2-dimensional geometries. Thus by the "plane" P we will mean either the surface of a unit 2-sphere, or the Lobachevsky plane [e.g., the upper half-plane $y > 0$ with the Poincaré metric $(dx^2 + dy^2)/y^2$], or the Euclidean plane. In different language, P is to be a complete, simply-connected, 2-dimensional Riemannian manifold of constant curvature $+1$, -1, or 0.

We recall some familiar facts. Given angles α, β, γ with $0<\alpha,\beta,\gamma<\pi$, there always exists a triangle T bounded by geodesics, in a suitably chosen plane P, with interior angles α, β, and γ. In fact P must be either spherical, hyperbolic, or Euclidean according as the difference $\alpha + \beta + \gamma - \pi$ is positive, negative, or zero. In the first two cases the area of the triangle T is precisely $|\alpha + \beta + \gamma - \pi|$, but in the Euclidean case the area of T can be arbitrary.

We are interested in a triangle with interior angles π/p, π/q, and π/r respectively, where $p, q, r \geq 2$ are fixed integers. Thus this triangle $T = T(p,q,r)$ lies either in the spherical, hyperbolic, or Euclidean plane according as the rational number $p^{-1} + q^{-1} + r^{-1} - 1$ is positive, negative, or zero.

DEFINITION. By the *full Schwarz triangle group* $\Sigma^* = \Sigma^*(p,q,r)$ we will mean the group of isometries of P which is generated by reflections $\sigma_1, \sigma_2, \sigma_3$ in the three edges of T(p, q, r). We will also be interested in the subgroup $\Sigma \subset \Sigma^*$ of index 2, consisting of all orientation preserving elements of Σ^*.

REMARK 2.1. Before studying these groups further, it may be helpful to briefly list the possibilities. Let us assume for convenience that $p \le q \le r$. In the *spherical case* $p^{-1} + q^{-1} + r^{-1} > 1$, it is easily seen that (p,q,r) must be one of the triples

$$(2,3,3),\ (2,3,4),\ (2,3,5);\ \text{or}\ (2,2,r)$$

for some $r \ge 2$. The corresponding group $\Sigma(p,q,r)$ of rotations of the sphere is respectively either the tetrahedral, octahedral, or icosahedral group; or a dihedral group of order 2r. The area of the associated triangle T can be any number of the form π/n with $n \ge 2$. In the *Euclidean case* $p^{-1} + q^{-1} + r^{-1} = 1$, the triple (p, q, r) must be either

$$(2,3,6),\ (2,4,4),\ \text{or}\ (3,3,3)\ .$$

For all of the infinitely many remaining triples, we are in the *hyperbolic case* $p^{-1} + q^{-1} + r^{-1} < 1$. The area of the hyperbolic triangle T can range from the minimum value of $(1 - 2^{-1} - 3^{-1} - 7^{-1})\pi = \pi/42$ to values arbitrarily close to π.

The structure of the full triangle group $\Sigma^* = \Sigma^*(p,q,r)$ is described in the following basic assertion. Recall that Σ^* is generated by reflections $\sigma_1, \sigma_2, \sigma_3$ in the three edges of a triangle $T \subset P$ whose interior angles are π/p, π/q, and π/r.

THEOREM 2.2 (Poincaré). *The triangle* T *itself serves as fundamental domain for the action of the group* Σ^* *on the "plane"* P. *In other words the various images* $\sigma(T)$ *with* $\sigma \in \Sigma^*$ *are mutually disjoint except for boundary points, and* *cover all of* P. *This group* Σ^* *has a presentation with generators* $\sigma_1, \sigma_2, \sigma_3$ *and relations*

$$\sigma_1^{\ 2} = \sigma_2^{\ 2} = \sigma_3^{\ 2} = 1$$

and

$$(\sigma_1\sigma_2)^p = (\sigma_2\sigma_3)^q = (\sigma_3\sigma_1)^r = 1 \ .$$

Here it is to be understood that the edges are numbered so that the first two edges e_1 and e_2 enclose the angle of π/p, while e_2 and e_3 enclose the angle of π/q, and e_3, e_1 enclose π/r.

Proof of 2.2. Inspection shows that the composition $\sigma_1\sigma_2$ is a rotation through the angle $2\pi/p$ about the first vertex of the triangle T, so the relation $(\sigma_1\sigma_2)^p = 1$ is certainly satisfied in the group Σ^*. The other five relations can be verified similarly.

Let $\hat{\Sigma}$ denote the abstract group which is defined by a presentation with generators $\hat{\sigma}_1, \hat{\sigma}_2, \hat{\sigma}_3$ and with relations $\hat{\sigma}_i^{\ 2} = 1$ and $(\hat{\sigma}_1\hat{\sigma}_2)^p = (\hat{\sigma}_2\hat{\sigma}_3)^q = (\hat{\sigma}_3\hat{\sigma}_1)^r = 1$. Thus there is a canonical homomorphism $\hat{\sigma} \mapsto \sigma$ from $\hat{\Sigma}$ onto Σ^*, and we must prove that this canonical homomorphism is actually an isomorphism.

Form a simplicial complex K as follows. Start with the product $\hat{\Sigma} \times T$, consisting of a union of disjoint triangles $\hat{\sigma} \times T$, one such triangle for each group element. Now for each $\hat{\sigma}$ and each $i = 1, 2, 3$ paste the i-th edge of $\hat{\sigma} \times T$ onto the i-th edge of $\hat{\sigma}\hat{\sigma}_i \times T$. More precisely, let K be the identification space of $\hat{\Sigma} \times T$ in which $(\hat{\sigma}, x)$ is identified with $(\hat{\sigma}\hat{\sigma}_i, x)$ for each $\hat{\sigma} \in \hat{\Sigma}$, for each $i = 1, 2, 3$, and for each $x \in e_i \subset T$. Using the relation $\hat{\sigma}_i^{\ 2} = 1$, we see that precisely two triangles are pasted together along each edge of K.

Consider the canonical mapping $\hat{\Sigma} \times T \to P$ which sends each pair $(\hat{\sigma}, x)$ to the image $\sigma(x)$ (using the homomorphism $\hat{\sigma} \mapsto \sigma$ from $\hat{\Sigma}$ to the group Σ^* of isometries of P). This mapping is compatible with the identification $(\hat{\sigma}, x) \equiv (\hat{\sigma}\hat{\sigma}_i, x)$ for $x \in e_i$ since the reflection σ_i fixes e_i. Hence there is an induced map

$$f : K \to P .$$

We must prove that f is actually a homeomorphism.

First consider the situation around a vertex $(\hat{\sigma}, v)$ of K. To fix our ideas, suppose that v is the vertex $e_1 \cap e_2$ of T. Using the identifications

$$(\hat{\sigma}, v) \equiv (\hat{\sigma}\hat{\sigma}_1, v) \equiv (\hat{\sigma}\hat{\sigma}_1\hat{\sigma}_2, v) \equiv (\hat{\sigma}\hat{\sigma}_1\hat{\sigma}_2\hat{\sigma}_1, v) \cdots ,$$

together with the relation $(\hat{\sigma}_1\hat{\sigma}_2)^p = 1$, we see that precisely $2p$ triangles of K fit cyclically around the vertex $(\hat{\sigma}, v)$. (These $2p$ triangles are distinct since the $2p$ elements $\hat{\sigma}_1, \hat{\sigma}_1\hat{\sigma}_2, \hat{\sigma}_1\hat{\sigma}_2\hat{\sigma}_1, \cdots, (\hat{\sigma}_1\hat{\sigma}_2)^p$ of $\hat{\Sigma}$ map to distinct elements of Σ^*.) *Now inspection shows that the star neighborhood, consisting of $2p$ triangles fitting around a vertex of K maps homeomorphically onto a neighborhood of the image point $\sigma(v)$ in P.* The image neighborhood is the union of $2p$ triangles in P, each with interior angle π/p at the common vertex $\sigma(v)$.

Thus the canonical map $f : K \to P$ is locally a homeomorphism. But it is not difficult to show that every path in P can be lifted to a path in K. Therefore f is a covering map. *Since P is simply connected, this implies that f is actually a homeomorphism.* The conclusions that $\hat{\Sigma}$ maps isomorphically to the group Σ^*, and that the various images $\sigma(T)$ cover P with only boundary points in common, now follow immediately. ■

REMARK 2.3. More generally, following Dyck, one can consider a convex n-sided polygon A with interior angles $\pi/p_1, \cdots, \pi/p_n$. Again A is the fundamental domain for a group $\Sigma^* = \Sigma^*(A)$ of isometries which is generated by the reflections $\sigma_1, \cdots, \sigma_n$ in the edges of A with relations

$$\sigma_i^2 = (\sigma_i \sigma_{i+1})^{p_i} = 1$$

for all i modulo n. In fact the above proof extends to this more general case without any essential change.

COROLLARY 2.4. *In the spherical case* $p^{-1} + q^{-1} + r^{-1} > 1$, *the full triangle group* $\Sigma^*(p,q,r)$ *is finite of order* $4/(p^{-1} + q^{-1} + r^{-1} - 1)$. *In the remaining cases* $p^{-1} + q^{-1} + r^{-1} \leq 1$, *the group* $\Sigma^*(p,q,r)$ *is infinite.*

Proof. Since the various images $\sigma(T)$ form a non-overlapping covering of P, the order of Σ^* can be computed as the area of P divided by the area of T. ∎

Recall that Σ denotes the subgroup of index 2 consisting of all orientation preserving isometries in the full triangle group Σ^*. Setting

$$\tau_1 = \sigma_1\sigma_2, \ \tau_2 = \sigma_2\sigma_3, \ \tau_3 = \sigma_3\sigma_1 \ ,$$

note that the product

$$\tau_1\tau_2\tau_3 = \sigma_1\sigma_2\sigma_2\sigma_3\sigma_3\sigma_1$$

is equal to 1.

COROLLARY 2.5. *The subgroup* $\Sigma(p,q,r)$ *has a presentation with generators* τ_1, τ_2, τ_3 *and relations* $\tau_1^p = \tau_2^q = \tau_3^r = \tau_1\tau_2\tau_3 = 1$.

Proof. This corollary can be derived, for example, by applying the Reidemeister-Schreier theorem.* [More generally, for the Dyck group described in 2.3 we obtain a presentation with generators $\tau_1, \cdots, \tau_n$ and relations

$$\tau_1^{p_1} = \cdots = \tau_n^{p_n} = \tau_1\tau_2 \cdots \tau_n = 1.]$$

Details will be left to the reader. ∎

We conclude with three remarks which further describe these groups Σ.

---

* See for example [Weir].

REMARK 2.6. Using 2.2, it is easy to show that an element of the group Σ has a fixed point in P if and only if it is conjugate to a power of τ_1, τ_2, or τ_3. *Hence every element of finite order in* Σ *is conjugate to a power of* τ_1, τ_2 *or* τ_3. Therefore the three integers p, q, r can be characterized as the orders of the three conjugate classes of maximal finite cyclic subgroups of Σ. (Caution: In the spherical case these three conjugate classes may not be distinct. In fact in the spherical case, since each vertex of our canonical triangulation of P is antipodal to some other vertex, it follows that each τ_i is conjugate to some τ_j^{-1} where j may be different from i.)

Here we have used the easily verified fact that every orientation preserving isometry of P of finite order has a fixed point.

THEOREM 2.7 (R. H. Fox). *The triangle group* $\Sigma(p, q, r)$ *contains a normal subgroup* N *of finite index which has no elements of finite order.*

[Fox] constructs two finite permutations of orders p and q so that the product permutation has order r. The subgroup N is then defined as the kernel of the evident homomorphism from Σ to the finite group generated by these two permutations. Using 2.6 we see that N has no elements of finite order. ■

Note also that N operates freely on P; that is, no non-trivial group element has a fixed point in P. Hence the quotient space $N \backslash P$ is a smooth compact Riemann surface which admits the finite group Σ/N as a group of conformal automorphisms. To compute the Euler characteristic $\chi(N \backslash P)$ of this Riemann surface, we count vertices, edges, and faces of the canonical triangulation of $N \backslash P$, induced from the triangulation of 2.2. This yields the formula

$$\chi(N \backslash P) = (p^{-1} + q^{-1} + r^{-1} - 1) \operatorname{order}(\Sigma/N) .$$

In the hyperbolic case $p^{-1} + q^{-1} + r^{-1} < 1$, it follows that the triangle group $\Sigma \supset N$ contains free non-abelian subgroups. For N is the fundamental group of a surface of genus $g \geq 2$, hence any subgroup of infinite index in N is the fundamental group of a non-compact surface and therefore is free.

Note that a given finite group Φ can occur as such a quotient Σ/N if and only if Φ is generated by two elements, and has order at least 3. For if Φ is generated by elements of order p and q, and if the product of these two generators has order r, then $\Sigma(p, q, r)$ maps onto Φ, and it follows from 2.6 that the kernel has no element of finite order. As an example, the triangle group $\Sigma(2,3,7)$ maps onto the simple group of order 168. (Compare [Klein and Fricke, pp. 109, 737] as well as [Klein, Entwicklung ⋯, p. 369].) Hence this simple group operates conformally on a Riemann surface $N\backslash P$ whose genus $g = 3$ can be computed from the equation $2 - 2g = 168(1 - 2^{-1} - 3^{-1} - 7^{-1})$.

More generally let Λ be any discrete group of isometries of P with compact fundamental domain. (That is, assume that there exists a compact set $K \subset P$ with non-vacuous interior so that the various translates of K by elements of Λ cover P, and have only boundary points in common.) *Then* Λ *also contains a normal subgroup* N *of finite index which operates freely on* P. (See [Fox] and [Bungaard, Nielsen]. A much more general theorem of this nature has been proved by [Selberg, Lemma 8].) Again the Euler characteristic $\chi(N\backslash P)$ of the smooth compact quotient surface is directly proportional to the index of N in Λ. In fact, the ratio $\chi(N\backslash P)/\text{order}(\Lambda/N)$ can be computed as a product $\chi(B_\Lambda)\chi(P)$ where the rational number $\chi(B_\Lambda)$ is the Euler characteristic of Λ in the sense of [Wall], and where $\chi(P) = \sum (-1)^n \text{ rank } H_n(P)$ is the usual Euler characteristic, equal to 1 or 2. Now assume that Λ preserves orientation.

The quotient $S = \Lambda\backslash P$ can itself be given the structure of a compact Riemann surface, even if Λ has elements of finite order. (Compare 6.3.) In general there will be finitely many ramification points, say $x_1, \cdots, x_k \in S$. Let $r_1, \cdots, r_k \geq 2$ be the corresponding ramification indices. Then classically the data $(S; x_1, \cdots, x_k; r_1, \cdots, r_k)$ provides a complete invariant for the group Λ. That is: a second such group Λ' is conjugate to Λ within the group of orientation preserving isometries of P if and only if the Riemann surface $S' = \Lambda' \backslash P$ is isomorphic to S under an isomorphism which preserves ramification points and ramification indices. The triangle group $\Sigma(p,q,r)$ corresponds to the special case where S has genus zero with three ramification points having ramification indices p, q, r.

REMARK 2.8. It is sometimes possible to deduce inclusion relations between the various groups $\Sigma(p,q,r)$ by noting that a triangle $T(p,q,r)$ can be decomposed into smaller triangles of the form $T(p',q',r')$. For example if $p = q$ one sees in this way that

$$\Sigma(p,p,r) \subset \Sigma(2,p,2r)$$

as a necessarily normal subgroup of index 2. Similarly, taking $p = r$ one sees that

$$\Sigma(2,p,2p) \subset \Sigma(2,3,2p)$$

as an abnormal subgroup of index 3. However, not all inclusions can be derived in this manner. A counterexample is provided by the inclusion $\Sigma(2,3,3) \subset \Sigma(2,3,5)$ of the alternating group on four letters into the alternating group on five letters.

§3. The centrally extended triangle group $\Gamma(p,q,r)$

As in the last section, let P denote either the Euclidean plane or the plane of spherical or hyperbolic geometry. Let $\overline{G}$ denote the connected Lie group consisting of all orientation preserving isometries of P. Then we can form the coset space $\overline{G}/\Sigma$ where

$$\Sigma = \Sigma(p,q,r) \subset \overline{G}$$

is the triangle group of Section 2. Clearly $\overline{G}/\Sigma$ is a compact 3-dimensional manifold. To compute the fundamental group $\pi_1(\overline{G}/\Sigma)$ it is convenient to pass to the universal covering group G of $\overline{G}$.

DEFINITION. The full inverse image in G of the subgroup $\Sigma \subset \overline{G}$ will be called the *centrally extended triangle group* $\Gamma = \Gamma(p, q, r)$.

Evidently the quotient manifold $\overline{G}/\Sigma$ can be identified with G/Γ, and hence has fundamental group $\pi_1(\overline{G}/\Sigma) \cong \Gamma$.

To describe the structure of Γ, let us start with the isomorphism $G/C \cong \overline{G}$ of Lie groups, where the discrete subgroup $C \cong \pi_1(\overline{G})$ is the center of G. In the spherical case, where $\overline{G}$ is the rotation group SO(3), it is well known that this fundamental group C is cyclic of order 2. In the Euclidean and hyperbolic cases we will see that C is free cyclic.

Evidently Γ, defined as the inverse image of Σ under the surjection $G \to \overline{G}$, contains C as a central subgroup with $\Gamma/C \cong \Sigma$. [In fact one can verify that C is precisely the center of Γ.] The main object of this section is to prove the following.

LEMMA 3.1. *The centrally extended triangle group* $\Gamma = \Gamma(p,q,r)$ *has a presentation with generators* $\gamma_1, \gamma_2, \gamma_3$ *and relations* $\gamma_1{}^p = \gamma_2{}^q = \gamma_3{}^r = \gamma_1\gamma_2\gamma_3$.

Proof. We will make use of the following construction. Choose some fixed orientation for the "plane" P. Given a basepoint x and a real number θ, let

$$\overline{r}_x(\theta) \in \overline{G}$$

denote the rotation through angle θ about the point x. Thus we obtain a homomorphism $\overline{r}_x : R \to \overline{G}$ which clearly lifts to a unique homomorphism

$$r_x : R \to G$$

into the universal covering group. Since $\overline{r}_x(2\pi)$ is the identity element of $\overline{G}$, it follows that the lifted element

$$r_x(2\pi) \in G$$

belongs to the central subgroup C. We will use the notation $c = r_x(2\pi) \in C$. *In fact* C *is a cyclic group generated by* c, as one easily verifies by studying the fibration

$$\overline{G} \to P \cong \overline{G}/S^1$$

defined by the formula $\overline{g} \mapsto \overline{g}(x)$. Here S^1 denotes the group $\overline{r}_x(R) \subset \overline{G}$ consisting of all rotations about x. In the Euclidean and hyperbolic cases, since P is contractible, it follows that the fundamental group $\pi_1(S^1) \cong Z$ maps isomorphically onto $\pi_1(\overline{G}) \cong C$.

Note that this element $r_x(2\pi) \in C$ depends continuously on x, and therefore is independent of the choice of x.

Now recall that the subgroup $\Sigma \subset \overline{G}$ is generated by the three rotations

$$\tau_1 = \overline{r}_{v_1}(2\pi/p),\ \tau_2 = \overline{r}_{v_2}(2\pi/q),\ \tau_3 = \overline{r}_{v_3}(2\pi/r)\ ,$$

where v_1, v_2, v_3 are the three vertices of T. It follows that the inverse image $\Gamma \subset G$ is generated by the three lifted rotations

$$\gamma_1 = r_{v_1}(2\pi/p),\ \gamma_2 = r_{v_2}(2\pi/q),\ \gamma_3 = r_{v_3}(2\pi/r)\ ,$$

together with the central element c. Clearly

$$\gamma_1{}^p = \gamma_2{}^q = \gamma_3{}^r = c\ .$$

Next consider the product $\gamma_1\gamma_2\gamma_3$. Since $\tau_1\tau_2\tau_3 = 1$, it is clear that $\gamma_1\gamma_2\gamma_3$ belongs to C, and hence is equal to c^k for some integer k. We must compute this unknown integer k.

It will be convenient to work with a more general triangle, with arbitrary angles. In fact, without complicating the argument, we can just as well consider an n-sided convex polygon $A \subset P$ with interior angles $\alpha_1, \cdots, \alpha_n$. Here we assume that $0 < \alpha_i < \pi$. If σ_i denotes the reflection in the i-th edge (suitably numbered), then $\sigma_i{}^2 = 1$, and therefore

$$(\sigma_1\sigma_2)(\sigma_2\sigma_3) \cdots (\sigma_{n-1}\sigma_n)(\sigma_n\sigma_1) = 1\ .$$

Lifting each rotation

$$\sigma_i \sigma_{i+1} = \bar{r}_{v_i}(2\alpha_i) \in \bar{G}$$

to the element

$$\gamma_i = r_{v_i}(2\alpha_i) \in G \ ,$$

it follows that the product $\gamma_1 \gamma_2 \cdots \gamma_n$ belongs to the central subgroup C. Now as we vary the polygon A continuously, this central element $\gamma_1 \cdots \gamma_n$ must also vary continuously. But C is a discrete group, so $\gamma_1 \cdots \gamma_n$ must remain constant.

In particular we can shrink the polygon A down towards a point x, in such manner that the angles $\alpha_1, \cdots, \alpha_n$ tend towards the angles $\beta_1, \cdots, \beta_n$ of some Euclidean n-sided polygon. Thus the element $\gamma_i = r_{v_i}(2\alpha_i) \in G$ tends towards the limit $r_x(2\beta_i)$, while the product $\gamma_1 \cdots \gamma_n$ tends towards the product $r_x(2\beta_1 + \cdots + 2\beta_n)$. Therefore, using the formula

$$\beta_1 + \cdots + \beta_n = (n-2)\pi$$

for the sum of the angles of a Euclidean polygon, we see that the constant product $\gamma_1 \cdots \gamma_n$ must be equal to

$$r_x((n-2)2\pi) = c^{n-2} \ .$$

Finally, specializing to the case $n = 3$, we obtain the required identity $\gamma_1 \gamma_2 \gamma_3 = c$.

Thus we have proved that Γ is generated by elements $\gamma_1, \gamma_2, \gamma_3$, and c which satisfy the relations

$$\gamma_1{}^p = \gamma_2{}^q = \gamma_3{}^r = \gamma_1 \gamma_2 \gamma_3 = c \ .$$

Conversely, if $\hat{\Gamma}$ denotes the group which is defined abstractly by generators $\hat{\gamma}_1, \hat{\gamma}_2, \hat{\gamma}_3, \hat{c}$ and corresponding relations, then certainly the element $\hat{c} \in \hat{\Gamma}$ generates a central subgroup $\hat{C}$, with quotient $\hat{\Gamma}/\hat{C}$ isomorphic to Σ by Section 2.5. Thus we obtain the commutative diagram

$$\begin{array}{ccccccccc} 1 & \longrightarrow & \hat{C} & \longrightarrow & \hat{\Gamma} & \longrightarrow & \Sigma & \longrightarrow & 1 \\ & & \downarrow \text{onto} & & \downarrow \text{onto} & & \downarrow \cong & & \\ 1 & \longrightarrow & C & \longrightarrow & \Gamma & \longrightarrow & \Sigma & \longrightarrow & 1 . \end{array}$$

In the Euclidean and hyperbolic cases, C is free cyclic, hence $\hat{C}$ maps isomorphically to C, and it follows that $\hat{\Gamma}$ maps isomorphically to Γ.

In the spherical case, since C is cyclic of order 2, we must prove that $\hat{c}^2 = 1$ in order to complete the proof. This relation can be verified by a case by case computation. (Compare [Coxeter].) There is an alternative argument which can be sketched as follows.

To prove that $\hat{c}^2 = 1$, it suffices to show that $\hat{c}^2$ maps to 1 in the abelianized group $\hat{\Gamma}/[\hat{\Gamma},\hat{\Gamma}]$. For clearly $\hat{\Gamma}$ is a central extension of the form

$$1 \to \hat{C}^2 \to \hat{\Gamma} \to \Gamma \to 1 .$$

Such a central extension is determined by a characteristic cohomology class in $H^2(\Gamma; \hat{C}^2)$. Consider the universal coefficient theorem

$$0 \to \mathrm{Ext}(H_1\Gamma, \hat{C}^2) \to H^2(\Gamma; \hat{C}^2) \to \mathrm{Hom}(H_2\Gamma, \hat{C}^2) \to 0$$

[Spanier, p. 243]. The group $H_2\Gamma$ is zero by Poincaré duality, since the finite group Γ is fundamental group of a closed 3-manifold. Therefore our extension is induced from an element of $\mathrm{Ext}(H_1\Gamma, \hat{C}^2)$, or in other words from an abelian group extension of the form

$$0 \to \hat{C}^2 \to A \to H_1\Gamma \to 0 .$$

Thus we obtain a commutative diagram

$$\begin{array}{ccccccccc} 1 & \longrightarrow & \hat{C}^2 & \longrightarrow & \hat{\Gamma} & \longrightarrow & \Gamma & \longrightarrow & 1 \\ & & \downarrow \cong & & \downarrow & & \downarrow & & \\ 0 & \longrightarrow & \hat{C}^2 & \longrightarrow & A & \longrightarrow & H_1\Gamma & \longrightarrow & 0 \end{array}$$

with A abelian. *Therefore, in the spherical case, the group* $\hat{C}^2$ *generated by* $\hat{c}^2$ *maps injectively into the abelianized group* $\hat{\Gamma}/[\hat{\Gamma},\hat{\Gamma}]$.

But a straightforward matrix computation shows that $\hat{c}$ maps to an element of order $m(p^{-1}+q^{-1}+r^{-1}-1)$ in this abelianized group, where m is the least common multiple of p, q, r. In all of the spherical cases this product is 1 or 2, so $\hat{c}^2 = 1$. ∎

REMARK. Similarly for the Dyck group of Section 2.3 one obtains a central extension with generators $\gamma_1, \cdots, \gamma_n$ and with relations

$$\gamma_1^{p_1} = \cdots = \gamma_n^{p_n} = c \quad \text{and} \quad \gamma_1 \cdots \gamma_n = c^{n-2} .$$

COROLLARY 3.2. *The abelianized group* $\Gamma/[\Gamma,\Gamma]$ *has order* $|qr+pr+pq-pqr| = pqr\,|p^{-1}+q^{-1}+r^{-1}-1|$.

Here we adopt the usual convention that an infinite group has "order" zero. Thus the commutator subgroup has finite index in Γ if and only if $p^{-1}+q^{-1}+r^{-1} \neq 1$. To prove this corollary, we apply the usual theorem that the order of an abelianized group with n generators and n relations is equal to the absolute value of the determinant of the $n \times n$ relation matrix. Taking the three relations to be $\gamma_1\gamma_2\gamma_3\gamma_1^{-p} = 1$, $\gamma_1\gamma_2\gamma_3\gamma_2^{-q} = 1$, $\gamma_1\gamma_2\gamma_3\gamma_3^{-r} = 1$, the relation matrix becomes

$$\begin{bmatrix} 1-p & 1 & 1 \\ 1 & 1-q & 1 \\ 1 & 1 & 1-r \end{bmatrix}$$

with determinant $qr + pr + pq - pqr$, as required. ∎

In the spherical case $p^{-1} + q^{-1} + r^{-1} > 1$, since Γ has order $4/(p^{-1}+q^{-1}+r^{-1}-1)$ as a consequence of 2.4, it follows that the commutator subgroup $[\Gamma,\Gamma]$ has order $4/(pqr\,(p^{-1}+q^{-1}+r^{-1}-1)^2)$.

One case of particular interest occurs when p, q, r are pairwise relatively prime. In this case the index $i = |qr+pr+pq-pqr|$ of $[\Gamma,\Gamma]$ in Γ is relatively prime to pqr. Therefore, using 2.6, it follows that for any element γ of Γ which has finite order modulo the center C there exists an element γ^i of $[\Gamma,\Gamma]$ having the same finite order modulo $[\Gamma,\Gamma] \cap C$. It then follows that the three integers p, q, r are invariants of the group $[\Gamma,\Gamma]$. Namely, they can be characterized as the orders of the maximal finite cyclic subgroups of $[\Gamma,\Gamma]$ modulo its center $[\Gamma,\Gamma] \cap C$.

§4. The spherical case $p^{-1} + q^{-1} + r^{-1} > 1$

This section gives a concrete description of the Brieskorn manifolds $M(p,q,r)$ in the spherical case. Since the conclusions are well known, the presentation is mainly intended as motivation for the analogous arguments in Section 6.

Let Γ be any finite subgroup of the group $SU(2)$ of unimodular 2×2 unitary matrices, acting by matrix multiplication on the complex coordinate space C^2. Note that $SU(2)$ acts simply transitively on each sphere centered at the origin.

DEFINITION. A complex polynomial $f(z) = f(z_1, z_2)$ is Γ-invariant if

$$f(\gamma(z)) = f(z)$$

for all $\gamma \in \Gamma$ and all $z \in C^2$. Let $H_\Gamma^{n,1}$ denote the finite dimensional vector space consisting of all homogeneous polynomials of degree n which are Γ-invariant. More generally, given any character of Γ, that is any homomorphism

$$\chi : \Gamma \to U(1) \subset C^{\cdot} = C - 0$$

from Γ to the unit circle, let $H_\Gamma^{n,\chi}$ denote the space of all homogeneous polynomials f of degree n which transform according to the rule

$$f(\gamma(z)) = \chi(\gamma) f(z) .$$

Note that the product of a polynomial in $H_\Gamma^{n,\chi}$ and a polynomial in $H_\Gamma^{m,\rho}$ belongs to the space $H_\Gamma^{n+m,\chi\rho}$. Thus the set of $H_\Gamma^{n,\chi}$ for all n and χ forms a bigraded algebra, which we denote briefly by the symbol $H_\Gamma^{*,*}$. This bigraded algebra possesses an identity element $1 \in H_\Gamma^{0,1}$.

LEMMA 4.1. *Let* $\Pi = [\Gamma, \Gamma]$ *be the commutator subgroup of* Γ. *Then the space* $H_{\Pi}^{n,1}$ *of* Π*-invariant homogeneous polynomials of degree* n *is equal to the direct sum of its subspaces* $H_{\Gamma}^{n,\chi}$ *as* χ *varies over all characters of* Γ.

Proof. Since every character of Γ annihilates Π, it follows that $H_{\Gamma}^{n,\chi} \subset H_{\Pi}^{n,1}$. On the other hand, since Π is normal in Γ, it follows that the quotient group Γ/Π operates linearly on $H_{\Pi}^{n,1}$. In fact, for each Π-invariant homogeneous polynomial f and each $\gamma \in \Gamma$ let $f\gamma$ denote the polynomial

$$z \mapsto f(\gamma(z)) .$$

(Thus Γ acts on the right.) This new polynomial is also Π-invariant since

$$(f\gamma)\pi = (f(\gamma \pi \gamma^{-1}))\gamma = f\gamma$$

for $\pi \in \Pi$. Clearly $f\gamma = f\gamma'$ whenever $\gamma \equiv \gamma' \bmod \Pi$. Since Γ/Π is finite and abelian, it follows that $H_{\Pi}^{n,1}$ splits as a direct sum of eigenspaces corresponding to the various characters of Γ/Π. ∎

Now consider a homogeneous polynomial $f \in H_{\Gamma}^{n,\chi}$ for some n and χ. According to the fundamental theorem of algebra, f must vanish along n (not necessarily distinct) lines $L_1, \cdots, L_n$ through the origin in $\mathbb{C}^2$. Given these lines, the polynomial f is uniquely determined up to a multiplicative constant. Evidently each element of the group Γ must permute these n lines. Conversely, given n lines through the origin which are permuted by Γ, the corresponding homogeneous polynomial $f(z)$ of degree n clearly has the property that the rotated polynomial $f(\gamma(z))$ is a scalar multiple of $f(z)$ for each group element γ. Setting

$$f(\gamma(z))/f(z) = \chi(\gamma)$$

we obtain a character χ of Γ so that $f \in H_{\Gamma}^{n,\chi}$.

Let us apply these constructions to the centrally extended triangle group $\Gamma = \Gamma(p, q, r)$ of Section 3; where $p^{-1} + q^{-1} + r^{-1} > 1$. To do this we must identify SU(2) with the universal covering group G of Section 3. In fact, SU(2) operates naturally on the projective space $P = P^1(C)$ of lines through the origin in C^2. Or rather, since the central element $-I$ carries each line to itself, the quotient group $\overline{G} = SU(2)/\{\pm I\}$ operates on P, which is topologically a 2-dimensional sphere. Choosing a $\overline{G}$-invariant metric, we see easily that P will serve as model for 2-dimensional spherical geometry, with $\overline{G}$ as group of orientation preserving isometries and $G = SU(2)$ as universal covering group.

Let $k = 2/(p^{-1}+q^{-1}+r^{-1}-1)$ denote the order of the quotient group $\Sigma = \Gamma/\{\pm I\}$. Then, by 2.6, nearly every orbit for the action of Σ on P contains k distinct points. The only exceptions are the three orbits containing the three vertices of the triangle T. These three exceptional orbits contain k/p, k/q, and k/r points respectively.

Let $f_1 \in H_\Gamma^{k/p,\chi_1}$, for appropriately chosen χ_1, be the polynomial which vanishes on the k/p lines through the origin corresponding to the orbit of the first vertex of T. Similarly construct the polynomials $f_2 \in H_\Gamma^{k/q,\chi_2}$ and $f_3 \in H_\Gamma^{k/r,\chi_3}$, each well defined up to a multiplicative constant. We will need some partial information about these three characters χ_1, χ_2, and χ_3.

LEMMA 4.2. *The three homomorphisms* $\chi_i : \Gamma \to U(1)$ *constructed in this way satisfy the relation* $\chi_1{}^p = \chi_2{}^q = \chi_3{}^r$.

Proof. Let $\gamma'_1, \cdots, \gamma'_k \in \Gamma$ be a set of representatives for the cosets of the subgroup $\{\pm I\} \subset \Gamma$. Then to each linear form $\ell(z) = a_1 z_1 + a_2 z_2$ we can associate the homogeneous polynomial

$$f(z) = \ell(\gamma'_1(z)) \cdots \ell(\gamma'_k(z))$$

of degree k. The argument above shows that $f \in H_\Gamma^{k,\chi_0}$ for some χ_0. Evidently this character χ_0 depends continuously on the linear form ℓ, and hence is independent of ℓ. Now specializing to the case where $\ell(z)$

vanishes at the line corresponding to one vertex of the triangle T, we see easily that $\chi_1{}^p = \chi_2{}^q = \chi_3{}^r = \chi_0$. ∎

REMARK. The characters χ_i themselves can be computed by the methods of Section 6.1. In fact, writing p_1, p_2, p_3 in place of p, q, r, the character $\chi_i(\gamma_j)$ is equal to $e(-k/2p_ip_j)$ for $i \neq j$ and to $e(1/p_j)e(-k/2p_jp_j)$ for $i = j$.

We are now ready to prove the following basic result.

LEMMA 4.3. *These three polynomials* f_1, f_2, f_3 *generate the bigraded algebra* $H_\Gamma^{*,*}$. *They satisfy a polynomial relation which, after multiplying each* f_i *by a suitable constant if necessary, takes the form* $f_1{}^p + f_2{}^q + f_3{}^r = 0$.

Proof. Let $f \in H_\Gamma^{n,\chi}$ be an arbitrary non-zero element of the bigraded algebra. Then f must have n zeros in $P = P^1(C)$. If one of these zeros lies at the i-th vertex of the triangle T, then clearly f is divisible by f_i. If f does not vanish at any vertex of T, then it must vanish at some point $x \in P$ which lies in an orbit with k distinct elements. Choose $\lambda \neq 0$ so that the linear combination $f_1{}^p + \lambda f_2{}^q \in H_\Gamma^{k,\chi_0}$ also vanishes at x, and hence vanishes precisely at the points of the orbit containing x. Then f is divisible by $f_1{}^p + \lambda f_2{}^q$. Now it follows easily by induction on the degree n that f can be expressed as a polynomial in the f_i.

A similar argument shows that the polynomial $f_3{}^r$ is divisible by $f_1{}^p + \lambda f_2{}^q$ for suitably chosen $\lambda \neq 0$, say

$$f_3{}^r = \lambda'(f_1{}^p + \lambda f_2{}^q) .$$

Multiplying each f_i by a suitable constant, we can put this relation in the required form $f_1{}^p + f_2{}^q + f_3{}^r = 0$. ∎

REMARK. More precisely, one can show that the ideal consisting of all polynomial relations between the f_i is actually generated by $f_1{}^p+f_2{}^q+f_3{}^r$. Compare 4.4 below.

Now let V denote the *Pham-Brieskorn variety* consisting of all triples $(v_1, v_2, v_3) \in C^3$ with $v_1{}^p + v_2{}^q + v_3{}^r = 0$. Evidently the correspondence

$$z \mapsto (f_1(z), f_2(z), f_3(z))$$

maps C^2 into V.

Let $\Pi = [\Gamma, \Gamma]$ denote the commutator subgroup of Γ. Since every character of Γ annihilates Π, we have $f_i(\pi(z)) = f_i(z)$ for $\pi \in \Pi$. Therefore (f_1, f_2, f_3) maps the orbit space $\Pi \backslash C^2$ into V.

LEMMA 4.4. *In fact, this correspondence* $\Pi z \mapsto (f_1(z), f_2(z), f_3(z))$ *maps the orbit space* $\Pi \backslash C^2$ *homeomorphically onto the Pham-Brieskorn variety* V.

Restricting to the unit sphere in C^2, we will prove the following statement at the same time.

THEOREM 4.5. *The quotient manifold* $\Pi \backslash S^3$ *or* $\Pi \backslash SU(2)$ *is diffeomorphic to the Brieskorn manifold* M(p, q, r).

The orbit space $\Pi \backslash S^3$ can be identified with the coset space $\Pi \backslash SU(2)$ since SU(2) operates simply transitively on S^3.

Proof. First consider two points z' and z'' which do not belong to the same Π-orbit. Choose a (not necessarily homogeneous) polynomial $g(z)$ which vanishes at z'', but does not vanish at any of the images $\pi(z')$. Setting

$$h(z) = g(\pi_1(z))\, g(\pi_2(z)) \cdots g(\pi_m(z))$$

where $\Pi = \{\pi_1, \cdots, \pi_m\}$, it follows that h is Π-invariant and $h(z') \neq h(z'')$. Expressing h as a sum of homogeneous polynomials and applying 4.1, we obtain a polynomial $f \in H_\Gamma^{n,\chi}$ for some n and χ satisfying the same condition $f(z') \neq f(z'')$. Finally, applying 4.3, we see that one of the f_i must satisfy $f_i(z') \neq f_i(z'')$. *Thus the mapping* (f_1, f_2, f_3) *embeds* $\Pi \backslash C^2$ *injectively into* V.

Note that each real half-line from the origin in C^2 maps to a curve

$$t \mapsto (t^{k/p} f_1(z), t^{k/q} f_2(z), t^{k/r} f_3(z))$$

in V which intersects the unit sphere of C^3 transversally and precisely once. Therefore we can map the unit sphere of C^2 into $M = M(p, q, r)$ by following each such image curve until it hits the unit sphere, and hence hits M. Thus we obtain a smooth one-to-one map from the quotient $\Pi \backslash S^3$ into M.

But a one-to-one map from a compact 3-manifold into a connected 3-manifold must necessarily be a homeomorphism. Therefore $\Pi \backslash S^3$ maps homeomorphically onto M. It follows easily that $\Pi \backslash C^2$ maps homeomorphically onto V, thus proving 4.4.

Now let us apply the theorem that a one-to-one holomorphic mapping between complex manifolds of the same dimension is necessarily a diffeomorphism. (See [Bochner and Martin, p. 179].) Since the complex manifold $\Pi \backslash C^2 - 0$ maps holomorphically onto $V - 0$, this mapping must have non-singular Jacobian everywhere. It then follows easily that the mapping $\Pi \backslash S^3 \to M$ is also a diffeomorphism. ■

§5. Automorphic differential forms of fractional degree

This section will develop some technical tools concerning functions of one complex variable which will be needed in the next section. Some of the concepts (e.g., "labeled" biholomorphic mappings) are non-standard.

It is common in the study of Riemann surfaces to consider abelian differentials (that is, expressions of the form $f(z)dz$) as well as quadratic differentials (expressions of the form $f(z)dz^2$). More generally, for any integer $k \geq 0$, a *differential* (= differential form) *of degree* k on an open set U of complex numbers can be defined as a complex valued function of two variables of the form

$$\phi(z, dz) = f(z)\,dz^k \ ,$$

where z varies over U and dz varies over $\mathbf{C}$.

To further explain this concept, one must specify how such a differential transforms under a change of coordinates. In fact, if $g: U \to U_1$ is a holomorphic map, and if $\phi_1(z_1, dz_1) = f_1(z_1)\,dz_1{}^k$ is a differential on U_1, then the *pull-back* $\phi = g^*(\phi_1)$ is defined to be the differential

$$\phi(z, dz) = \phi_1(g(z), dg(z)) = f_1(g(z))\,\dot{g}(z)^k dz^k$$

on U. Here $\dot{g}(z)$ denotes the derivative $dg(z)/dz$. This pull-back operation carries sums into sums and products into products.

We will need to generalize these constructions, replacing the integer k by an arbitrary rational number α. There are two closely related difficulties: If α is not an integer, then the fractional power dz^α is not uniquely defined, and similarly the fractional power $\dot{g}(z)^\alpha$ is not uniquely defined.

To get around the first difficulty we agree that the symbol dz is to vary, not over the complex numbers, but rather over the universal covering group $\widetilde{\mathbf{C}}^{\cdot}$ of the multiplicative group $\mathbf{C}^{\cdot}$ of non-zero complex numbers. Since every element of $\widetilde{\mathbf{C}}^{\cdot}$ has a unique n-th root for all n, it follows that the fractional power dz^α is always well defined in $\widetilde{\mathbf{C}}^{\cdot}$.

REMARK. This universal covering group $\widetilde{\mathbf{C}}^{\cdot}$ is of course canonically isomorphic to the additive group of complex numbers. In fact, the exponential homomorphism $e(z) = \exp(2\pi i z)$ from $\mathbf{C}$ to $\mathbf{C}^{\cdot}$ lifts uniquely to an isomorphism

$$\tilde{e}: \mathbf{C} \to \widetilde{\mathbf{C}}^{\cdot}$$

of complex Lie groups. The kernel of the projection homomorphism $\widetilde{\mathbf{C}}^{\cdot} \to \mathbf{C}^{\cdot}$ is evidently generated by the image $\tilde{e}(1)$.

We are now ready to describe our basic objects.

DEFINITION. A *differential* (= differential form) *of degree* α on an open set $U \subset \mathbf{C}$ is a complex valued function of the form

$$\phi(z, dz) = f(z)dz^{\alpha}$$

where z varies over U and dz varies over $\widetilde{\mathbf{C}}^{\cdot}$. Here it is understood that the fractional power dz^{α} is to be evaluated in $\widetilde{\mathbf{C}}^{\cdot}$ and then projected into $\mathbf{C}^{\cdot}$ to be multiplied by $f(z)$. In practice we will always assume that f is holomorphic, so that ϕ is holomorphic as a function of two variables. Note that the product of two holomorphic differentials of degrees α and β is a holomorphic differential of degree $\alpha + \beta$.

In order to define the pull-back $g^{*}(\phi)$ of a differential of fractional degree, we must impose some additional structure on the map g.

DEFINITION. By a *labeled holomorphic map* g from U to U_1 will be meant a holomorphic map $z \mapsto g(z)$ with nowhere vanishing derivative, together with a continuous lifting $\dot{g}$ of the derivative from $\mathbf{C}^{\cdot}$ to $\widetilde{\mathbf{C}}^{\cdot}$. More precisely,

$$\dot{g} : U \to \widetilde{\mathbf{C}}^{\cdot}$$

must be a holomorphic function whose projection into $\mathbf{C}^{\cdot}$ is precisely the derivative $dg(z)/dz$. (Alternatively, a labeling could be defined as a choice of one single valued branch of the many valued function $\log dg(z)/dz$ on U.) Given two labeled holomorphic maps

$$g : U \to U_1 \quad \text{and} \quad g_1 : U_1 \to U_2 \ ,$$

the *composition* $g_1 g : U \to U_2$ has a unique labeling which is determined by the requirement that the chain law identity

$$(g_1 g)^{\cdot}(z) = \dot{g}(z)\dot{g}_1(g(z))$$

should be valid in $\widetilde{\mathbf{C}}^{\cdot}$.

Now consider a labeled holomorphic map $g : U \to U_1$ together with a differential

$$\phi_1(z_1, dz_1) = f_1(z_1)dz_1{}^{\alpha}$$

on U_1. The *pull-back* $g^*(\phi_1)$ is defined to be the differential

$$\phi(z, dz) = \phi_1(g(z), \dot{g}(z)dz)$$

on U. Note that this pull-back operation carries sums into sums and products into products. Furthermore, given any composition

$$U \xrightarrow{g} U_1 \xrightarrow{g_1} U_2$$

of labeled holomorphic maps, the pull-back $(g_1 g)^*(\phi_2)$ of a differential on U_2 is clearly equal to the iterated pull-back $g^*(g_1{}^*(\phi_2))$.

Let Γ be a discrete group of labeled biholomorphic maps of U onto itself.

DEFINITION. A holomorphic differential form $\phi(z, dz) = f(z)dz^a$ on U is *Γ-automorphic* if it satisfies

$$\gamma^*(\phi) = \phi$$

for every $\gamma \in \Gamma$. More generally, given any character $\chi : \Gamma \to U(1) \subset C^{\cdot}$, the form ϕ is called *χ-automorphic* if

$$\gamma^*(\phi) = \chi(\gamma)\phi$$

for every γ. (Thus the Γ-automorphic forms correspond to the special case $\chi = 1$.) Note that a form $\phi(z, dz) = f(z)dz^a$ is χ-automorphic if and only if f satisfies the identity

$$f(\gamma(z))\dot{\gamma}(z)^a = \chi(\gamma)f(z)$$

for all $\gamma \in \Gamma$ and all $z \in U$.

Evidently the χ-automorphic forms of degree a on U form a complex vector space which we denote by the symbol $A_\Gamma^{a,\chi}$. In this way we obtain a bigraded algebra $A_\Gamma^{*,*}$, where the first index a ranges over the additive group of rational numbers and the second index χ ranges over the multiplicative group $\mathrm{Hom}(\Gamma, U(1))$ of characters. This algebra possesses an identity element $1 \in A_\Gamma^{0,1}$. It is associative, commutative, and has no zero-divisors so long as the open set U is connected.

REMARK. The classical theory of automorphic forms of non-integer degree is due to [Petersson]. (Compare [Gunning], [Lehner].) It is based on definitions which superficially look rather different.

Suppose that we are given a normal subgroup of Γ.

LEMMA 5.1. *If* $N \subset \Gamma$ *is a normal subgroup, then the quotient* Γ/N *operates as a group of automorphisms of the algebra* $A_N^{*,1}$ *with fixed point set* $A_\Gamma^{*,1}$. *If the quotient group* Γ/N *is finite abelian of order* m, *then each* $A_N^{\alpha,1}$ *splits as the direct sum of its subspaces* $A_\Gamma^{\alpha,\chi}$ *as* χ *varies over the* m *characters of* Γ *which annihilate* N.

The proof is easily supplied. (Compare 4.1.) ∎

COROLLARY 5.2. *If* $N \subset \Gamma$ *is a normal subgroup of finite index* m, *then every* $\phi \in A_N^{\alpha,1}$ *has a well defined "norm"* $(\gamma_1{}^*\phi)\cdots(\gamma_m{}^*\phi) \in A_\Gamma^{m\alpha,1}$. *Here* $\gamma_1,\cdots,\gamma_m$ *are to be representatives for the cosets of* N *in* Γ.

Again the proof is easily supplied. ∎

It will be important in Section 6 to be able to extract n-th roots of automorphic forms.

LEMMA 5.3. *Let* $\phi(z,dz) = f(z)dz^\alpha$ *be a* χ*-automorphic form. If* f *possesses an* n*-th root,* $f(z) = f_1(z)^n$ *where* f_1 *is holomorphic, then the form* $\phi_1(z,dz) = f_1(z)dz^{\alpha/n}$ *is itself* χ_1*-automorphic for some character* χ_1 *of* Γ *satisfying* $\chi_1{}^n = \chi$.

Proof. For any group element γ, since the holomorphic forms ϕ_1 and $\gamma^*(\phi_1)$ both have degree α/n, the quotient $\gamma^*(\phi_1)/\phi_1$ is a well defined meromorphic function on U. Raising this function to the n-th power we obtain the constant function $\gamma^*(\phi)/\phi = \chi(\gamma)$. Therefore $\gamma^*(\phi_1)/\phi_1$ must itself be a constant function. Setting its value equal to $\chi_1(\gamma)$, it is easy to check that χ_1 is a character of Γ with $\chi_1{}^n = \chi$. ■

As open set U, let us take the upper half-plane P consisting of all $z = x + iy$ with $y > 0$. Then every biholomorphic map from U to itself has the form

$$z \mapsto z' = (g_{11}z + g_{12})/(g_{21}z + g_{22})$$

where

$$\begin{bmatrix} g_{11} & g_{12} \\ g_{21} & g_{22} \end{bmatrix}$$

is an element, well defined up to sign, of the group $SL(2, \mathbf{R})$ of 2×2 real unimodular matrices. The derivative dz'/dz is equal to $(g_{21}z + g_{22})^{-2}$.

It follows easily that the group G consisting of all labeled biholomorphic maps from P to itself can be identified with the universal covering group of $SL(2, \mathbf{R})$. This group G contains an infinite cyclic central subgroup C consisting of group elements which act trivially on P. The generator c of C is characterized by the formulas

$$c(z) = z, \quad \dot{c}(z) = \widetilde{e}(1), \quad \dot{c}(z)^{a} = \widetilde{e}(a) \mapsto e^{2\pi i a} \text{ in } \mathbf{C} .$$

A group $\Sigma \subset \overline{G}$ of conformal automorphisms of P is said to have *compact fundamental domain* if there exists a compact subset $K \subset P$ with non-vacuous interior so that the various images $\sigma(K)$ cover P, and are mutually disjoint except for boundary points. We will be interested in subgroups of G whose images in $\overline{G} = G/C$ satisfy this hypothesis.

LEMMA 5.4. *Let $\Gamma \subset G$ be such that the image $\overline{\Gamma} = \Gamma/(\Gamma \cap C)$ in $\overline{G}$ operates on the upper half-plane P with compact fundamental domain. Then Γ is discrete as a subgroup of the Lie group G, and the coset space $\Gamma\backslash G$ is compact. This group Γ necessarily intersects the center C non-trivially.*

Proof. As noted in Section 2.7 there exists a normal subgroup $N \subset \Gamma$ of finite index so that $\bar{N} = N/N \cap C$ operates freely on P. The orbit space under this action, denoted briefly by the symbol $\bar{N}\backslash P$, is then a smooth compact surface S of genus $g \geq 2$ with fundamental group $\pi_1(S) \cong \bar{N} \cong NC/C$. Here NC denotes the subgroup of G generated by N and C.

Since the group G/C operates simply transitively on the unit tangent bundle $T_1(P)$ of P, it follows easily that the coset space $(NC)\backslash G$ can be identified with the unit tangent bundle $T_1(S)$ of the quotient surface $\bar{N}\backslash P$. In particular this coset space is compact, with fundamental group

$$NC \cong \pi_1(T_1(S)) .$$

Hence the abelianized group $NC/[NC, NC] = NC/[N, N]$ can be identified with the homology group $H_1(T_1(S))$.

It follows that N must intersect C non-trivially. For otherwise NC would split as a cartesian product $N \times C$ with $N = N/N \cap C \cong \pi_1(S)$. Hence $T_1(S)$ would have first Betti number $2g + 1$, rather than its actual value of $2g$.

(Carrying out this argument in more detail and using the Gysin sequence of the tangent circle bundle (see [Spanier, p. 260] as well as [Milnor and Stasheff, pp. 143, 130]), one finds that the kernel of the natural homomorphism from $H_1(T_1(S))$ onto $H_1(S)$ is cyclic, with order equal to the absolute value of the Euler characteristic $\chi(S) = 2 - 2g$. Identifying these two groups with $NC/[N, N]$ and $\bar{N}/[\bar{N}, \bar{N}] \cong NC/[N, N]C$ respectively, we see that this kernel can be identified with $C/[N, N] \cap C$. *Therefore the element* c^{2-2g} *of* C *necessarily belongs to the commutator subgroup* $[N, N] \subset N$.)

Thus N has finite index in NC, so $N\backslash G$ is also compact, and it follows that $\Gamma\backslash G$ is compact. ∎

REMARK. Conversely, if $\Gamma \subset G$ is any discrete subgroup with compact quotient, then one can show that the hypothesis of 5.4 is necessarily satisfied. Such subgroups Γ can be partially classified as follows. Recall from Section 2.7 that the image $\bar{\Gamma} = \Gamma C/C$ is completely classified by the quotient Riemann surface $\bar{\Gamma}\backslash P$ together with a specification of ramification points and ramification indices. But Γ has index at most $2g-2$ in the full inverse image ΓC of $\bar{\Gamma}$. Therefore, for each fixed $\bar{\Gamma}$ there are only finitely many possible choices for Γ.

To show that automorphic forms really exist, we can proceed as follows. Again let Γ satisfy the hypothesis of 5.4 and let $N \subset \Gamma$ be normal of finite index m, with $N/N \cap C$ operating freely on P.

LEMMA 5.5. *If* α *is a multiple of* m, *then the space* $A_\Gamma^{\alpha,1}$ *is non-zero. In fact, this space contains a form* ϕ *which does not vanish throughout any prescribed finite* (*or even countable*) *subset of* P.

Proof. Recall that $A_N^{1,1}$ can be identified with the space of holomorphic abelian differentials $f(z)dz$ on the quotient surface $S = \bar{N}\backslash P$ of genus $g \geq 2$. By a classical theorem, this space has dimension g. Furthermore, using the Riemann-Roch theorem, the space of abelian differentials vanishing at some specified point of S has dimension $g-1$. (Compare [Springer, pp. 252, 270].) Clearly we can choose an element ψ of this g-dimensional vector space so as to avoid any countable collection of hyperplanes. Now the norm $\phi = \gamma_1{}^*(\psi) \cdots \gamma_m{}^*(\psi) \in A_\Gamma^{m,1}$ of Section 5.2 will be non-zero at any specified countable collection of points. Setting $\alpha = km$, it follows that $\phi^k \in A_\Gamma^{\alpha,1}$ has the same properties. ∎

The density of zeros of an automorphic form can be computed as follows. We will think of the upper half-plane P as a model for the Lobachevsky plane, using the Poincaré metric $(dx^2+dy^2)/y^2$, and its associated area element $dxdy/y^2$.

Again let $\Gamma/\Gamma \cap C$ operate on P with compact fundamental domain. Let $\chi : \Gamma \to U(1)$ be a character of finite order. (The hypothesis that χ has finite order is not essential. It is made only to simplify the proof.)

LEMMA 5.6. *If* $\phi \in A_\Gamma^{\alpha,\chi}$ *is a non-zero automorphic form, then the density of zeros of* ϕ *is* $\alpha/2\pi$. *More explicitly: the number of zeros of* ϕ *in a large disk of Lobachevsky area* a, *each zero being counted with its appropriate multiplicity, tends asymptotically to* $a\alpha/2\pi$ *as* $a \to \infty$.

In particular it follows that $\alpha \geq 0$.

Proof. Again we may choose a normal subgroup N of finite index so that $\bar{N} = N/N \cap C$ operates freely on P. Furthermore, after raising ϕ to some power if necessary, we may assume that the character χ is trivial and that the degree $\alpha = k$ is an integer. By a classical theorem, the number of zeros of a k-th degree differential in a compact Riemann surface $\bar{N}\backslash P$ of genus $g \geq 2$ is equal to $(2g-2)k$, where k is necessarily non-negative. (For the case of an abelian differential $f(z)dz$, see for example [Springer, pp. 252, 267]. Given such a fixed abelian differential, any k-th degree differential on $\bar{N}\backslash P$ can be written uniquely as $h(z)f(z)^k dz^k$ where h is meromorphic on $\bar{N}\backslash P$, and hence has just as many zeros as poles.)

Since the quotient $\bar{N}\backslash P$ has area $(2g-2)2\pi$ by the Gauss-Bonnet theorem, it follows that the ratio of number of zeros to area is $k/2\pi$, as asserted. ■

REMARK. If $\phi \neq 0$ is a form in $A_\Gamma^{\alpha,\chi} \subset A_N^{\alpha,\chi|N}$, then it follows that the number of zeros of ϕ in $\bar{N}\backslash P$ is equal to $(2g-2)\alpha$. In particular, $(2g-2)\alpha$ is an integer. Thus we obtain a uniform common denominator for the rational numbers α which actually occur as degrees.

The algebra of N-automorphic forms can be described rather explicitly as follows. Let k be the order of the finite cyclic group $C/N \cap C$.

LEMMA 5.7. *If the rational number* α *is a multiple of* $1/k$, *then,*

$$\dim A_N^{\alpha,1} \geq (2g-2)\alpha + 1 - g ,$$

with equality whenever $\alpha > 1$. *In particular, this vector space is non-zero whenever* $\alpha > \frac{1}{2}$. *On the other hand, if* α *is not a multiple of* $1/k$, *then* $A_N^{\alpha,1} = 0$.

It follows incidentally that $(2g-2)/k$ is necessarily an integer. The following will be proved at the same time.

LEMMA 5.8. *If* α *is a multiple of* $1/k$ *and* $\alpha > g/(g-1)$, *then given two distinct points of* $\overline{N}\backslash P$ *there exists a form in* $A_N^{\alpha,1}$ *which vanishes at the first point but not at the second.*

Proof. For any form ϕ of degree α the identity

$$c^*(\phi) = e(\alpha)\phi$$

is easily verified. Thus if ϕ is N-automorphic and non-zero, with $c^k \in N$, then it follows that $e(k\alpha) = 1$. Hence α must be a multiple of $1/k$.

Conversely, if α is a multiple of $1/k$, then it is not difficult to construct a complex analytic line bundle ξ^α over the surface $S = \overline{N}\backslash P$ so that the holomorphic sections of ξ^α can be identified with the elements of $A_N^{\alpha,1}$. For example, the total space of ξ^α can be obtained as the quotient of $P \times C$ under the group $N/N \cap C$ which operates freely by the rule $\nu : (z, w) \mapsto (\nu(z), \dot{\nu}(z)^{-\alpha} w)$. Every holomorphic section $z \mapsto f(z)$ of the resulting bundle must satisfy the identity $f(\nu(z)) = \dot{\nu}(z)^{-\alpha} f(z)$ appropriate to N-automorphic forms of degree α. Note that the tensor product $\xi^\alpha \otimes \xi^\beta$ can be identified with $\xi^{\alpha+\beta}$.

To compute the Chern class $c_1(\xi^\alpha)$ we raise to the k-th tensor power so that holomorphic cross-sections exist as in 5.5, and then count the number of zeros of a holomorphic section as in 5.6. In this way we obtain the formula

$$c_1(\xi^\alpha)[S] = (2g-2)\alpha\ .$$

Now let us apply the Riemann-Roch theorem as stated in [Hirzebruch, p. 144]: For any analytic line bundle ξ over S,

$$\dim(\text{space of holomorphic sections}) \geq c_1(\xi)[S] + 1 - g \ .$$

Taking $\xi = \xi^a$ this yields

$$\dim A_N^{a,1} \geq (2g-2)a + 1 - g$$

as asserted.

To decide when equality holds, and to prove 5.8, it is perhaps easier to use the older form of the Riemann-Roch theorem, as described in [Springer] or [Hirzebruch, p. 4]. Choosing some fixed $\phi \neq 0$ in $A_N^{a,1}$, any element of $A_N^{a,1}$ can be obtained by multiplying ϕ by a meromorphic function h on $\overline{N} \setminus P$ which has poles at most on the $(2g-2)a$ zeros of ϕ. More precisely the divisors (h) and (ϕ) of h and ϕ must satisfy $(h) \geq (\phi)^{-1}$. According to Riemann-Roch, the number of linearly independent h satisfying this condition is $\geq \deg(\phi) + 1 - g$, with equality whenever the degree $(2g-2)a$ of (ϕ) is greater than the degree $2g-2$ of the divisor of an abelian differential. This proves 5.7.

If we want this form $h\phi$ to vanish at z' [or at both z' and z''], then we must use the divisor $(\phi)^{-1}z'$ [respectively $(\phi)^{-1}z'z''$] in place of $(\phi)^{-1}$. A brief computation then shows the following. If the degree $(2g-2)a - 2$ of the divisor $(\phi)z'^{-1}z''^{-1}$ satisfies

$$(2g-2)a - 2 > 2g - 2 \ ,$$

or in other words if $a > g/(g-1)$, then the space of forms in $A_N^{a,1}$ which vanish at z' [respectively at z' and z''] is equal to $(2g-2)a-g$ [respectively $(2g-2)a-1-g$]. Since these two dimensions are different, there is a form which vanishes at z' but not z''. ■

REMARK. More generally consider the vector space $A_N^{a,\rho}$ where ρ is an arbitrary character of N. Suppose that $\gamma = c^j$ is an element of the intersection $N \cap C$. Then the appropriate equation

$$f(\gamma(z))\dot{\gamma}(z)^a = f(z)\rho(\gamma)$$

takes the form $f(z)e(j\alpha) = f(z)\rho(c^j)$. Evidently there can be a solution $f(z) \neq 0$ only if the rational number α and the character ρ satisfy the relation

$$e(j\alpha) = \rho(c^j)$$

for every c^j in $N \cap C$. Conversely, if this condition is satisfied, then the argument above can easily be modified so as to show that

$$\dim A_N^{\alpha,\rho} \geq (2g-2)\alpha + 1 - g ,$$

with equality whenever $\alpha > 1$.

In the next section we will need a sharp estimate which says that "enough" automorphic forms exist. To state it we must think of an automorphic form ϕ explicitly as a function

$$\phi(z,w) = f(z)w^{\alpha}$$

of two variables, where $z \in P$ and $w \in \widetilde{C}^{\cdot}$. Let the groups $\Gamma \subset G$ operate freely on $P \times \widetilde{C}^{\cdot}$ by the rule

$$g(z,w) = (g(z), \dot{g}(z)w) .$$

With this notation, the statement that ϕ is Γ-automorphic can be expressed by the equation

$$\phi(\gamma(z,w)) = \phi(z,w)$$

for all $\gamma \in \Gamma$, $z \in P$, and $w \in \widetilde{C}^{\cdot}$.

THEOREM 5.9. *With Γ as in 5.4, two points (z', w') and (z'', w'') of $P \times \widetilde{C}^{\cdot}$ belong to the same Γ-orbit if and only if $\phi(z',w') = \phi(z'',w'')$ for every Γ-automorphic form ϕ.*

Proof. First consider the corresponding statement for the normal subgroup $N \subset \Gamma$ of Section 2.7. If $\phi(z',w') = \phi(z'',w'')$ for every $\phi \in A_N^{*,1}$ note that z' and z'' belong to the same N-orbit. For otherwise by 5.8 there would exist a form $\phi \in A_N^{3,1}$ which vanishes at z' but not z''.

Thus there exists $\nu \in N$ with $\nu(z'') = z'$. Note that

$$\phi(z', w') = \phi(z'', w'') = \phi(\nu(z'', w''))$$

for every N-automorphic form $\phi \in A_N^{\alpha,1}$. Defining the element $\tilde{e}(u) \in \tilde{C}^{\cdot}$ by the equation $\nu(z'', w'') = (z', w'\tilde{e}(u))$, note that

$$\phi(z', w'\tilde{e}(u)) = \phi(z', w')e(\alpha u) .$$

Setting this equal to $\phi(z', w')$, we see that $e(\alpha u) = 1$ whenever ϕ is non-zero at z'. By 5.8, α can be any sufficiently large multiple of $1/k$. Therefore u must be a multiple of k, say $u = nk$. Hence the corresponding power c^u is in N; completing the proof that (z', w') and (z'', w'') belong to the same N-orbit. In fact $c^{-u}\nu(z'', w'') = (z', w')$.

To prove the corresponding assertion for Γ we will make temporary use of inhomogeneous automorphic forms, that is, elements of the direct sum $\bigoplus A_N^{\alpha,1}$, to be summed over α. Given points (z', w') and (z'', w'') not in the same Γ-orbit, consider the m images $\gamma_j(z', z'')$ where $\gamma_1, \cdots, \gamma_m$ represent the cosets of N in Γ. The above argument constructs forms $\phi_j \in A_N^{*,1}$ with

$$\phi_j(z'', w'') \neq \phi_j(\gamma_j(z', w')) .$$

Subtracting the constant $\phi_j(z'', w'') \in A_N^{0,1} \cong C$ from each ϕ_j, we obtain an *inhomogeneous* form which vanishes at (z'', w') but not at $\gamma_j(z', w')$. Now almost any linear combination ϕ of $\phi_1, \cdots, \phi_m$ will vanish at (z'', w'') but not anywhere in the Γ-orbit of (z', w'). Hence the norm

$$\psi = \gamma_1{}^*(\phi) \cdots \gamma_m{}^*(\phi) \in \bigoplus A_\Gamma^{\alpha,1}$$

of Section 5.2 will vanish at (z'', w'') but not at (z', w'). Expressing ψ as the sum of its homogeneous constituents, clearly at least one must take distinct values at (z'', w'') and (z', w'). ■

§6. The hyperbolic case $p^{-1} + q^{-1} + r^{-1} < 1$

The computations in this section will be formally very similar to those of Section 4. However, automorphic forms will be used in place of homogeneous polynomials.

Let Γ be the extended triangle group $\Gamma(p,q,r)$ of Section 3, with $p^{-1} + q^{-1} + r^{-1} < 1$ so that Γ can be considered as a group of labeled biholomorphic maps of the upper half-plane P. Recall that Γ has generators $\gamma_1, \gamma_2, \gamma_3$ which represent rotations about the three vertices of the triangle $T \subset P$. With this choice of Γ, the characters χ which actually occur for non-zero χ-automorphic forms can be described as follows. We continue to use the abbreviation $e(\alpha) = e^{2\pi i \alpha}$.

LEMMA 6.1. *Let χ be a character of the extended triangle group Γ. If $\phi \neq 0$ is a χ-automorphic form of degree α, then*

$$\chi(\gamma_1) = e((k+\alpha)/p)$$

where k is the order of the zero of ϕ at the first vertex of the triangle T. The values $\chi(\gamma_2)$ and $\chi(\gamma_3)$ can be computed similarly.

In particular, if ϕ does not vanish at the first vertex of T, then $\chi(\gamma_1) = e(\alpha/p)$.

Proof. Since $\gamma_1 = r_{v_1}(2\pi/p)$ is a lifted rotation through the angle $2\pi/p$, the derivative $\dot{\gamma}_1(v_1)$ equals $\tilde{e}(1/p)$, hence the fractional power $\dot{\gamma}_1(v_1)^\alpha$ in $\tilde{C}^{\cdot}$ projects to the complex number $e^{2\pi i \alpha/p} = e(\alpha/p)$. Setting $\phi(z,dz) = f(z)\,dz^\alpha$, and substituting the Taylor expansion

$$f(z) = a(z-v_1)^k + b(z-v_1)^{k+1} + \cdots$$

in the identity

$$f(\gamma_1(z))\dot{\gamma}_1(z)^\alpha = \chi(\gamma_1)f(z)$$

we obtain

$$a(e(1/p)(z-v_1))^k e(\alpha/p) + \cdots = \chi(\gamma_1) a(z-v_1)^k + \cdots .$$

Hence $e(k/p)e(\alpha/p) = \chi(\gamma_1)$ as asserted. ■

Define a rational number s by the formula $s^{-1} = 1-p^{-1}-q^{-1}-r^{-1}$. Thus π/s is the Lobachevsky area of the base triangle T. Define a character χ_0 of Γ by the formulas

$$\chi_0(\gamma_1) = e(s/p), \ \chi_0(\gamma_2) = e(s/q), \ \chi_0(\gamma_3) = e(s/r) .$$

The necessary identities

$$\chi_0(\gamma_1{}^p) = \chi_0(\gamma_2{}^q) = \chi_0(\gamma_3{}^r) = \chi_0(\gamma_1\gamma_2\gamma_3)$$

are easily verified.

COROLLARY 6.2. *If the automorphic form* $\phi \in A_\Gamma^{\alpha,\chi}$ *does not vanish at any vertex of the triangle* T, *then the degree* α *must be a multiple of* s, *and the character* χ *must be equal to* $\chi_0{}^{\alpha/s}$.

Proof. By 6.1 we have $\chi(\gamma_1) = e(\alpha/p)$, $\chi(\gamma_2) = e(\alpha/q)$, $\chi(\gamma_3) = e(\alpha/r)$. Hence the relations

$$\gamma_1{}^p = \gamma_2{}^q = \gamma_3{}^r = \gamma_1\gamma_2\gamma_3$$

of Section 3.1 imply that $\chi(\gamma_1)^p = \chi(\gamma_2)^q = \chi(\gamma_3)^r = e(\alpha)$ must be equal to

$$\chi(\gamma_1)\chi(\gamma_2)\chi(\gamma_3) = e((p^{-1}+q^{-1}+r^{-1})\alpha) = e((1-s^{-1})\alpha) .$$

Therefore $e(\alpha/s) = 1$, or in other words α must be a multiple of s. The equation $\chi = \chi_0{}^{\alpha/s}$ clearly follows. ■

Now we can begin to describe the algebra $A_\Gamma^{*,*}$ more explicitly.

LEMMA 6.3. *With* Γ, s, *and* χ_0 *as above, the complex vector space* A_Γ^{s,χ_0} *has dimension* 2. *This space contains one and (up to a constant multiple) only one automorphic form which vanishes at any given point of* P.

Proof. We begin with the basic existence Lemma 5.5. For some α there exists a form $\phi \in A_\Gamma^{\alpha,1}$ which is non-zero throughout any specified finite subset of P. In particular we can choose ϕ to be non-zero on the three vertices of T. By 6.2, the degree α of this ϕ must be a multiple of s, say $\alpha = ks$.

Let us count the number of zeros of ϕ. Since the triangle T has Lobachevsky area $(1-p^{-1}-q^{-1}-r^{-1})\pi = \pi/s$, it follows that a fundamental domain $T \cup \sigma(T)$ for the action of $\Gamma/\Gamma \cap C$ on P has Lobachevsky area $2\pi/s$. But the number of zeros of ϕ per unit area is $ks/2\pi$ by 5.6. Therefore the number of zeros of ϕ in the fundamental domain $T \cup \sigma(T)$ is precisely equal to k. [Here each pair of zeros z and $\gamma(z)$ on the boundary of the fundamental domain must of course be counted as a single zero. Note that ϕ does not vanish at the corners of the fundamental domain.] In other words there are precisely k (not necessarily distinct) zeros of ϕ in the quotient space $\bar{\Gamma}\backslash P$.

Next note that this quotient space $\bar{\Gamma}\backslash P$ can be given the structure of a smooth Riemann surface. If we stay away from the three exceptional orbits, this is of course clear. To describe the situation near the vertex v_1 it is convenient to choose a biholomorphic map h from P onto the unit disk satisfying $h(v_1) = 0$. Then the coordinate $w = h(z)$ can be used as a local uniformizing parameter near v_1. Since the rotation γ_1 of P about v_1 corresponds to the rotation

$$h\gamma_1 h^{-1}(w) = e(1/p)w$$

of the unit disk about the origin, it follows that a locally defined holomorphic function of w is invariant under this rotation if and only if it is actually a holomorphic function of w^p. Hence w^p can be used as local uniformizing parameter for the quotient surface $\bar{\Gamma}\backslash P$ about the image of v_1. The other two vertices are handled similarly. Note that a meromorphic function on $\bar{\Gamma}\backslash P$ having a simple zero at the image of v_1 corresponds to a Γ-invariant meromorphic function on P having a p-fold zero at each point of the exceptional orbit Γv_1.

Topologically, this quotient $\bar{\Gamma}\backslash P$ can be identified with the "double" of the triangle T. Hence it is a surface of genus zero. More explicitly, following Schwarz, $\bar{\Gamma}\backslash P$ can be identified biholomorphically with the unit 2-sphere by using the Riemann mapping theorem to map T onto a hemisphere and then applying the reflection principle.

Since $\bar{\Gamma}\backslash P$ is a compact Riemann surface of genus zero, it possesses a meromorphic function with k arbitrarily placed zeros and k arbitrarily placed poles. Starting with the non-zero form $\phi \in A_\Gamma^{ks,1}$ constructed above, we can multiply by a Γ-invariant meromorphic function h which has poles precisely at the k zeros of ϕ, and thus obtain a new form $\psi = h\phi \in A_\Gamma^{ks,1}$ whose k zeros can be prescribed arbitrarily in $\bar{\Gamma}\backslash P$. In particular we can choose ψ so as to have a k-fold zero at one point of $\bar{\Gamma}\backslash P$, and no other zeros. (To avoid confusion, let us choose this point to be distinct from the three ramification points.) Then by 5.3 this form has a k-th root $\psi_1 \in A_\Gamma^{s,\chi}$ for some character χ, and by 6.2 the character χ must be precisely χ_0. Evidently the form ψ_1 has a simple zero at just one point of $\bar{\Gamma}\backslash P$.

Similarly we can choose $\psi_2 \in A_\Gamma^{s,\chi_0}$ which vanishes at a different point of $\bar{\Gamma}\backslash P$. Then ψ_1 and ψ_2 are linearly independent. A completely arbitrary element $\psi \neq 0$ of A_Γ^{s,χ_0} must have precisely one simple zero in $\bar{\Gamma}\backslash P$, using 5.6. Choosing a linear combination $\lambda_1\psi_1 + \lambda_2\psi_2$ which vanishes at this zero, we see that the ratio $\psi/(\lambda_1\psi_1+\lambda_2\psi_2) \in A_\Gamma^{0,1}$ represents a holomorphic function defined throughout $\bar{\Gamma}\backslash P$, hence a constant. Thus ψ_1 and ψ_2 form a basis for A_Γ^{s,χ_0}, and this space contains precisely one 1-dimensional subspace consisting of forms which vanish at any prescribed point of $\bar{\Gamma}\backslash P$. ∎

The structure of $A_\Gamma^{*,*}$ can now be described as follows

LEMMA 6.4. *With* $\Gamma = \Gamma(p,q,r)$ *as above, the bigraded algebra* $A_\Gamma^{*,*}$ *is generated by three forms*

$$\phi_1 \in A_\Gamma^{s/p,\chi_1}, \quad \phi_2 \in A_\Gamma^{s/q,\chi_2}, \quad \phi_3 \in A_\Gamma^{s/r,\chi_3}$$

where χ_1, χ_2, χ_3 *are characters satisfying*

$$\chi_1{}^p = \chi_2{}^q = \chi_3{}^r = \chi_0 \ .$$

The automorphic form ϕ_i *has a simple zero at each point of the orbit* Γv_i, *and no other zeros. These three forms satisfy a polynomial relation* $\phi_1{}^p + \phi_2{}^q + \phi_3{}^r = 0$.

REMARK. The meromorphic function $-\phi_1{}^p/\phi_3{}^r$ is the Schwarz triangle function, which maps the quotient Riemann surface $\bar{\Gamma}\backslash P$ biholomorphically onto the extended complex plane, sending the three vertices of T to $0, 1$ and ∞ respectively.

Proof of 6.4. To construct ϕ_1 we use 6.3 to construct a form ϕ in A_Γ^{s,χ_0} which vanishes only along the orbit of v_1. This form must have a p-fold zero at v_1 by 5.6 or by the proof of 6.3. Since P is simply connected, it follows that ϕ possesses a holomorphic p-th root ϕ_1. Then ϕ_1 is itself an automorphic form by 5.3. The rest of the proof is completely analogous to the proof of 4.3. ■

Let Π denote the commutator subgroup of $\Gamma(p,q,r)$. Then by 3.2, 5.1 and 6.4 the graded algebra $A_\Pi^{*,1}$ is generated by the three forms ϕ_1, ϕ_2, ϕ_3.

COROLLARY 6.5. *The coset space* $\Pi\backslash G$ *is diffeomorphic to the Brieskorn manifold* $M(p,q,r)$.

Proof. Let $V \subset \mathbf{C}^3$ be the Pham-Brieskorn variety $z_1{}^p + z_2{}^q + z_3{}^r = 0$, singular only at the origin. Since the three functions ϕ_1, ϕ_2, ϕ_3 on $P \times \widetilde{\mathbf{C}}^{\cdot}$ satisfy the relation $\phi_1{}^p + \phi_2{}^q + \phi_3{}^r = 0$, and are never simultaneously zero, they together constitute a holomorphic mapping $(\phi_1, \phi_2, \phi_3) : P \times \widetilde{\mathbf{C}}^{\cdot} \to V - 0 \subset \mathbf{C}^3$ between complex 2-dimensional manifolds.

Recall from Section 5.9 that the groups $\Pi \subset G$ operate freely on $P \times \widetilde{\mathbf{C}}^{\cdot}$ by the rule $g : (z, w) \mapsto (g(z), \dot{g}(z) w)$. Setting $z = x + iy$ and identifying w with dz, this action preserves the Poincaré metric $|dz|^2/y^2 = |w|^2/y^2$. In fact, G operates simply transitively on each 3-dimensional manifold $|w|/y = \text{constant}$. Since Π is a discrete subgroup of G, it follows that the quotient $\Pi \backslash (P \times \widetilde{\mathbf{C}}^{\cdot})$ is again a complex 2-dimensional manifold.

Since each ϕ_i is Π-automorphic, the triple ϕ_1, ϕ_2, ϕ_3 give rise to a holomorphic mapping

$$\Phi : \Pi \backslash (P \times \widetilde{\mathbf{C}}^{\cdot}) \to V - 0$$

on the quotient manifold. By 5.9, since the ϕ_i generate $A_{\Pi}^{*,1}$, this mapping Φ is one-to-one. Hence by [Bochner and Martin, p. 179], Φ *maps* $\Pi \backslash (P \times \mathbf{C}^{\cdot})$ *biholomorphically onto an open subset of* $V - 0$. (It will follow in a moment that the image of Φ is actually all of $V - 0$.)

Choosing a base point $(z_0, 1)$ in $P \times \widetilde{\mathbf{C}}^{\cdot}$, map the coset space $\Pi \backslash G$ into the Brieskorn manifold $M(p, q, r) = V \cap S^5$ as follows. For each coset Πg the image $\Phi(\Pi g(z_0, 1))$ is a well defined point (z_1, z_2, z_3) of $V - 0$. Consider the curve

$$t \mapsto (t^{1/p} z_1, t^{1/q} z_2, t^{1/r} z_3) = \Phi(\Pi g(z_0, t^{1/s}))$$

through this point in $V - 0$, where $t > 0$. Intersecting this curve with the unit sphere, we obtain the required point $\Psi(\Pi g)$ of $M(p, q, r)$. It is easily verified that Ψ is smooth, well defined, one-to-one, and that its derivative has maximal rank everywhere. Since $\Pi \backslash G$ is compact while $M(p, q, r)$ is connected, it follows that Ψ is a diffeomorphism. ∎

COROLLARY 6.6. *The Brieskorn manifold* $M(p, q, r)$ *has a finite covering manifold diffeomorphic to a non-trivial circle bundle over a surface.*

Proof. Choosing $N \subset \Pi$ as in 2.7, it is easily verified that $N\backslash G$ fibers as a circle bundle over the surface $\bar{N}\backslash P$. ■

This corollary remains true in the spherical and nilmanifold cases.

CONCLUDING REMARKS. It is natural to ask whether there is a generalization of 6.5 in which the group Π is replaced by an arbitrary discrete subgroup of $G = \widetilde{SL}(2,R)$ with compact quotient. It seems likely that such a generalization exists:

CONJECTURE. For any discrete subgroup $\Gamma \subset G$ with compact quotient, the algebra $A_\Gamma^{*,1}$ of Γ-automorphic forms is finitely generated.*

Choosing generators $\phi_1, \cdots, \phi_k$ for this algebra, it would then follow from 5.9 that the k-tuple $(\phi_1, \cdots, \phi_k)$ embeds the complex 2-manifold $\Gamma\backslash(P \times \widetilde{C}^{\cdot})$ into the complex coordinate space C^k. It is conjectured that the image in C^k is of the form $V-0$ where $V = V_\Gamma$ is an irreducible algebraic surface, singular only at the origin. Intersecting this image V_Γ with a sphere centered at the origin, we then obtain a 3-manifold diffeomorphic to $\Gamma\backslash G$.

In general it is not claimed that V_Γ embeds as a hypersurface. Presumably V_Γ can be embedded in C^3 only if the algebra $A_\Gamma^{*,1}$ happens to be generated by three elements.

Note that this surface V_Γ is *weighted homogeneous*. That is, if each variable z_j is assigned a weight equal to the degree of ϕ_j, then V_Γ can be defined by polynomial equations $f(z_1, \cdots, z_k) = 0$ which are homogeneous in these weighted variables.

Not every weighted homogeneous algebraic surface can be obtained in this way. Here is an interesting class of examples. Start with an algebraic curve S of genus $g \geq 2$ together with a complex analytic line bundle ξ over S with Chern number $c_1 < 0$. Let $V(\xi)$ be obtained from the total space $E(\xi)$ by collapsing the zero-section to a point. Applying the Riemann-Roch theorem to negative tensor powers ξ^{-n} one can presumably construct enough holomorphic mappings $V(\xi) \to C$ to embed $V(\xi)$ as a weighted homogeneous algebraic surface in some C^k.

---

* Added in proof: See [Serre, p. 20-13].

CONJECTURE. The algebraic surface $V(\xi)$ obtained in this way is isomorphic to V_Γ for some discrete $\Gamma \subset G$ if and only if some tensor power $\xi^k = \xi \otimes \cdots \otimes \xi$ is isomorphic to the tangent bundle $\tau(S)$.

For each fixed S there are uncountably many line bundles ξ with negative Chern number. Only finitely many of these (the precise number is k^{2g} for each k dividing $2g-2$) satisfy the condition that the k-th tensor power is isomorphic to $\tau(S)$.

§7. A fibration criterion

In this section p, q, r may be any integers ≥ 2.

LEMMA 7.1. *If the least common multiples of* (p, q) *of* (p, r) *and of* (q, r) *are all equal*:

$$\ell.c.m.(p, q) = \ell.c.m.(p, r) = \ell.c.m.(q, r) \ ,$$

then the Brieskorn manifold $M(p, q, r)$ *fibers smoothly as a principal circle bundle over an orientable surface.* ∎

The precise surface B and the precise circle bundle will be determined below.

At the same time we show that the complement of the origin in the Pham-Brieskorn variety $z_1{}^p + z_2{}^q + z_3{}^r = 0$ fibers complex analytically as a principal $C^\cdot$-bundle over the Riemann surface B. In other words this variety V can be obtained from a complex analytic line bundle ξ over B by collapsing the zero cross-section to a point.

One special case is particularly transparent. If $p = q = r$, then the hypothesis of 7.1 is certainly satisfied. The equation $z_1{}^p + z_2{}^p + z_3{}^p = 0$ is then homogeneous, and hence defines an algebraic curve B in the complex projective plane $P^2(C)$. Clearly the mapping $(z_1, z_2, z_3) \mapsto (z_1 : z_2 : z_3)$ fibers $M(p, q, r)$ as a circle bundle over B.

Proof of 7.1. Starting with any values of p, q, r, let m denote the least common multiple of p, q, and r. Then the group $C^\cdot$ of non-zero complex numbers operates on the variety $z_1{}^p + z_2{}^q + z_3{}^r = 0$ by the correspondence

$$u : (z_1, z_2, z_3) \mapsto (u^{m/p}z_1, u^{m/q}z_2, u^{m/r}z_3)$$

for $u \neq 0$. Restricting to the unit circle $|u| = 1$ and the unit sphere $|z_1|^2 + |z_2|^2 + |z_3|^2 = 1$, we obtain a circle action on $M = M(p, q, r)$.

Let us determine whether any group elements have fixed points in M or in $V - 0$. If

$$(u^{m/p}z_1, u^{m/q}z_2, u^{m/r}z_3) = (z_1, z_2, z_3) \in V - 0 \ ,$$

then at least two of the complex numbers z_1, z_2, z_3 must be non-zero, hence at least two of the numbers $u^{m/p}, u^{m/q}, u^{m/r}$ must equal 1. If the three integers $m/p, m/q, m/r$ happen to be pairwise relatively prime, then it clearly follows that $u = 1$.

Thus, if $m/p, m/q, m/r$ are pairwise relatively prime, we obtain a smooth free $C^{\cdot}$ action on $V - 0$ restricting to a smooth free circle action on $M = M(p, q, r)$. Evidently M fibers as a smooth circle bundle over the quotient space $S^1 \backslash M = B$, which must be a compact, orientable, 2-dimensional manifold. In fact, using the alternative description

$$B = C^{\cdot} \backslash (V - 0)$$

we see that B has the structure of a complex analytic 1-manifold. (The two quotient spaces can be identified since every $C^{\cdot}$-orbit intersects the unit sphere precisely in a circle orbit.)

Since an elementary number theoretic argument shows that $m/p, m/q, m/r$ are pairwise relatively prime if and only if the hypothesis of 7.1 is satisfied, this completes the proof. ■

To compute the genus of the surface $B = S^1 \backslash M = C^{\cdot} \backslash (V - 0)$, we describe it as a branched covering of the 2-sphere $P^1(C)$ by means of the holomorphic mapping

$$f : (u^{m/p}z_1, u^{m/q}z_2, u^{m/r}z_3) \mapsto (z_1{}^p : z_2{}^q) \ .$$

Clearly f is well defined. A counting argument, which will be left to the reader, shows that the pre-image of a general point of $P^1(C)$ consists of precisely pqr/m points of B. Thus f is a map of degree pqr/m from B to $P^1(C)$.

There are just three branch points in $P^1(C)$, corresponding to the possibilities $z_1 = 0$, $z_2 = 0$, and $z_3 = 0$ respectively. The preimage of a branch point contains qr/m, or pr/m, or pq/m points respectively. Again the count will be left to the reader.

Now choose a triangulation of $P^1(C)$ with the three branch points $(0:1)$, $(1:0)$, and $(-1:1)$ as vertices. Counting the numbers of vertices, edges, and faces in the induced triangulation of B, we easily obtain the following.

LEMMA 7.2. *Let* p, q, r *be as in 7.1, with least common multiple* m. *Then the surface* $B = S^1 \backslash M$ *has Euler characteristic* $\chi(B) = (qr + pr + pq - pqr)/m = pqr(p^{-1} + q^{-1} + r^{-1} - 1)/m$.

In particular the sign of $\chi(B)$ is equal to the sign of $p^{-1} + q^{-1} + r^{-1} - 1$. The genus g can now be recovered from the usual formula $\chi = 2 - 2g$. Note that the genus satisfies $g \geq 2$, except in the four special cases $(2,2,2)$, $(2,3,6)$, $(2,4,4)$, and $(3,3,3)$. (Compare Section 2.1.)

To determine the precise circle bundle in question, we must compute the Chern class

$$c_1 = c_1(\xi) \in H^2(B; Z)$$

or equivalently the Chern number $c_1(\xi)[B]$ of the associated complex line bundle ξ. (The Chern class c_1 can also be described as the Euler class of ξ. Using the Gysin sequence ([Spanier, pp. 260-261], [Milnor and Stasheff, p. 143]), one sees that $H_1(M; Z)$ is the direct sum of a free abelian group of rank 2g and a cyclic group of order $|c_1(\xi)[B]|$.)

To compute c_1 we consider the map

$$F : (z_1, z_2, z_3) \mapsto (z_1{}^p, z_2{}^q)$$

from $V-0$ to C^2-0. Thus we obtain a commutative diagram

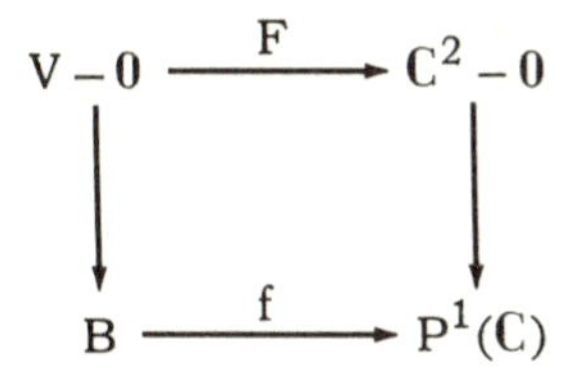

where the right hand vertical arrow is the canonical fibration $(z_1, z_2) \mapsto (z_1 : z_2)$ with Chern number -1, associated with the Hopf fibration $S^3 \to S^2$.

This map F is not quite a bundle map, since inspection shows that each fiber of the left hand fibration covers the corresponding fiber on the right m times. To correct this situation we must factor $V-0$ by the action of the subgroup

$$\Omega \subset C^{\cdot}$$

consisting of all m-th roots of unity. Thus we identify (z_1, z_2, z_3) with $\omega^{m/p}z_1, \omega^{m/q}z_2, \omega^{m/r}z_3$ for each $\omega^m = 1$, obtaining a new commutative diagram

$$\begin{array}{ccc} \Omega\backslash(V-0) & \xrightarrow{\overline{F}} & C^2-0 \\ \downarrow & & \downarrow \\ B & \xrightarrow{f} & P^1(C) \end{array}$$

where $\overline{F}$ is now a bundle map. Since f has degree pqr/m, it follows that the new $C^{\cdot}$-bundle $\Omega\backslash(V-0) \to B$ has Chern number $-pqr/m$. But this new bundle can be described as the $C^{\cdot}$-bundle associated with the m-fold tensor product $\xi \otimes \cdots \otimes \xi$ of the original complex line bundle ξ. Therefore ξ has Chern number $c_1(\xi)[B] = -pqr/m^2$.

Recapitulating, we have proved the following.

THEOREM 7.3. *If the hypothesis*

$$m = \ell.c.m.(p,q) = \ell.c.m.(p,r) = \ell.c.m.(q,r)$$

of 7.1 is satisfied, then the Brieskorn manifold $M(p,q,r)$ *fibers as a smooth circle bundle with Chern number* $-pqr/m^2$ *over a Riemann surface of Euler characteristic* $pqr(p^{-1}+q^{-1}+r^{-1}-1)/m$.

The number pqr/m^2 can be described more simply as the greatest common divisor of p, q, r.

The negative sign of the Chern number has no particular topological significance, but is meaningful in the complex analytic context, since ξ is a complex analytic line bundle with no non-zero holomorphic cross-sections.

Note that the Euler characteristic of B is always a multiple of the Chern number of ξ. In general it is a large multiple, for it is not difficult to show that the ratio satisfies

$$\chi(B)/c_1(\xi)[B] = m(1-p^{-1}-q^{-1}-r^{-1}) \geq m/6$$

in the hyperbolic case. Hence this ratio tends to infinity with m. Therefore the genus of B also tends to infinity with m.

Here are two examples to illustrate 7.3.

EXAMPLE 1. For any $g \geq 0$, the manifold $M(2, 2(g+1), 2(g+1))$ fibers as a circle bundle with Chern number -2 over a surface of genus g. Similarly, for any $g \geq 1$, the manifold $M(2, 2g+1, 2(2g+1))$ fibers as a circle bundle with Chern number -1 over a surface of genus g.

EXAMPLE 2. The Brieskorn manifolds $M(p,q,r)$ are not all distinct. For example, $M(2,9,18)$ and $M(3,5,15)$ are diffeomorphic, since each fibers as a circle bundle with Chern number -1 over a surface of genus 4.

CONCLUDING REMARK. If it is known that $M(p,q,r)$ fibers as a circle bundle over a surface, does it follow that the hypothesis of 7.1 must be satisfied? The lens spaces $M(2,2,r)$ with $r \geq 3$ provide counter-examples. These fiber as circle bundles with Chern number $\pm r$ over a surface of genus zero. (Presumably there is no associated analytic fibration of $V-0$?) However, these are the only counter-examples. In the cases $p^{-1}+q^{-1}+r^{-1}\geq 1$, this can be verified by inspection. Thus we need only prove the following.

LEMMA 7.4. *In the hyperbolic case* $p^{-1}+q^{-1}+r^{-1}<1$, *if* $M(p,q,r)$ *has the fundamental group of a principal circle bundle over an orientable surface, then the hypothesis of 7.1 must be satisfied.*

The proof can be sketched as follows. First note that the fundamental group of a principal circle bundle over an orientable surface, modulo its center, has no elements of finite order. Now consider the fundamental group $\Pi=\Pi(p,q,r)$ of Section 6. The center of Π is precisely equal to $\Pi\cap C$. As noted in 2.6, an element of $\Gamma/C\supset\Pi/\Pi\cap C$ has finite order if and only if it is conjugate to a power of γ_1,γ_2, or γ_3 modulo C. To decide which powers of say γ_1 belong to Π, we carry out a matrix computation in the abelianized group Γ/Π. (Compare Section 3.2.) Setting $\mu=\ell.\text{c.m.}(q,r)$, it turns out that the order k of γ_1 modulo Π is given by

$$k = p\mu(1-p^{-1}-q^{-1}-r^{-1}) \equiv -\mu(\text{mod } p)\ .$$

Evidently the element ${\gamma_1}^k$ of Π belongs to $\Pi\cap C$ if and only if k is a multiple of p, or in other words if and only if μ is a multiple of p. Thus $\Pi/\Pi\cap C$ has no elements of finite order if and only if

$$\ell.\text{c.m.}(q,r)\equiv 0 \qquad (\text{mod } p)\ ,$$

and similarly

$$\ell.\text{c.m.}(p,r)\equiv 0 \qquad (\text{mod } q)\ ,$$
$$\ell.\text{c.m.}(p,q)\equiv 0 \qquad (\text{mod } r)\ .$$

Clearly these conditions are equivalent to the hypothesis of 7.1. ∎

§8. The nil-manifold case $p^{-1} + q^{-1} + r^{-1} = 1$

As noted in 2.1, we are concerned only with three particular cases. The triple (p, q, r), suitably ordered, must be either $(2,3,6)$ or $(2,4,4)$ or $(3,3,3)$. Clearly each of these triples satisfies the hypothesis of 7.1. Hence by 7.3 the corresponding manifold $M = M(p, q, r)$ is a circle bundle over a torus. The absolute value of the Chern number of this circle bundle is the greatest common divisor of p, q, r which is either 1, or 2, or 3 respectively.

But any non-trivial circle bundle over a torus can also be described as a quotient manifold N/N_k as follows. Let N be the nilpotent Lie group consisting of all real matrices of the form

$$A = \begin{bmatrix} 1 & a & c \\ 0 & 1 & b \\ 0 & 0 & 1 \end{bmatrix},$$

and let N_k be the discrete subgroup consisting of all such matrices for which a, b, and c are integers divisible by k. (Here k should be a positive integer.) Then the correspondence

$$A \mapsto (a \bmod k,\ b \bmod k)$$

maps N/N_k to the torus with a circle as fiber. The first homology group

$$H_1(N/N_k; \mathbf{Z}) \cong N_k/[N_k, N_k]$$

is isomorphic to $\mathbf{Z} \oplus \mathbf{Z} \oplus (\mathbf{Z}/k)$, so the Chern number of this fibration must be equal to $\pm k$. Thus we obtain the following three diffeomorphisms

$$M(2,3,6) \cong N/N_1$$
$$M(2,4,4) \cong N/N_2$$
$$M(3,3,3) \cong N/N_3 .$$

It must be admitted that this proof is rather ad hoc. I do not know whether there exists a more natural construction of these diffeomorphisms.

THE INSTITUTE FOR ADVANCED STUDY

REFERENCES

Arnol'd, V. I., Normal forms for functions near degenerate critical points, Weyl groups A_k, D_k, E_k and lagrangian singularities. Functional Anal. and Appl. 6 (1972), 254-272.

__________, Critical points of smooth functions. Proc. Int. Congr. Math. Vancouver, to appear.

Bochner, S., and Martin, W. T., Several Complex Variables. Princeton Univ. Press 1948.

Brieskorn, E., Beispiele zur Differentialtopologie von Singularitäten. Invent. Math. 2 (1966), 1-14.

__________, Rationale Singularitäten komplexer Flachen. Invent. Math. 4 (1967-68), 336-358.

Bundgaard, S., and Nielsen, J., On normal subgroups with finite index in F-groups. Mat. Tidsskrift B 1951, 56-58.

Carathéodory, C., Theory of Functions of a Complex Variable, 2. Verlag Birkhäuser 1950, Chelsea 1954, 1960.

Conner, P., and Raymond, F., Injective operations of the toral groups. Topology 10 (1971), 283-296.

Coxeter, H. S. M., The binary polyhedral groups and other generalizations of the quaternion group. Duke Math. J. 7 (1940), 367-379.

Dolgačev, I. V., Conic quotient singularities of complex surfaces. Functional Anal. and Appl. 8:2 (1974), 160-161.

Dyck, W., Gruppentheoretische Studien. Math. Ann. 20 (1882), 1-45.

Fox, R. H., On Fenchel's conjecture about F-groups. Mat. Tidsskrift B 1952, 61-65.

Gunning, R., The structure of factors of automorphy. Amer. J. Math. 78 (1956), 357-382.

Hirzebruch, F., Topological Methods in Algebraic Geometry. Grundlehren 131, Springer Verlag 1966.

Klein, F., Lectures on the Icosahedron and the Solution of Equations of the Fifth Degree. Teubner 1884, Dover 1956.

__________, Vorlesungen über die Entwicklung der Mathematik im 19. Jahrhundert, 1. Springer 1926.

Klein, F., and Fricke, R., Vorlesungen über die Theorie der elliptischen Modulfunctionen, 1. Teubner 1890.

Lehner, J., Discontinuous Groups and Automorphic Functions. Math. Surveys 18, Amer. Math. Soc. 1964.

Magnus, W., Noneuclidean Tessalations and their Groups. Academic Press 1974.

Milnor, J., Singular Points of Complex Hypersurfaces. Annals of Math. Studies 61, Princeton Univ. Press 1968.

————, Isolated critical points of complex functions, in Proc. Symp. Pure Math. 27, Amer. Math. Soc. 1974.

Milnor, J., and Stasheff, J., Characteristic Classes. Annals of Math. Studies 76, Princeton Univ. Press 1974.

Orlik, P., Weighted homogeneous polynomials and fundamental groups. Topology 9 (1970), 267-273.

————, Seifert Manifolds. Springer Lecture Notes in Math. 291, 1972.

Orlik, P., and Wagreich, P., Isolated singularities of algebraic surfaces with C^* action. Annals of Math. 93 (1971), 205-228.

Petersson, H., Theorie der automorphen Formen beliebiger reeller Dimension und ihre Darstellung durch eine neue Art Poincaréscher Reihen. Math. Ann. 103 (1930), 369-436.

Pham, F., Formules de Picard-Lefschetz généralisees et ramification des intégrales. Bull. Soc. Math. France 93 (1965), 333-367.

Prill, D., Local classification of quotients of complex manifolds by discontinuous groups. Duke Math. J. 34 (1967), 375-386.

Raymond, F., and Vasquez, A., Closed 3-manifolds whose universal covering is a Lie group, to appear.

Saito, K., Einfach elliptische Singularitäten. Invent. math. 23 (1974), 289-325.

Schwarz, H. A., Ueber diejenigen Fälle in welchen die Gaussische hypergeometrische Reihe eine algebraische Function ihres vierten Elementes darstellt. Math. Abh. 2, Springer Verlag 1890, 211-259.

Seifert, H., Topologie dreidimensionaler gefaserter Räume. Acta Math. 60 (1932), 147-238.

Seifert, H., and Threlfall, W., Topologische Untersuchungen der Diskontinuitäts-bereiche endlicher Bewegungsgruppen des dreidimensionalen sphärischen Raumes. Math. Ann. 104 (1931), 1-70 and 107 (1933), 543-596.

Selberg, A., On discontinuous groups in higher dimensional symmetric spaces, Contributions to Function Theory. Tata Institute, Bombay 1960, 147-164.

Siegel, C. L., Topics in Complex Function Theory, 2. Wiley 1969.

Siersma, D., Classification and deformation of singularities. Thesis, Amsterdam 1974.

Spanier, E. H., Algebraic Topology. McGraw-Hill 1966.

Springer, G., Introduction to Riemann Surfaces. Addison-Wesley 1957.

Wagreich, P., Singularities of complex surfaces with solvable local fundamental group. Topology 11 (1972), 51-72.

Wall, C. T. C., Rational Euler characteristics. Proc. Cambr. Phil. Soc. 57 (1961), 182-183.

Weir, A. J., The Reidemeister-Schreier and Kuroš subgroup theorems. Mathematika 3 (1956), 47-55.

Added in Proof:

Durfee, A., and Kauffman, L., Periodicity of branched cyclic covers, to appear.

Neumann, W., Cyclic suspension of knots and periodicity of signature for singularities. Bull. Amer. Math. Soc. 80 (1974), 977-981.

Serre, J.-P., Fonctions automorphes, Exp. 20 in Seminaire Cartan 1953/54, Benjamin 1967.

Note: I am informed by V. Arnold that all of the essential results of this paper have been obtained independently by I. V. Dolgačev. Parts of Dolgačev's work are described in his paper cited above, and in his paper "Automorphic forms and quasihomogeneous singularities," Funct. anal. 9:2 (1975), 67-68.

SURGERY ON LINKS AND DOUBLE BRANCHED COVERS OF S^3

José M. Montesinos

§0. *Introduction*

This paper deals with the relationship between 2-fold cyclic coverings of S^3 branched over a link and closed, orientable 3-manifolds which are obtained by doing surgery on a link in S^3. In Theorem 1 it is shown that every 2-fold cyclic branched covering of S^3 can be obtained by doing surgery on a "strongly invertible" link, that is, a link L which has the property that there is an orientation preserving involution of S^3 which induces in each component of L an involution with two fixed points. This result has some interesting consequences. Let K be a non-trivial knot in S^3. Then Theorem 1, which is a constructive result, allows us to obtain a link L in S^3 such that the 2-fold covering space $\tilde{K}$ of S^3 branched over K can be obtained by doing surgery on L. Note that if L has property P, then $\tilde{K}$ cannot be a counterexample to Poincaré Conjecture because $\pi(\tilde{K}) \neq 1$. Thus, every simply connected 2-fold cyclic covering of S^3 is S^3 iff every strongly invertible link has property P (Corollary 1). As a second consequence of Theorem 1 we obtain a new proof of a result established earlier by Viro [25] and also by Birman and Hilden [2], that every closed, orientable 3-manifold of Heegaard genus ≤ 2 is a 2-fold cyclic branched covering of S^3 (Corollary 2). In Corollary 3 we will sharpen Theorem 1 showing that every 2-fold cyclic branched covering of S^3 can be obtained by doing surgery on a member of a special family of strongly invertible links in S^3.

Let L be a link such that there is an orientation preserving involution of S^3 with fixed points which induces an involution in each component of L. Let M be a manifold that is obtained by doing surgery on L. We will see in Theorem 2 that M is a 2-fold cyclic covering of a manifold that is obtained by doing surgery on a link in S^3. As an application of Theorem 2, it is shown that each manifold that is obtained by doing surgery on a noninvertible pretzel knot or on the noninvertible "borromeans rings" is a 2-fold cyclic branched covering of a 2-fold cyclic branched covering of S^3. This yields some insight into the answer to a question (Question 3) raised by Birman and Hilden.

The construction of the link L in Theorem 1 uses some knot modifications, defined by Wendt, which have the effect of changing K into the trivial knot. Having in mind the purpose of finding, for a given knot K, if $\pi(\tilde{K})$ is or is not trivial, we define in Section 3 some modifications of a knot which generalize Wendt's modifications. These modifications have the effect of exhibiting $\tilde{K}$ as a manifold which is obtained by doing "generalized surgery" on a link in S^3, that is, removing n disjoint solid tori from S^3 and replacing each torus with a special "graph-manifold" which is bounded by a torus. The advantage of this is that if a link has property P, then a counterexample to the Poincaré conjecture cannot be obtained by doing generalized surgery on it (Theorem 4).

This fact allows us, in Section 4, to establish that there cannot be a counterexample to the Poincaré Conjecture among the 2-fold cyclic coverings of S^3 which are branched over the knots of Kinoshita-Terasaka (Section 4.1), or over Conway's 11-crossing knot with Alexander polynomial 1 (see Section 4.2), or over a special class of closed 3-braids (see Section 4.3) first studied by Birman and Hilden.

In Section 5 it is established that graph-manifolds are in the Poincaré Category. This fact was used earlier in the paper, in the proof of Theorem 4.

Acknowledgement. I would like to express my deep gratitude to Professor J. S. Birman for her valuable suggestions and comments in reading my manuscript.

§1. *Statement of the problems*

In this section we will discuss several interesting questions which have been posed by Ralph Fox and others about the Poincaré Conjecture and related matters. These questions will serve to motivate the main results of this paper, which are given in Sections 2, 3, 4 and 5, below.

Let L denote a link in S^3, and let $\tilde{L}$ denote the 2-fold cyclic covering space of S^3 branched over L. Since 2-fold branched covering spaces are in many ways especially simple (see [2, 5, 6, 15, 25, 27]), one might like to know how they are related to the class of *all* closed, orientable 3-manifolds? Ralph Fox has proved [6] that the 3-dimensional torus $S^1 \times S^1 \times S^1$ is not a 2-fold cyclic branched cover of S^3. However he has given a conjecture [6, Conjecture A′] that implies an affirmative answer to the question:

Question 1. Is every closed, orientable, simply-connected 3-manifold a 2-fold cyclic branched cover of S^3?

This appears to be a deep and difficult question, and, as will be seen below, it may even be equivalent to the Poincaré Conjecture.

Now, in [17], [18] it was shown that there are Seifert fiber spaces, different from $S^1 \times S^1 \times S^1$, which are not 2-fold cyclic coverings of S^3. However, all of them, are 2-fold cyclic coverings branched over a 3-sphere with handles [18].

Question 2. Is every closed, orientable 3-manifold a 2-fold cyclic covering branched over a 3-sphere with handles?

If Question 2 has an affirmative answer, then *each closed, orientable 3-manifold* M *with* $H_1(M)$ *finite is a 2-fold cyclic covering of* S^3, because the lift to M of a non-separating 2-sphere (in S^3 with $g > 0$ handles) must be a non-separating closed, orientable surface in M. Thus $H_2(M)$ and $H_1(M)$ are infinite. Then, an affirmative answer to Question 2 implies an affirmative answer to Question 1.

Note that a 3-sphere with $g > 0$ handles is a 2-fold cyclic branched covering of S^3. Joan S. Birman and Hugh M. Hilden have suggested that it is reasonable to ask the following question, which looks like a weaker question than Question 2.

Question 3. Is every closed, orientable 3-manifold a 2-fold branched cyclic covering of a 2-fold branched cyclic covering of $\cdots$ of a 2-fold branched cyclic covering of S^3?

It was observed by Birman and Hilden that if the answer to Question 3 is affirmative, then Fox's argument [5] implies that *if a counterexample exist to the Poincaré Conjecture, then there is also a counterexample which is a 2-fold branched cyclic covering of* S^3.

Thus an affirmative answer to one of the three above questions would reduce the investigation of the Poincaré Conjecture, to the case of 2-fold cyclic coverings of S^3.

Now, the trivial knot is the only knot which has S^3 as associated 2-fold cyclic covering branched over it [27]. On the other hand, if L has more than one component, then $H_1(\tilde{L}) \neq 0$ [6] and if $L = L_1 \# L_2$ is a composite knot, then $\pi(\tilde{L}) = \pi(\tilde{L}_1) * \pi(\tilde{L}_2)$ [15, Theorem V.5.3.]. Thus, one is led to consider the following Conjecture (see [15, Conjecture I.1.1.]):

CONJECTURE 1. If N is a non-trivial prime knot, then $\pi(\tilde{N}) \neq 1$.

If one searches for a counterexample to Questions 2, 3, then one need not consider Seifert fiber spaces or closed *graph-manifolds* ("Graphen-mannigfaltigkeiten," see [26]) because all of them are 2-fold cyclic coverings of S^3 with handles.[1] I suggest looking for M among the closed, orientable 3-manifolds obtained by doing surgery on a knot in S^3.

---

[1] In [18] this was proved for Seifert manifolds and for graph-manifolds M represented by a graph $A(M)$. Of course, this can be extended to each closed graph-manifold according to [26, Satz 6.3, p. 88] and [15, Teorema V.5.3.] and [25; 3.10].

Therefore, in this paper, we explore the relationship between 2-fold cyclic coverings of S^3 branched over a link and closed, orientable 3-manifolds which are obtained by doing surgery on a link in S^3.

§2. *Surgery on links and double branched covers of* S^3

Let L be a link in S^3. L is called *strongly-invertible* if there is an orientation-preserving involution of S^3 which induces in each component of L an involution with two fixed points. Every strongly-invertible link L is invertible, but I do not know if every invertible link is a strongly-invertible link.

THEOREM 1. *Let* M *be a closed, orientable* 3-*manifold that is obtained by doing surgery on a strongly-invertible link* L *of* n *components. Then* M *is a* 2-*fold cyclic covering of* S^3 *branched over a link of at most* n+1 *components. Conversely, every* 2-*fold cyclic branched covering of* S^3 *can be obtained in this fashion.*

Proof of Theorem 1. Let S^3 be represented as Euclidean space with an ideal point at infinity. It can be supposed without loss of generality [27], that there is an axis E in S^3 such that the axial symmetry u with respect to E induces in each component of L an involution with two fixed points. For the sake of brevity, the first part of Theorem 1 will be proved for a knot N in S^3.

Let U(N) be a regular neighborhood of N such that u induces an involution in U(N) (a typical case is illustrated in Figure 1a). Let V be the solid torus, as represented in Figure 1b, and let u′ be the symmetry with respect to the axis E′. There is a homeomorphism ψ of $\partial U(N)$ onto ∂V such that $(u'|\partial V)\psi = \psi(u|\partial U(N))$.

Let ϕ now be a homeomorphism of ∂V onto $\partial U(N)$. Then $\psi\phi$ is an autohomeomorphism of ∂V and it can be supposed (by composing ϕ, if necessary, with an isotopy) that

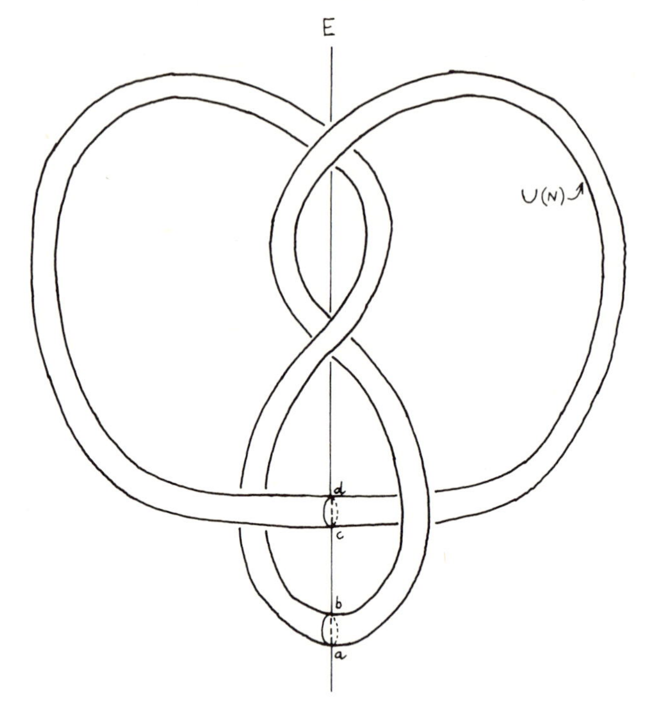

Fig. 1a.

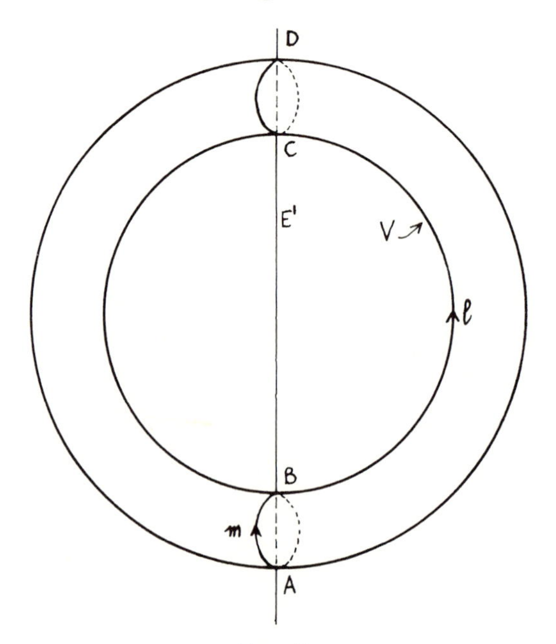

Fig. 1b.

$$(\psi\phi)(u'|\partial V) = (u'|\partial V)(\psi\phi) .$$ [2]

Thus $\phi(u'|\partial V) = \psi^{-1}(u'|\partial V)\psi\phi = (u|\partial U(N))\phi$.

Then, the space M obtained by pasting V to $S^3 - U(N)$ by means of ϕ is compatible with the involutions u and u′, and admits an involution u″, induced by u and u′. The orbit-space of $(S^3 - U(N)) \cup V$ under u″ can be obtained by adjoining the orbit space of V under u′ (which is a ball) to the orbit-space of $S^3 - U(N)$ under u, which is S^3 minus a ball (see in Figure 1c a fundamental set for the action of u on U(N)). Then M is a 2-fold cyclic covering of S^3, branched over the image of $E - (ab + cd) + (AB + CD)$ (see Figures 1a and 1b). This is a link in S^3 which has, at most, two components.

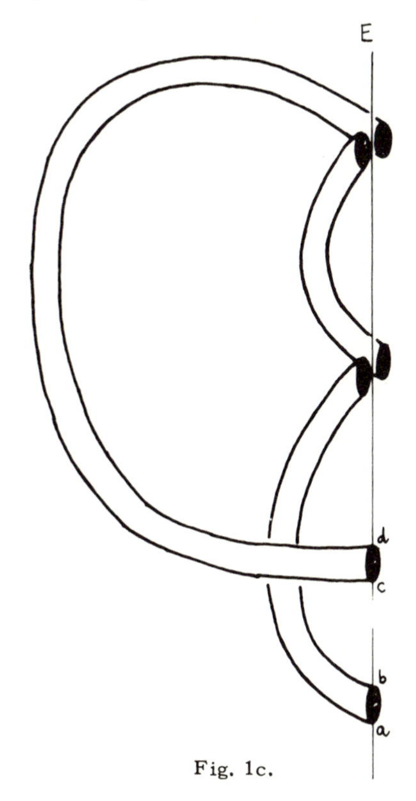

Fig. 1c.

Conversely, suppose that M is a 2-fold cyclic covering of S^3, branched over a link L. We consider two ways to modify this link, by removing certain solid balls from S^3 and sewing them back differently. First, it is possible, by applying modifications of type W_1 (see Figure 2a), to change a given link L in S^3 into a knot K in S^3. Then, by

---

[2] This result is contained implicitly in [3], and is proved in [2], [25] and [18]. In [2] and [25] this result has been generalized for orientable surfaces of genus $g = 2$. For $g > 2$ this generalization is not true in general (see [6] and [17]).

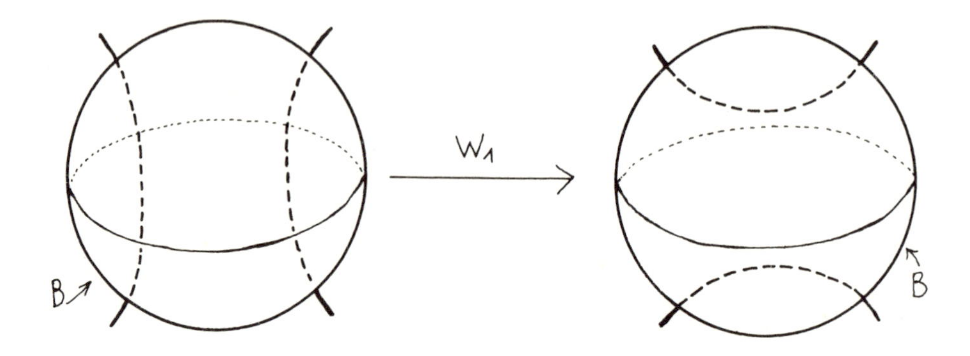

Fig. 2a.

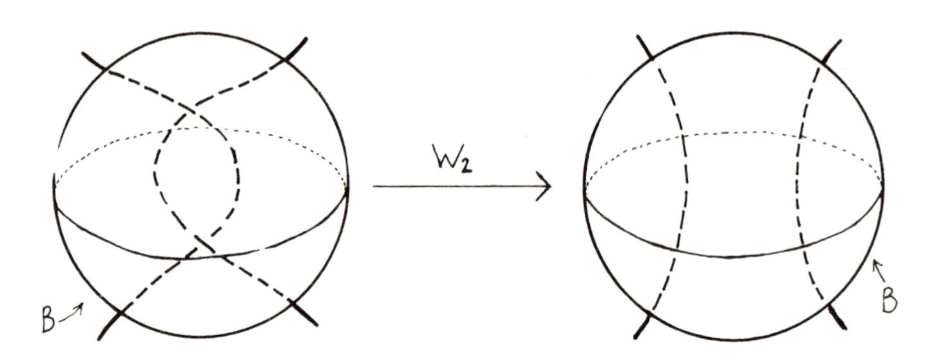

Fig. 2b.

applying further modifications of type W_2 (see Figure 2b), it is possible to change the knot K into the trivial knot T (see [28]). Let n be the minimal number of modifications of type W_1, W_2 that are necessary in order to change the given link L into the trivial knot T.

It may be supposed that these modifications are set up in the inner of n disjointed balls $B_1, \cdots, B_n$ of S^3 (see Figure 2). Note that the 2-fold cyclic coverings of B_i branched over $B_i \cap T$ are solid tori. Thus, in order to build up $\tilde{L}$ it is sufficient to do surgery along n solid tori in $\tilde{T} = S^3$.

Let $\tilde{B}_i$ be the 2-fold cyclic covering of B_i, branched over $B_i \cap T$. Then $\bigcup_{i=1}^{n} \tilde{B}_i$ can be interpreted as a regular neighborhood of a strongly-invertible link in S^3. Thus, $\tilde{L}$ can be obtained by doing surgery on a strongly-invertible link in S^3 which has, at most, n components. □

Recall that a link L in S^3 has *property* P when it is not possible to obtain a counterexample to the Poincaré Conjecture by doing surgery on it.

COROLLARY 1. *Conjecture 1 is true iff every strongly invertible link has property* P. □

As property P is known to be true for many links [1], [8], [23], Corollary 1 implies that Conjecture 1 can be established for a large family of knots. In Section 4 we will apply Theorem 1 in this way to establish that there cannot be a counterexample to the Poincaré Conjecture among the 2-fold coverings of S^3 which are branched over the knots of Kinoshita-Terasaka (see Section 4.1), or over Conway's 11-crossing knot with Alexander polynomial 1 (see Section 4.2), or over a special class of closed 3-braids (see Section 4.3).

We now give a different application of Theorem 1. Let $g \geq 1$ be an integer. Let L be a link in $R^3 = S^3 -$ (one point) made up of a disjoint union of circles, each being one of the following: (i) a circle of radius < 1, center at $(2n+1, 0, 0)$ where $0 \leq n \leq g$, and lying in the x, z plane, or (ii) a circle of radius < 1, center at $(2n, 0, 0)$ where $1 \leq n \leq g$, and lying in the x, y plane, or (iii) a circle of radius ≤ 2, center at $(2n, 2, 0)$ where $1 < n < g$, and lying in a parallel plane P_n to the y, z plane. We assume also that the annulus determinated by two concentric components of L must be cut by some other component in exactly one point. Let $\mathcal{L}_g$ be the family of links defined in this way, for a given g. It was proved by Lickorish [13] that every closed, orientable 3-manifold of genus g may be obtained by doing surgery on a link in the class $\mathcal{L}_g$. Let $\mathcal{L}'_g$ be the subfamily of $\mathcal{L}_g$ consisting of those links whose components in P_n have radius 2. Note that a link in $\mathcal{L}'_g$ is strongly-invertible.

Since $\mathcal{L}'_g = \mathcal{L}_g$ for $g \leq 2$, then according to Theorem 1, we obtain another proof of the following result by Viro [25] and Birman and Hilden [2]:

COROLLARY 2. *Every closed, orientable 3-manifold of genus* ≤ 2 *is a 2-fold cyclic branched covering of* S^3. □

COROLLARY 3. *Each 2-fold cyclic covering branched over* S^3 *can be obtained doing surgery on a link in* $\mathcal{L}'_g$, *for some* $g \geq 1$.

Proof of Corollary 3. First, we recall the definition of a "plat on 2m strings." If we represent S^3 as $R^3 + \infty$, then the x, y plane separates S^3 in two balls B_1 and B_2, B_1 containing the positive part of axis z. Let C be a collection of m circles in the x, z plane of radius 2 and centers at points $(1+5i, 0, 0)$, where $0 \leq i \leq m-1$. Let f be any orientation-preserving autohomeomorphism of ∂B_1 which keeps the set $C \cap \partial B_1$ fixed as a set. Since f is isotopic to the identity map in ∂B_1, there is an autohomeomorphism $F': \partial B_1 \times [0,1] \to \partial B_1 \times [0,1]$ such that $F'(x,t) = (x',t)$, $F'(x,1) = (x,1)$ and $F'(x,0) = (fx,0)$. Then F' is extended by the identity map outside $\partial B_1 \times [0,1]$ to an autohomeomorphism F of B_1. The subset $L = F(C \cap B_1) \cup (C \cap B_2)$, which is a link in S^3, is called a *plat on* 2m *strings* (for further details, see [2]). It is a known result (see, for instance, [2]) that every link type is represented by at least one plat. Note that $F(C \cap (\partial B_1 \times [0,1]))$ is a geometric braid on 2m strings. Thus a plat on 2m strings can be exhibited as a geometric braid on 2m strings by joining the initial points in pairs, and also the terminal points in pairs.

The proof of Corollary 3 may be illustrated by the following example (the general case is left to the reader). Let us consider the plat on 8 strings of Figure 3a. It is possible to change L into the trivial knot by removing ten solid balls B_i $(i = 1, \cdots, 10)$ from S^3 and sewing them back differently (see Figures 3a, 3b, 4a). Note that the 2-fold covering of B_i branched over $B_i \cap L$ or $B_i \cap T$ are solid tori. It is clear that we can

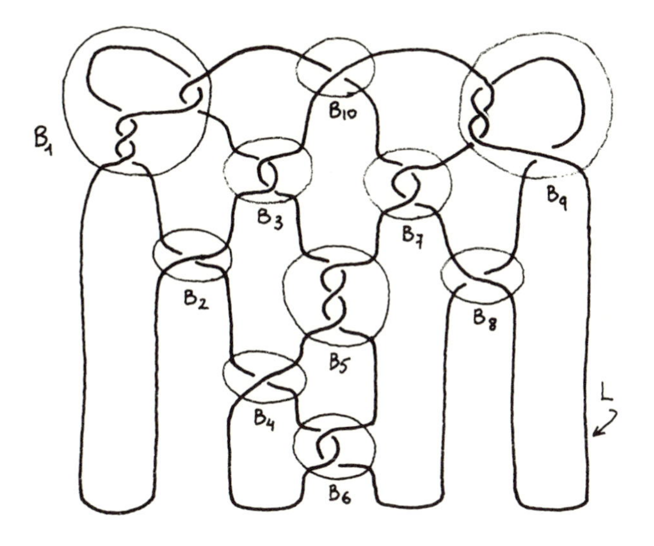

Fig. 3a.

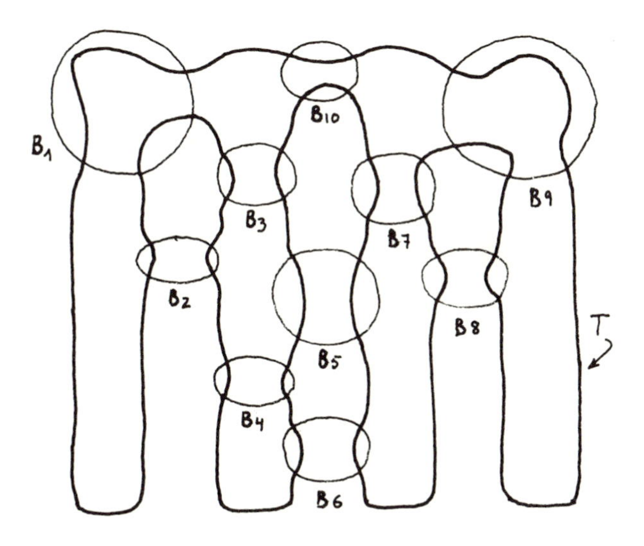

Fig. 3b.

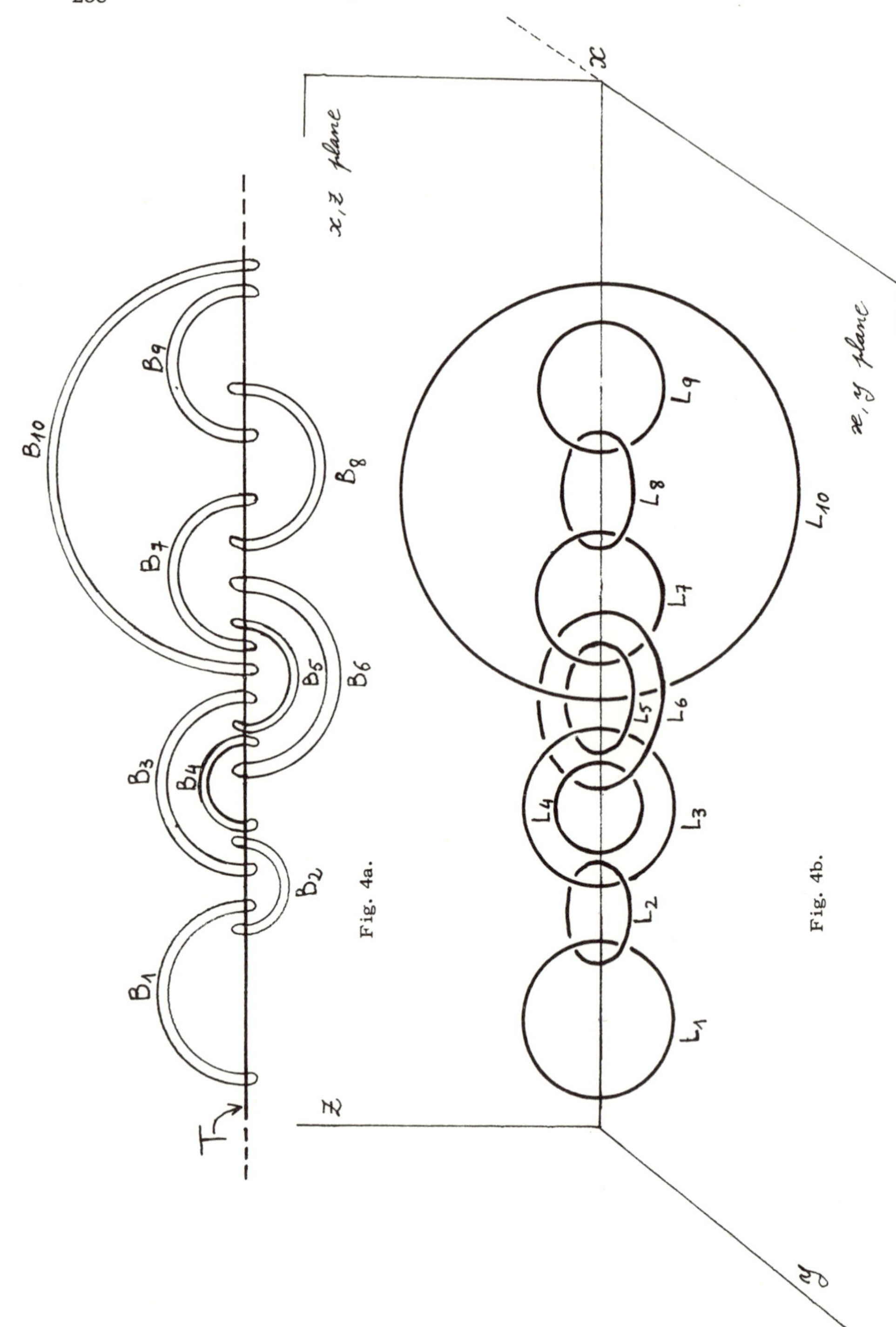

T
B1
B2
B3
B4
B5
B6
B7
B8
B9
B10
x
y
z
x, z plane
x, y plane
L1
L2
L3
L4
L5
L6
L7
L8
L9
L10

Fig. 4a.

Fig. 4b.

obtain $\tilde{L}$ by doing surgery on the link in $\mathcal{L}'_3$ of Figure 4b. In general, if L is a plat on $2m$ strings then $\tilde{L}$ can be obtained by doing surgery on a link in $\mathcal{L}'_{m-1}$.[3] □

As a consequence of Corollary 3 we have:

COROLLARY 4. *Conjecture 1 is equivalent to the Conjecture that each member of* $\mathcal{L}'_g$, $g \geq 1$, *has property* P.

To explore further the implications of Theorem 1, observe that if there is a closed, orientable 3-manifold M which gives a negative answer to Questions 2 or 3, it must be obtained by doing surgery on a link which is not strongly invertible. This suggests that one study Questions 2 or 3 by studying the 3-manifolds obtained by doing surgery on a non-invertible link.

Let L be a link in S^3 and let suppose that there is an orientation-preserving involution u in S^3, with fixed points, which induces an involution in each component of L. Let L' be the link consisting of those components of L for which the number of fixed points of u is different from two. Let $p: S^3 \to S^3$ the 2-fold cyclic branched covering of S^3 defined by u.

THEOREM 2. *Every manifold obtained by doing surgery on a link* L *is a 2-fold cyclic covering branched over a manifold obtained by doing surgery on* $p(L')$.

REMARK. Theorem 1 is a special case of Theorem 2.

---

[3] J. S. Birman has pointed out to me that it is interesting to note that the class of 3-manifolds which are obtained by doing surgery on links in $\mathcal{L}'_g$ are exactly the class of 3-manifolds which are "2-symmetric" in the notation of [2].

Proof of Theorem 2. For the sake of brevity, suppose that L has only one component and that either u is without fixed points in L or leaves each point of L fixed. Let $U(L)$ be a regular neighborhood of L, such that u induces in $U(L)$ an involution. Let $u' = u|\partial U(L)$.

Let V be a solid torus (see Figure 5) whose core C is a circle in the x, y plane with center 0 and radius one. Let z (resp. v) be the involution of V induced by the symmetry with respect to axis OZ (resp. C). There is a homeomorphism ψ of $\partial U(L)$ onto ∂V such that $z\psi = \psi u'$. Let $p = \psi^{-1}P$ and $m = \psi^{-1}M$ be a pair of simple oriented curves in $\partial U(L)$ (see Figure 5).

We now paste V to $S^3 - U(L)$ in the way that M is homologous to $\alpha m + \beta p$, where α and β are coprime integers. It is easy to see that there is a homeomorphism ϕ of ∂V onto $\partial U(L)$ such that $\phi(M) \sim \alpha m + \beta p$ and $\phi^{-1}\psi^{-1}z\psi\phi$, that is $\phi^{-1}u'\phi$, is equal to z if α is odd, or is equal to v if α is even.

Let W be the space obtained by pasting $S^3 - U(L)$ to V by ϕ. The map ϕ is compatible with the involutions u and z (or v, as the case may be). Thus, there is an involution u'' of W, the orbit-space of which is obtained by adjoining the orbit-space of u (that is S^3 minus a solid torus) with the orbit-space of z (or v, as the case may be), which is a solid torus. □

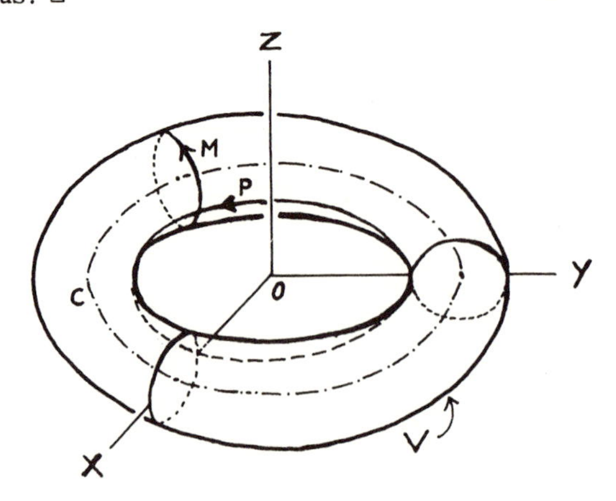

Fig. 5

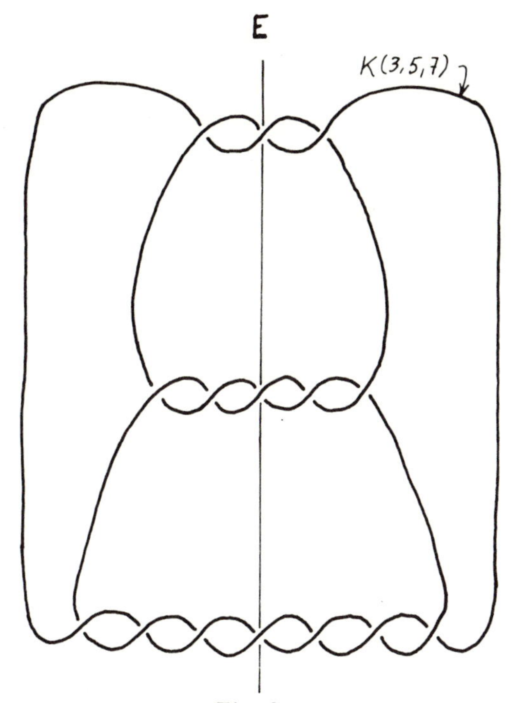

Fig. 6a.

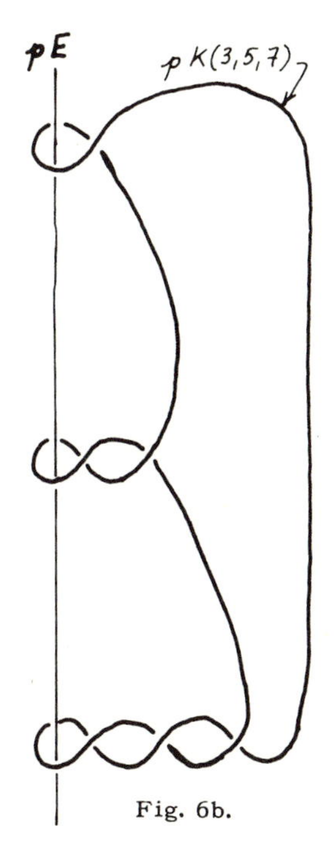

Fig. 6b.

As an application of Theorem 2 consider the pretzel knot $K(p,q,r)$ (see [24]). If any of the numbers p,q,r is even, it is clear that $K(p,q,r)$ is a strongly-invertible link. Thus, one obtains a 2-fold cyclic covering branched over S^3 by doing surgery on $K(p,q,r)$. If the numbers p,q,r are all odd, then there is an involution u of S^3 which induces in the knot $K(p,q,r)$ an involution without fixed points. (A typical case is illustrated in Figure 6a). Thus, every manifold that is obtained by doing surgery on $K(p,q,r)$ is a 2-fold cyclic covering branched over a manifold that is obtained by doing surgery on the trivial knot $p(K(p,q,r))$, where p is the covering defined by u (see Figure 6b). As the trivial knot is strongly-invertible it follows that the manifold obtained by doing surgery on $K(p,q,r)$, $(p,q,r$ odd), is a 2-fold cyclic covering of a 2-fold cyclic

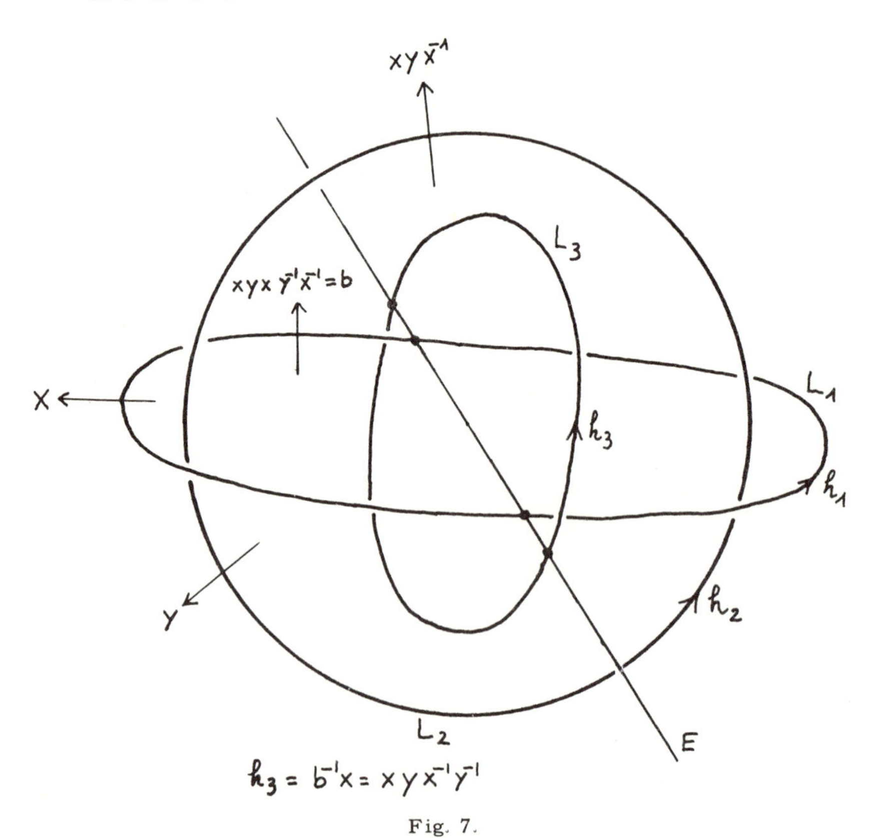

Fig. 7.

covering of S^3. This confirms Question 3. Note that Trotter has shown that $K(p, q, r)$ is non-invertible if p, q, r are distinct odd integers, each to greater than one. The author does not know whether the manifolds obtained by doing surgery on these knots are also representable as 2-fold cyclic branched covering of S^3.

As a second application, consider the manifold obtained by doing surgery on the "borromeans rings," B, illustrated in Figure 7. If we remove the solid tori $U(L_1)$, $U(L_2)$, $U(L_3)$ from S^3 and sew them back in such a way that the curves h_1, h_2, h_3 are identified with meridians, then we obtain $S^1 \times S^1 \times S^1$ [12], which is *not* a 2-fold branched cyclic covering of S^3 [6]. This shows that B is not a strongly invertible link. But the axial symmetry with respect to axis E (see Figure 7) induces in each component of B an involution. Then, by Theorem 2, every manifold that is obtained by doing surgery on B is a 2-fold cyclic branched covering of a manifold that is obtained by doing surgery on the trivial knot and this confirms Question 3. For instance, $S^1 \times S^1 \times S^1$ is a 2-fold cyclic branched covering of $S^1 \times S^2$.

It is interesting to note that not only is B non-invertible,[4] but also there is no orientation-preserving involution of S^3 which induces an involution in each component of B and which keeps fixed exactly two points of B.

---

[4] To the author's knowledge, this fact has not been established elsewhere in the literature. To prove it, let $F_2 = \{x, y / -\}$ be the group of the link formed by the components L_1, L_2. The group F_2 is a free group on two generators and the element $xyx^{-1}y^{-1}$ is represented by the loop h_3. If ϕ is an automorphism of F_2, then by [14, Theorem 3.9, p. 165], $\phi(xyx^{-1}y^{-1}) = w(xyx^{-1}y^{-1})^{\varepsilon}w^{-1}$, where w is a word in x, y which can be assumed to be reduced. Now, let us assume that B is an invertible link. Then there is an automorphism ϕ of F_2 that carries x to a conjugate of its inverse, carries y to a conjugate of its inverse and carries $xyx^{-1}y^{-1}$ to its inverse (compare [29]). The abelianizing homomorphism λ maps F_2 onto the abelian group $Z \oplus Z$, and ϕ induces an automorphism ϕ' of $Z \oplus Z$. It is easy to see that ε is equal to the determinant of the matrix of ϕ' with respect to λx, λy. Therefore, it follows that $w(xyx^{-1}y^{-1})w^{-1} = yxy^{-1}x^{-1}$. But induction on the length of w shows that this is impossible. Thus B is a non-invertible link. The same argument implies that there is not an orientation-preserving involution of S^3 which induces an involution in each component of B and which keeps fixed exactly two points of B.

At this point, it may be useful to remark there is a possibility of existence of a knot N such that there is no orientation-preserving involution of S^3 which induces an involution on N. One of these possible knots seems to be 8_{17} (see [4] and [19]).

§3. *Generalized surgery on links*

In this section we will define modifications of the projection of a link L that generalize the modifications W_1, W_2 introduced earlier and also the ones defined in [10]. These modifications have the effect of exhibiting $\tilde{L}$ as a manifold which is obtained by doing generalized surgery on a link in S^3, that is, removing n disjoint solid tori from S^3 and replacing each torus with a special "graph-manifold" which is bounded by a torus. The advantage of this is that if a link has property P, then it will be shown that a counterexample to the Poincaré Conjecture cannot be obtained by doing generalized surgery on it (Theorem 4). This fact will allow us to establish Conjecture 1 for a large set of knots (see Section 4).

Let R be a finite tree with a distinguished vertex v(R) (the *origin of* R). The tree is to be valued as follows: each vertex of R is labeled either with a hyphen, or with an arbitrary integer, in such a way that each vertex labeled with a hyphen belong to exactly one edge, and the origin v(R) is always labeled with an integer. Each edge of R is labeled with a pair of coprime integers (α, β) where $0 \leq \beta \leq \alpha$. We call R a *valued tree*.

We will describe a procedure for assigning to each valued tree R a manifold W(R), such that $\partial W(R)$ is a torus with a fixed oriented fiber, and moreover such that W(R) is a 2-fold cyclic covering of a 3-ball B, which is branched over a system of curves L(R) such that $\partial L(R)$ is the set $\{a, b, c, d\}$ of Figure 8. To do this, we need some definitions.

Let $M(s, \bar{m})$ be a manifold obtained as follows. Let M be the S^1-bundle over S^2 which admits a section, and let H be a fiber of M. Suppose that S^2 and H have a fixed orientation. We remove $m + 2$ fibered solid tori V_i from $M, i = -1, 0, 1, \cdots, m$. Then, S^2 cuts ∂V_i

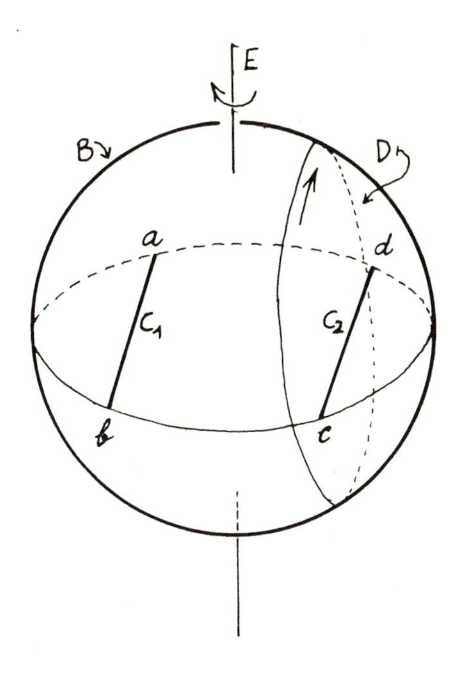

Fig. 8.

in a meridian curve m_i of V_i and we give to m_i the orientation induced by $S^2 - \operatorname{int} V_i$. Let us take in ∂V_i a fiber h_i, with the orientation inherited from H. In order to obtain M(s, m) we now paste a solid torus in such a way that its meridian curve is homologous to $m_i + sh_i$, $i = -1$. The boundary of M(s, m) is $\bigcup_{i=0}^{m} \partial V_i$, and m_i, h_i are fixed oriented curves in ∂V_i. M(s, m) is a 2-fold cyclic covering of $B - \operatorname{int}(B_1 \cup \cdots \cup B_m)$ branched over the curves L(s, m) of Figure 9 (for further details on the construction, see [18, Section 2 and Section 3]).

Let B be the ball of Figure 8. We define an autohomeomorphism t of ∂B as the composition of a rotation, of angle $\pi/2$ about the axis E which transforms a to d, and a symmetry with respect to the equatorial plane (see Figure 8). We define an autohomeomorphism v of ∂B as follows. Let D be a disc in ∂B which contains in its interior the points c, d and is disjoint from a, b (see Figure 8). Then, $v|D$ is

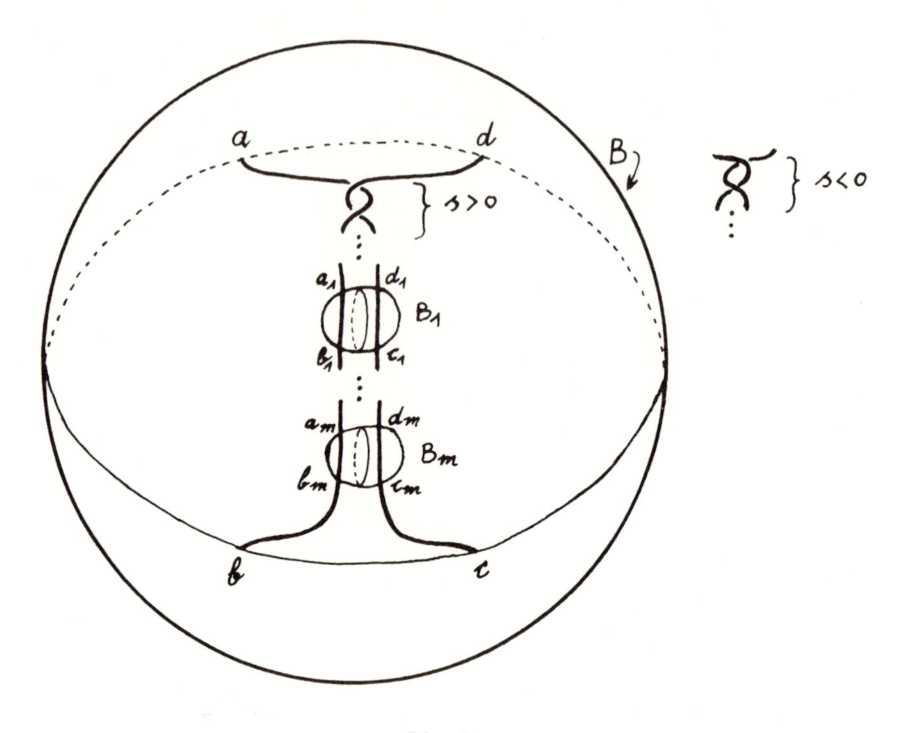

Fig. 9.

defined to be a "twist," holding D fixed, in the direction that is indicated in Figure 8, in order to move c to d. Now v is extended by the identity map outside D.

Now let α, β be two coprime integers. If α/β is the continued fraction

$$n + \frac{1}{m+} \cdots \frac{1}{+j} \, \frac{1}{+i} \;,$$

we define an autohomeomorphism $g(\alpha, \beta)$ of ∂B as the composition $g(\alpha, \beta) = v^n t v^m t \cdots t v^j t v^i$, where v^0 is the identity map. Let $f(\alpha, \beta) = g(\alpha, \beta)t$. Extend the homeomorphisms t, v to B. Then, $g(\alpha, \beta)$ and $f(\alpha, \beta)$ admit an extension to B, which we denote with the same symbols $g(\alpha, \beta)$, $f(\alpha, \beta)$.

We are now ready to define W(R) and L(R) by induction on the number n of vertices of R which are labeled with an integer.

Let $v(R)$ be labeled with the integer s. Let us suppose also that $v(R)$ belongs to m edges $t_1,\cdots,t_m$ and that t_i is labeled with (α_i,β_i). Let u_i denote another vertex of t_i and assume that u_i, where $1 \le i \le r$, is labeled with a hyphen and that u_j, where $r+1 \le j \le m$, is labeled with an integer. Then, u_j, $r+1 \le j \le m$, is the "beginning" of a valued tree R_j.

Let $v(R_j) = u_j$. Note that the number of vertices of R_j which are labeled with an integer is $< n$. $W(R)$ is defined inductively, pasting the r solid torus $V_1,\cdots,V_r$ and the $m-r$ manifolds $W(R_j)$, $r+1 \le j \le m$, to $M(s,m)$ in such a way that a meridian curve of V_i is homologous to $\alpha_i m_i + \beta_i h_i$, and the oriented fiber, fixed in $\partial W(R_j)$, is homologous to $\alpha_j m_j + \beta_j h_j$. Note that in $\partial W(R) = \partial V_0$, the oriented fiber h_0 remains fixed. Then $L(R)$ is obtained replacing $f(\alpha_i,\beta_i)(L(s,0) \cap B_i)$, where $1 \le i \le r$, by $L(s,0) \cap B_i$ and replacing $g(\alpha_j,\beta_j)L(R_j)$, where $r+1 \le j \le m$, by $L(s,0) \cap B_j$ (see Figure 9). As an illustration of this process see the example of Figure 10.

Let L be a link in S^3 having m components $N_1,\cdots,N_m$. We will say that a 3-manifold M is obtained by doing *general surgery* m times on L if M is obtained by removing from S^3 a regular neighborhood $U(N_i)$ of N_i, $1 \le i \le m$, and replacing it with $W(R_i)$, where R_i is some valued tree, by pasting $\partial W(R_i)$ to $\partial(S^3 - U(N_i))$.

Let L be a link in S^3 and let us suppose that there is a ball B in S^3 such that $\partial(B \cap L)$ is the set $\{a,b,c,d\}$ (see Figure 8) and $B \cap L$ is a system of curves $g(\alpha,\beta)L(R)$, where R is an arbitrary valued tree and α,β are an arbitrary pair of coprime integers. We will say that has made a *general modification* on L, if we replace $B \cap L$ for the pair of curves C_1, C_2 of Figure 8. Let m be the minimum number of general modifications which have to be applied to L in order to change L into the trivial knot. It is clear that $\tilde{L}$ has been obtained by doing general surgery on a strongly-invertible link in S^3 of m components.

The following theorem is proved in the same way as Theorem 1:

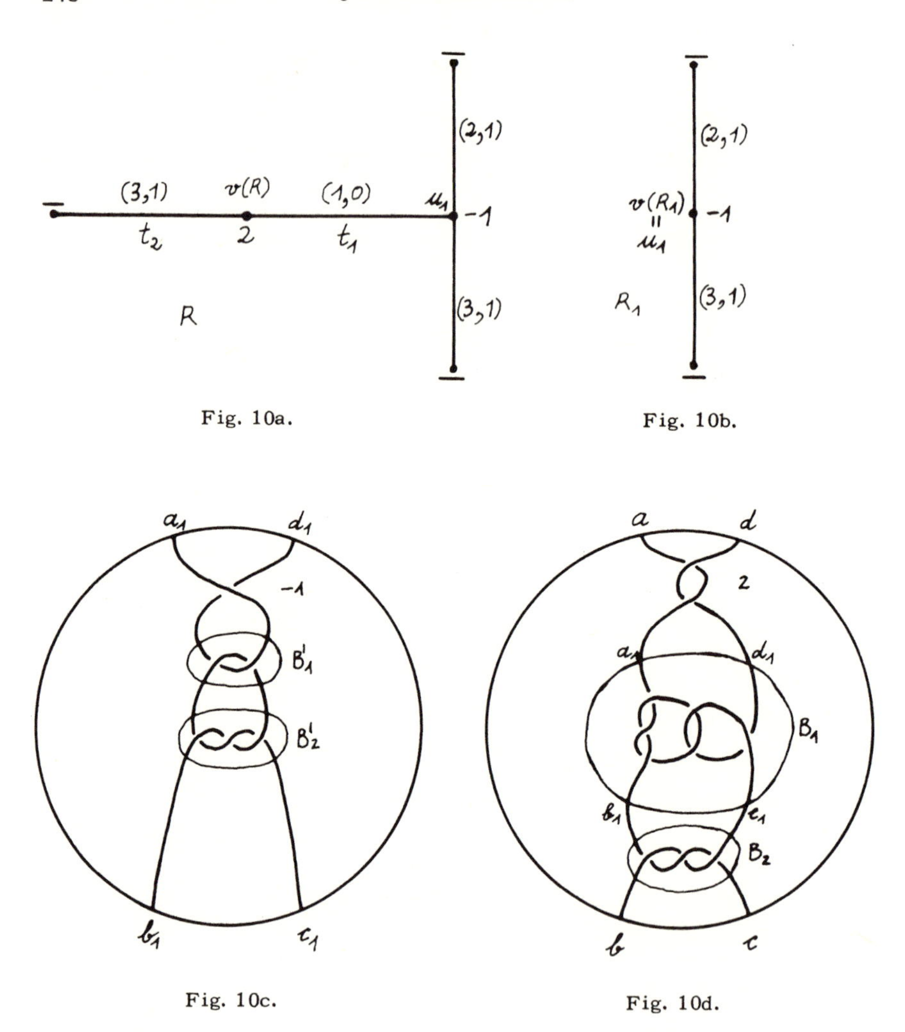

Fig. 10a.

Fig. 10b.

Fig. 10c.

Fig. 10d.

THEOREM 3. *Every manifold that is obtained by doing general surgery on a strongly-invertible link is a 2-fold cyclic branched covering of* S^3.

The following theorem indicates a useful application of general surgery.

THEOREM 4. *If* M *is a simply-connected 3-manifold that is obtained by doing general surgery on a link* L *with property* P, *then* $M = S^3$.

In order to prove Theorem 4 we first need the following Lemma:

LEMMA 1. *Every homotopy* 3-*ball that lies in a graph-manifold is a* 3-*ball.*

We defer the proof of this lemma until Section 5.

Proof of Theorem 4. We are going to demonstrate the theorem by induction on the number n of graph-manifolds distinct from a solid torus which are introduced by surgery. If $n = 0$, there is nothing to prove, thus let $n > 0$. Let L_1 be a component of L such that a regular neighborhood, $U(L_1)$, of L_1 has been replaced by a graph-manifold $W(R)$ which is not a solid torus. If $\pi(M) = 1$, then $\partial U(L_1)$ bounds in M a homotopy solid torus ([1] and [8; Lemma 5.1]). If $W(R)$ were a homotopy solid torus, it would be a solid torus (by Lemma 1), hence $M - \text{int } W(R)$ is a homotopy solid torus. Then, $\pi(M - \text{int } W(R))$ is an infinite cyclic group with one generator which is represented by a simple curve C in $\partial(M - \text{int } W(R))$. We paste a solid torus to $M - \text{int } W(R)$ in such a way that C is a meridian curve of it. Thus we have built a manifold M', with $\pi(M') = 1$, which is obtained from S^3 by doing surgery on the link L, and replacing $n-1$ components of L by $n-1$ graph-manifolds which are not solid tori. By the induction hypothesis, $M' = S^3$ and thus $M - \text{int } W(R)$ is a solid torus. Therefore, M is a graph-manifold. Making use of the result of Lemma 1 we conclude that Theorem 4 is true. □

With the purpose of justifying the definitions of general modifications and general surgery, we make the following remarks. Let K be a non-trivial knot in S^3. If we wish to check Conjecture 1 for K, we can, for instance, apply m modifications of type W_2 in order to change K into the trivial knot. Then, $\tilde{K}$ is a manifold that is obtained by doing surgery on a strongly invertible link in S^3 of m components. By doing this in all possible ways, we obtain a family $\mathfrak{L}(K)$ of links in S^3 such that $\tilde{K}$ can be exhibited as a manifold obtained by doing surgery on an arbitrary member of $\mathfrak{L}(K)$. Let $m(K)$ be the minimal number of modifications of type W_2 which we have to apply to K in order to change K into the trivial knot. We define $\mathcal{L}'(K)$, $m'(K)$, in the same way as $\mathcal{L}(K)$ and $m(K)$, but replacing modifications of type W_2 for general modifications. Thus $\tilde{K}$ can be exhibited as a manifold obtained doing *general* surgery on an arbitrary member of $\mathcal{L}'(K)$.

As a consequence of Theorem 4, if a member of $\mathcal{L}'(K)$ has property P, then $\pi(\tilde{K}) \neq 1$. On the one hand $m'(K) \leq m(K)$ and this makes it easier to check Conjecture 1 for K in many cases, especially when $m'(K) = 1$, because property P has been intensively studied for knots. On the other hand, $\mathcal{L}(K) \subset \mathcal{L}'(K)$ and this increases our possibilities of finding a link with property P such that $\tilde{K}$ is obtained by doing general surgery on it.

It could happen that $m'(K) = 1$, for every non-trivial knot K. If this was so, then every 2-fold cyclic covering branched over a knot of S^3, would be obtained by doing general surgery on a strongly-invertible knot of S^3. Then, Conjecture 1 would be equivalent to the conjecture that every strongly-invertible knot has property P.

§4. *Applications*

If one seeks a counterexample to the Poincaré Conjecture among the 2-fold branched coverings of S^3, it is natural to examine covering spaces which are branched over knots which share deep properties with the trivial knot. One such property is that the trivial knot has Alexander polynomial $\Delta(t) = 1$. Note that if a knot N has Alexander polynomial $\Delta(t) = 1$, then $\tilde{N}$ is a homology 3-sphere.

1. *Kinoshita-Terasaka knots*

Let us consider the knots of Kinoshita-Terasaka [11, p. 149] $k(p,2n)$ ($k(3,6)$ is illustrated in Figure 11a or 11b). All of them have Alexander polynomial $\Delta(t) = 1$. Note that $k(3,6)$ can be obtained from the link of Figure 11c by substituting B_i for C_i $(i = 1,2,3)$. Thus [18] $\tilde{k}(3,6)$ is the graph-manifold that is represented (in Waldhausen's notation) by the graph of Figure 12, where $p = 3$, $n = 2$. In general, for $k(p,2n)$, $\tilde{k}(p,2n)$

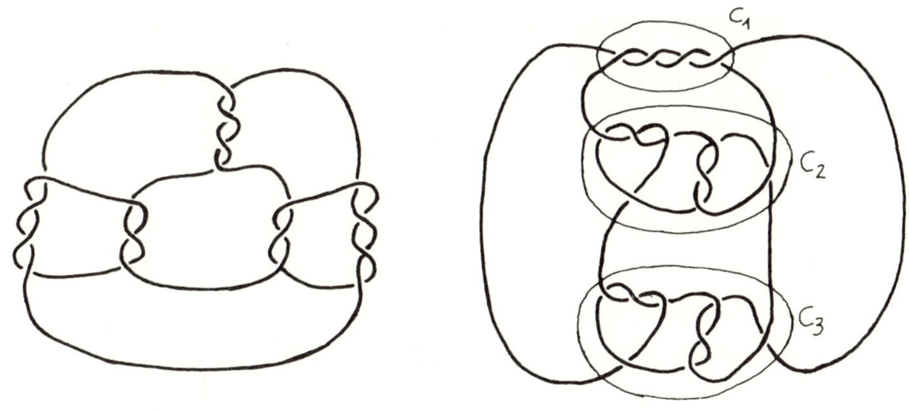

Fig. 11a. Fig. 11b.

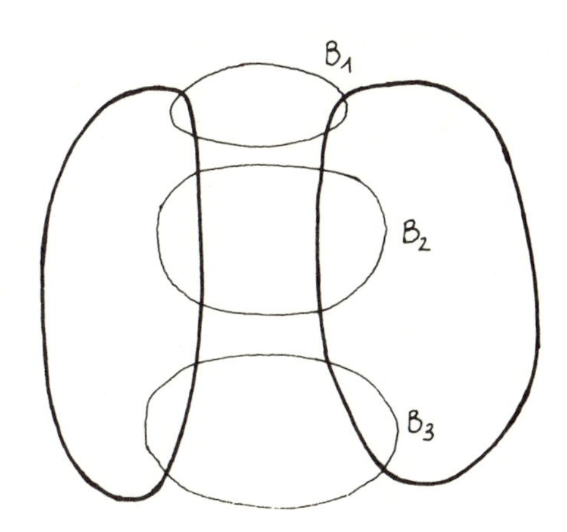

Fig. 11c.

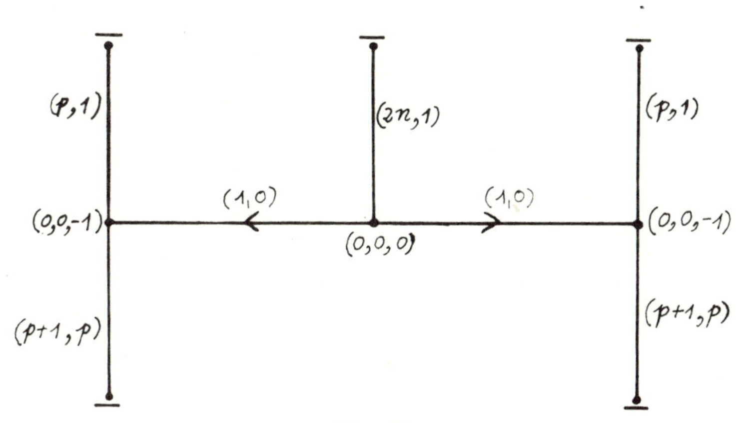

Fig. 12.

is the graph-manifold represented by the graph of Figure 12. Thus, by Lemma 1, k(p, 2n) cannot give a counterexample to the Poincaré Conjecture.

2. *Conway's 11-crossing knot*

Let L be the knot, with Alexander polynomial $\Delta(t) = 1$, of Figure 13a, which was discovered by J. Conway in his enumeration of the non-alternating 11-crossing knots [3] (see also [20, p. 615]).

The trivial knot T can be obtained by doing one general modification in L (see Figure 13a, b). The 2-fold cyclic covering $\tilde{B}$ (resp $\tilde{C}$) of the ball B (resp. C) branched over $B \cap L$ (resp. $C \cap L$) is a solid torus. Then, $\tilde{L}$ can be obtained by removing $\tilde{C}$ from $\tilde{T} = S^3$ and sewing it back differently. The position of the ball C with respect to the trivial knot T is shown in Figure 14a. Then, $\tilde{C}$ is a regular neighborhood of the square knot (Figure 14b). Thus, $\tilde{L}$ can be obtained by doing surgery on the square knot, hence $\pi(\tilde{L}) \neq 1$, because a composite knot has property P ([1], [8]).

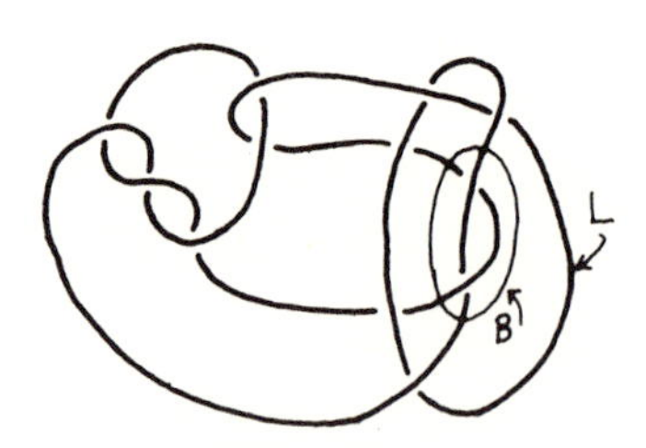

Fig. 13a.

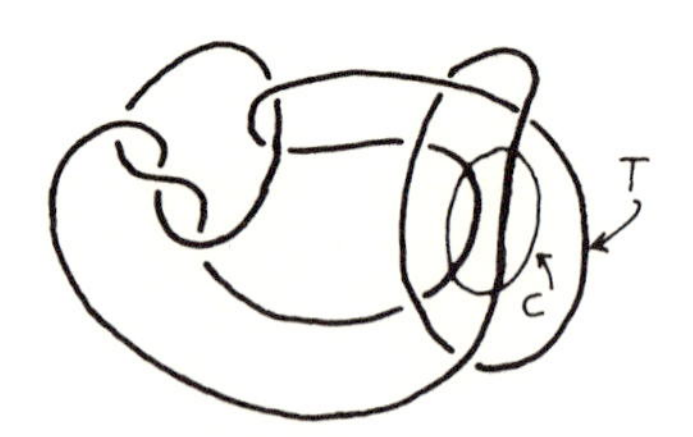

Fig. 13b.

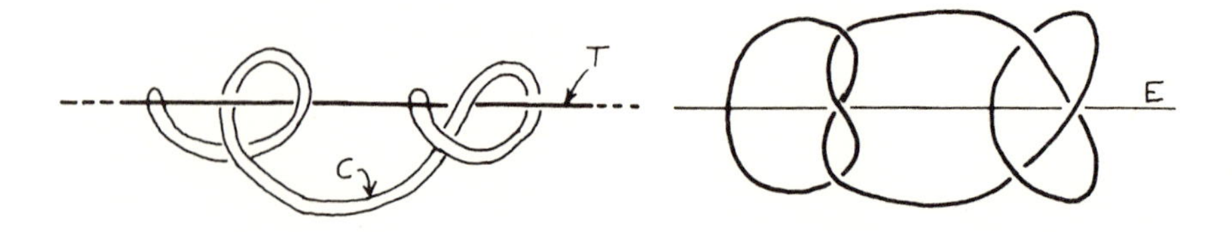

Fig. 14a. Fig. 14b.

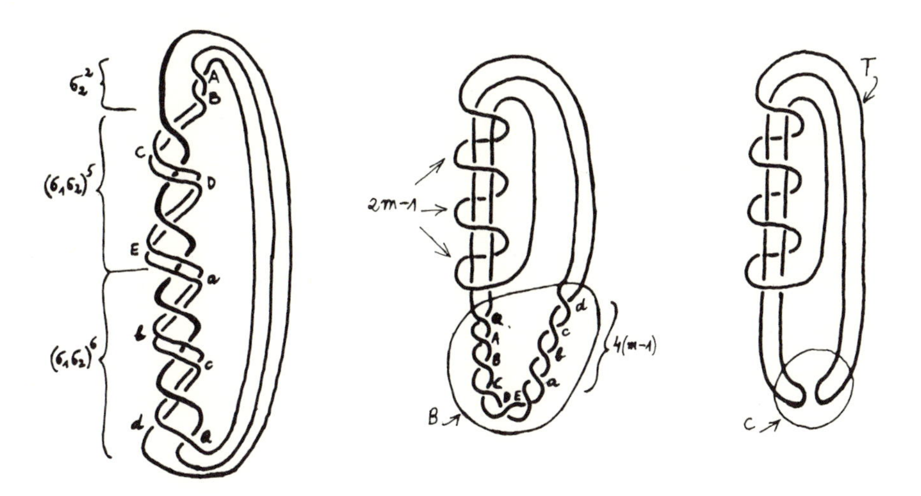

Fig. 15a. Fig. 15b. Fig. 15c.

3. *The 3-braid knots* $(\sigma_2\sigma_1^{-1})(\sigma_1\sigma_2)^{6m}$, $m \geq 1$

In [2] is is proved by Birman and Hilden that if Conjecture 1 is true for the knots $(\sigma_2\sigma_1^{-1})(\sigma_1\sigma_2)^{6m}$, $m \geq 1$ then Conjecture 1 is true for every 3-braid knot. We prove now that Conjecture 1 is true for the knots $(\sigma_2\sigma_1^{-1})(\sigma_1\sigma_2)^{6m}$, $m \geq 1$. For the sake of brevity, let L be the knot $(\sigma_2\sigma_1^{-1})(\sigma_1\sigma_2)^{12}$ of Figure 15a, b. The trivial knot T can be obtained by doing one general modification in L (see Figure 15b, c). The 2-fold cyclic covering $\tilde{B}$ (resp. $\tilde{C}$) of the ball B (resp. C) branched over $B \cap L$ (resp. $C \cap L$) is a solid torus. The position of the ball C with respect to the trivial knot T is shown in Figure 16a. Then, $\tilde{C}$ is a regular neighborhood of the twist knot T_3 (Figure 16b). Hence $\tilde{L}$ can be obtained by doing surgery on the twist knot T_3, hence $\pi(\tilde{L}) \neq 1$ because a twist knot has property P ([1], [8]).

A similar argument applies to the case where m is arbitrary. In general, the 2-fold cyclic covering branched over the 3-braid knot $(\sigma_2\sigma_1^{-1})(\sigma_1\sigma_2)^{6m}$ can be obtained by surgery on the twist knot T_{2m-1}.

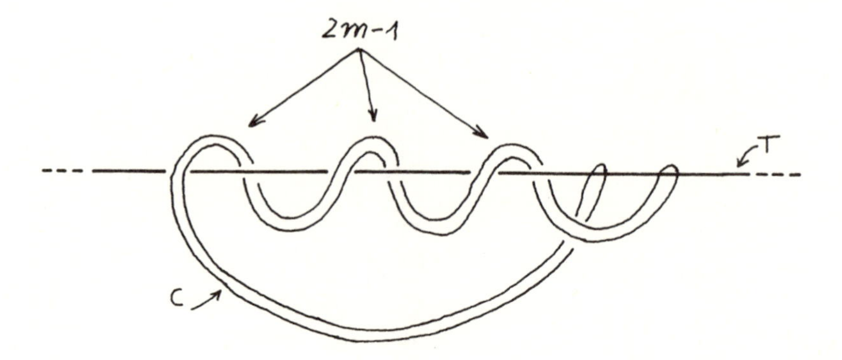

Fig. 16a.

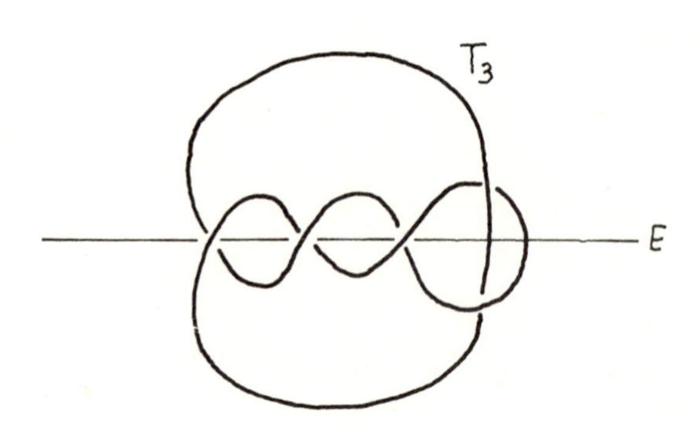

Fig. 16b.

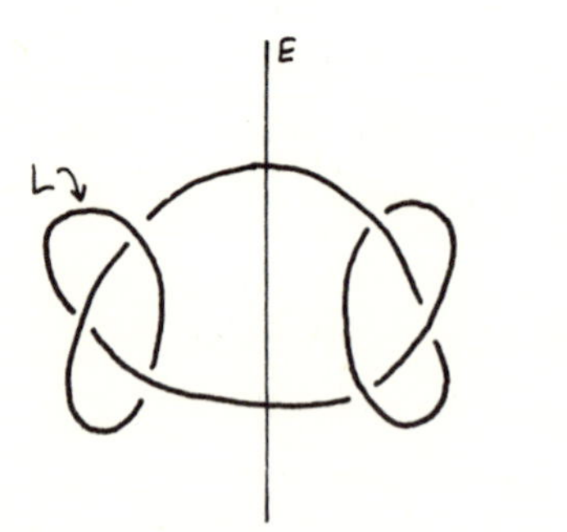

Fig. 17a.

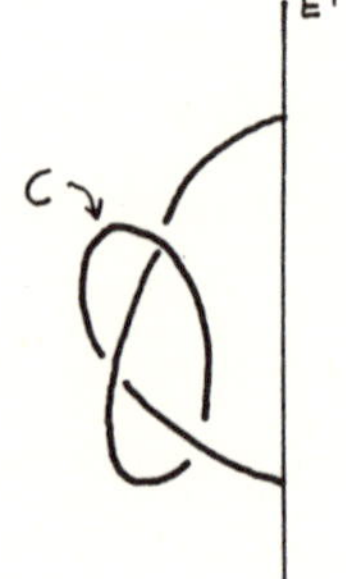

Fig. 17b.

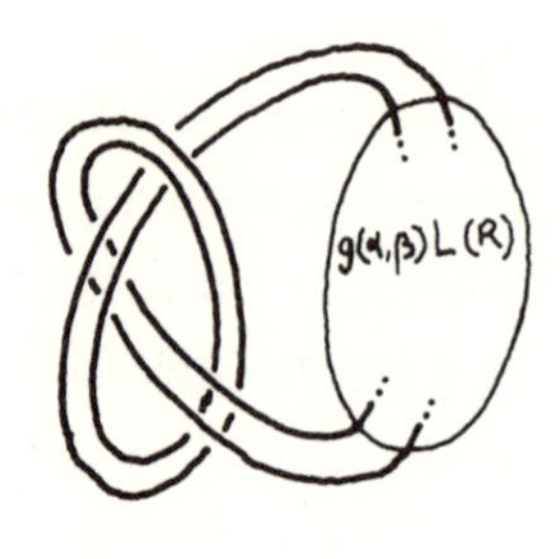

Fig. 17c.

4. *Generalized doubled knots*

Let L be the knot of Figure 17a. L is a strongly-invertible knot because the symmetry u with respect to the axis E leaves L invariant. Let $p: S^3 \to S^3$ be the 2-fold cyclic branched covering induced by u. Then, p(L) is the path C of Figure 17b. As a composite knot has property P ([1], [8]) then Conjecture 1 is true for the family of links of Figure 17c, where R is an arbitrary valued tree and where α, β are an arbitrary pair of coprime integers. As the same argument can be applied to an arbitrary strongly-invertible composite knot, we obtain in particular, that Conjecture 1 is true for every doubled knot (a fact proved by algebraic methods by Giffen [7]).

The same method can be applied to an arbitrary strongly-invertible link with property P (examples of these can be found in [1], [8] and [23]).

5. The idea illustrated in the following example may be useful. Let N be the knot of Figure 18 and let us consider a plane P with cuts N in the set $\{a, b, c, d\}$. Thus P divides S^3 into two balls A, B. The 2-fold cyclic covering $\tilde{A}$ (resp. $\tilde{B}$) of A (resp. B), branched over $A \cap N$ (resp. $B \cap N$) is the complement of a regular neighborhood of a non-trivial knot in S^3 (see Section 4.4.). Then, $\tilde{N}$ can be obtained by pasting $\partial\tilde{A}$ to $\partial\tilde{B}$. According to [1] and [8; Lemma 5.1] $\pi(\tilde{N}) \neq 1$.

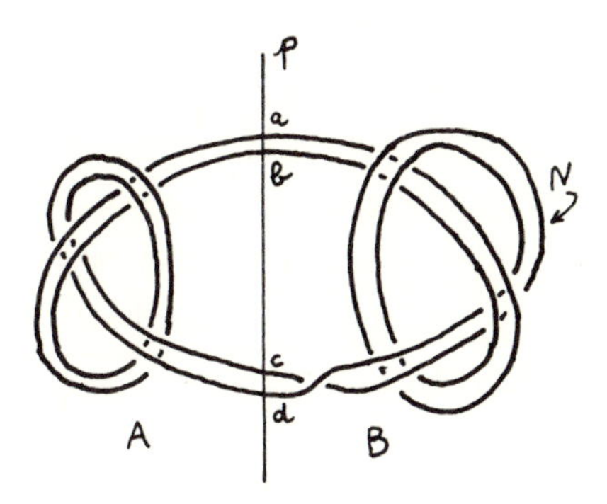

Fig. 18.

§5. *Demonstration of Lemma 1* *

Recall that the only simply connected Seifert manifold is S^3 [22].

On the one hand, every graph-manifold with boundary is a submanifold of a graph-manifold without boundary. On the other hand, every graph-manifold without boundary is [26, Satz 6.3] a connected sum of lens-spaces and *reduced graph-manifolds* ("Reduzierte Graphenmannigfaltigkeiten," see [26, 6.2]). Then, according to [9] and [26, Satz 7.1], Lemma 1 will be proved if we can show that a simply-connected, reduced, closed graph-manifold is S^3.

A reduced graph-manifold is either defined by a graph A(M) (see [26; 9]), or is a torus-bundle over S^1, or is a Seifert manifold over S^2 with three exceptional fibers. Thus, according to [9], it is sufficient to prove Lemma 1 for closed, reduced graph-manifolds M defined by a graph A(M). All of them [18; 7.5] are 2-fold cyclic coverings branched over a 3-sphere with g handles. If the graph A(M) is not a tree, or if any of the vertices of A(M) are valued with a triple $(g_j, 0, s_j)$, $g_j > 0$, then $g > 0$, hence $H_1(M) \neq 0$. If the graph A(M) is a tree with its vertices valued with triples $(g_j, 0, s_j)$, $g_j \leq 0$, then M is a 2-fold cyclic covering branched over a link L of S^3 [18; 7.3]. This link L has more than one component if $g_j < 0$ for any j [18; §3]. In this case, we have $H_1(M) \neq 0$. Then, let M be represented by a tree A(M) whose vertices are valued with triples $(0, 0, s_j)$. For [26, 9.2.3., 9.2.4.a), b) and c)] the vertices of A(M) either are of order ≥ 3, or are valued with a hyphen but there is always a vertex of order ≥ 3. We are going to prove Lemma 1, for those manifolds, by induction on the number m of vertices of order ≥ 3. If $m = 1$, M is a Seifert manifold and there is nothing to prove. Assume that $m > 1$. Then, there is a torus in M that splits M into two reduced graph-manifolds, M_1, M_2, corresponding to the graphs $A(M_1)$, $A(M_2)$ respectively. In order to build $A(M_1)$, $A(M_2)$ it is sufficient to

---

* In this section we will follow the notation of Waldhausen in [26].

remove from A(M) an edge which joins two vertices of order ≥ 3 and to value these vertices again with $(0,1,-)$. Then, $A(M_i)$, $i = 1,2$, has at least one vertex of order ≥ 2, valued with $(0,1,-)$.

According to [1], [8; Lemma 5.1], if $\pi(M) = 1$, then either M_1 or M_2 is a homotopy solid torus. We may assume that M_1 is a homotopy solid torus. Then, M_1 may be considered as a submanifold of either a Seifert manifold with three exceptional fibers, or a graph-manifold that is represented by a graph with $n < m$ vertices of order ≥ 3. Thus, by the induction hypothesis and according to [9; 2.2], M_1 is a solid torus. But then, [26; Satz, 9.4] $A(M_1)$ is a graph which has exactly one vertex of order zero, valued with $(0,1,-)$. This is a contradiction, hence $\pi(M) \neq 1$. □

Therefore, a simply connected graph-manifold is S^3.

UNIVERSIDAD COMPLUTENSE, MADRID

REFERENCES

[1] Bing, R. H., and Martin, M., "Cubes with knotted holes." Trans. Am. Math. Soc. 155 (1971), 217-231.

[2] Birman, J. S., and Hilden, H. M., "Heegaard splittings of branched covering of S^3" to appear in Trans. Am. Math. Soc.

[3] Conway, J. H., "An Enumeration of knots and links, and some of Their Algebraic properties." Computational Problems in Abstract Algebra, Pergamon Press, New York (1970), 329-358.

[4] Fox, R. H., "Some problems in knot theory." Topology of 3-manifolds (M. K. Fort, Ed.) 168-176. Prentice Hall, New York, 1962.

[5] ———, "Two Theorems about periodics transformations of the 3-sphere." Michigan Math. J. 14 (1967), 331-334.

[6] ———, "A note on branched cyclic coverings of spheres." Rev. Mat. Hisp.-Amer. 32 (1972), 158-166.

[7] Giffen, C. H., "Cyclic branched coverings of doubled curves in 3-manifolds." Illinois J. Math. 11 (1967), 644-646.

[8] González-Acuña, F., "On homology spheres." Princeton Ph.D. Thesis. (1969).

[9] Gross, J. L., "Manifolds in which the Poincaré Conjecture is true." Trans. Am. Math. Soc. 142 (1969), 177-189.

[10] Kinoshita, S., "On Wendt Theorem of Knots." Osaka Math. J. 9 (1957), 61-66.

[11] Kinoshita, S., and Terasaka, H., "On union of knots." Osaka Math. J. 9 (1959), 131-153.

[12] Laudenbach, F., and Roussarie, R., "Un example de feuilletage sur S^3." Topology 9 (1970), 63-70.

[13] Lickorish, W. B. R., "A foliation for 3-manifolds." Ann. of Math. 82 (1965), 414-420.

[14] Magnus, W., Karras, A., and Solitar, D., *"Combinatorial Group Theory": Presentations of Groups in Terms of Generators and Relations*. Pure and Appl. Maths., vol. 13, Interscience, New York, 1966.

[15] Montesinos, J. M., "Sobre la Conjetura de Poincaré y los recubridores ramificados sobre un nudo." Tesis doctoral. Madrid, 1971.

[16] ————————, "Reduccion de la Conjetura de Poincaré a otras Conjeturas Geométricas." Rev. Mat. Hisp.-Amer. 32 (1972), 33-51.

[17] ————————, "Una familia infinita de nudos representados no separables." Rev. Mat. Hisp.-Amer. 33 (1973), 32-35.

[18] ————————, "Variedades de Seifert que son recubridores cíclicos ramificados de dos hojas." Boletín Soc. Mat. Mexicana. 18 (1973), 1-32.

[19] Murasugi, K., "On periodics knots." Comment. Math. Helv. 46 (1971), 162-174.

[20] Riley, R., "Homomorphisms of Knot Groups on Finite Groups." Math. Compt. 25 (1971), 603-619.

[21] Schubert, H., "Knoten mit zwei Brücken." Math. Z., 65 (1956), 133-170.

[22] Seifert, H., "Topologie dreidimensionaler gefaserter Räume." Acta Math. 60 (1933), 147-238.

[23] Simon, J., "Some class of knots with property P." Topology of manifolds. Markham, Chicago, Ill. (1970), 195-199.

[24] Trotter, H. F., "Non-invertible knot exist." Topology 2 (1964), 275-280.

[25] Viro, O. Ja, "Linkings, 2-sheeted branched coverings, and braids." Math. USSR, Sbornik, 16 (1972), No. 2, 223-236 (English translation).

[26] Waldhausen, F., "Eine Klasse von 3-dimensionalen Mannigfaltigkeiten, I and II." Invent. Math. 3 (1967), 308-333 y 4 (1967), 87-117.

[27] Waldhausen, F., "Über Involutionen der 3-sphäre." Topology 8 (1969), 81-91.

[28] Wendt, H., "Die Gordische Auflösung von Knoten." Math. Z. 42 (1937), 680-696.

[29] Whiten, W. C., JR., "A pair of noninvertible links." Duke Math. J. 36 (1969), 695-698.

PLANAR REGULAR COVERINGS OF ORIENTABLE CLOSED SURFACES

C. D. Papakyriakopoulos

§1. Introduction

In 1963 this author reduced the Poincaré conjecture to two other conjectures ([7], p. 251). The first of those conjectures is a special case of the group theoretic Conjecture 1 of ([8], p. 205). A special case of Conjecture 1 was proved by Karrass, Magnus and Solitar ([4], p. 57). Elvira Rapaport proved Conjecture 1 in full generality ([9], p. 506). In a recent paper Eldon Dyer and Vasquez ([1], pp. 348-349) proved the algebraic topological and stronger Conjecture 1′ of ([8], p. 205).

However, Maskit ([6], p. 342, *ll.* 2-7) gave a new proof and a simpler statement of our key theorem ([7], p. 290), so that only the second of the two conjectures ([7], p. 251) is needed for the reduction of the Poincaré conjecture. That second conjecture leads us to the following problem.

PROBLEM. Let N be an orientable closed surface of genus at least two. Let g be an element of the fundamental group F of N, and let $\hat{N}$ be the regular covering of N corresponding to the normal closure G of g in F. *Is* $\hat{N}$ *planar?*

The Planarity theorem of Maskit ([6], p. 351) is a theorem of structure, and describes a way of obtaining any planar regular covering of any compact surface closed or not. However, that theorem does not seem to be directly applicable to our problem. So we will try to find another way of solving our problem. We observe that $\hat{N}$ is planar if and only if the intersection number of any two loops on $\hat{N}$ is zero, see No. 11 of this paper. We also observe that $\hat{N}$ and the loops on it depend on g. Thus, we need a formula which will give us the intersection number of any two loops on $\hat{N}$ by means of g. The equation with right hand side zero and left hand side the right hand side of the formula will give the solution of our problem.

We find such a formula, and actually in the case where G is the normal closure in F of a finite or infinite sequence $g_1, g_2, \cdots$ of elements of F, see Theorems 10.13 and 11.1 of this paper. Thus, the problem posed above is solved in theory. However, the result provided by the solution is not sufficient to solve the second conjecture of ([7], p. 251), see Section 5 at the end of this paper.

The main theorems of this paper are Theorems 10.13 and 11.1. The first of those theorems provides us with the *intersection* and *expansion formulas*, and the second provides us with a more explicit expansion formula and the necessary and sufficient conditions that $\hat{N}$ be planar. The core of those formulas is the operator Λ, which is defined by means of Fox-derivatives ([3], p. 550). The definition of Λ and the way we obtain the formulas are involved, and in the sequel of this Introduction we will try to explain things briefly.

In Section 2, which is preparatory, we first define the notion of conjugation in No. 1. The conjugate of a group ring element is obtained by replacing every element of the group by its inverse. In No. 2 we define the notion of inner product in a group ring, the natural way. In No. 3 antiderivations are defined. The conjugate of an anti-derivation is a derivation. In No. 4 biderivations are introduced. A biderivation is a map on two variables, it is a derivation with respect to the left variable and antiderivation with respect to the right variable. In No. 5 biderivations in a free group ring are examined and proved to be of a special form, see Theorem 5.3. In No. 6 some propositions concerning biderivations are proved, Theorem 6.3 will be needed in Section 4.

In Section 3 we study intersection theory on $\hat{N}$, and we obtain the intersection formula (8.5). This is an indispensable formula for Section 4, which expresses the intersection number of two 1-chains on $\hat{N}$ as a sum of inner products of their coefficients. Section 3 was inspired by Reidemeister's paper [10]. However, we emphasize that he neither obtains a formula nor does he introduce the inner product. The notion of inner product, though so simple and natural, is of decisive importance in obtaining the formulas needed to solve our problem.

The following Section 4 is the main part of our paper. In No. 9 we introduce the operator Λ, and in No. 10 after elaborate work we obtain the intersection formula (10.11) and the expansion formula (10.12). Those formulas are expressed with the help of the operator Λ and the inner product. We then obtain the first main Theorem 10.13, the proof of which is based on formulas (10.11) and (10.12). Finally in No. 11, we obtain the second main Theorem 11.1, which provides a more explicit expansion formula, and the necessary and sufficient conditions that $\hat{N}$ be planar. Thus, the second main theorem provides us with a solution of our problem. The proof of the second main theorem is based on the first one and Theorem 6.3 of Section 2.

We finish our paper with Section 5, where we formulate a conjecture. The solution of that conjecture will provide us with a proof of the Poincaré conjecture.

The paper is dedicated to the memory of Ralph Fox. A small sign of the gratitude the author feels to Ralph, for helping him to come to Princeton and stay here.

§2. Operations in Group Rings

1. *Conjugation.*

Let Γ be a finite or infinite denumerable group with elements g_i, $i = 1, 2, \cdots$. We denote by $Z[\Gamma]$ the integral group ring of Γ. Thus any element r of $Z[\Gamma]$ is of the form

$$r = \sum_i m_i g_i, \quad m_i \in Z, \quad g_i \in \Gamma, \quad i = 1, 2, \cdots$$

where only a finite number of m_i are different from zero. We write

$$\bar{r} = \sum_i m_i \bar{g}_i, \quad \bar{g}_i = g_i^{-1}$$

and we call it the *conjugate* of r. We also denote $(\overline{\cdots})$ by $(\cdots)^-$, i.e., $(\cdots)^- = (\overline{\cdots})$. The following hold.

(1.1) $\bar{m} = m$, for any element m of Z.

(1.2) $\bar{\bar{r}} = r$, for any element r of $Z[\Gamma]$.

(1.3) $\overline{r+s} = \bar{r} + \bar{s}$, for any two elements r, s of $Z[\Gamma]$.

(1.4) $\overline{rs} = \bar{s}\bar{r}$.

For any finite number of elements r_j, $j = 1, \cdots, n$ of $Z[\Gamma]$ we have

$$(1.5) \qquad \Big(\sum_j r_j\Big)^- = \sum_j \bar{r}_j, \qquad \Big(\prod_j r_j\Big)^- = \prod_k \bar{r}_k$$

where $k = n, \cdots, 1$.

Let $\psi : \Gamma \to \Gamma'$ be a homomorphism of the group Γ into another finite or infinite denumerable group Γ'. This induces a ring-homomorphism $\psi : Z[\Gamma] \to Z[\Gamma']$. (N.B. for the sake of simplicity, instead of denoting the ring-homomorphism by $\tilde{\psi}$, say, we denote it simply by ψ, see ([3], p. 548).) The following holds.

(1.6) $\overline{\psi(r)} = \psi(\bar{r})$, for any element r of $Z[\Gamma]$.

We define r^0 to be the sum of the coefficients of r, where r is given at the starting of this No. 1, i.e.,

$$r^0 = \sum_i m_i$$

see ([3], p. 549, $\ell\ell$. 18-20). A final remark is that the conjugation was introduced by Reidemeister ([11], p. 23, ℓ.8).

2. *Inner product*

Let g_i and g_j be two elements of Γ, we define the *inner product* of them by

(2.1) $$g_i \circ g_j = \delta_{ij}, \quad i,j = 1,2,\cdots$$

δ_{ij} being the Kronecker index. Let now r and s be two elements of $Z[\Gamma]$,

$$r = \sum_i m_i g_i, \qquad s = \sum_j n_j g_j, \qquad i,j = 1,2,\cdots .$$

We then define the *inner product* of them by

$$r \circ s = \sum_{i,j} m_i n_j (g_i \circ g_j) = \sum_i m_i n_i .$$

This is an element of Z, i.e., integer. We observe that the operation $\circ$ is a map $\circ : Z[\Gamma] \times Z[\Gamma] \to Z$, and that, generally, it does not behave naturally under a homomorphism of Γ. However, the following hold, where r, s and t are elements of $Z[\Gamma]$.

(2.2) The inner product is bilinear, i.e.,

$$(r+t) \circ s = r \circ s + t \circ s$$
$$r \circ (s+t) = r \circ s + r \circ t .$$

(2.3) The inner product is symmetric, i.e.,

$$r \circ s = s \circ r .$$

(2.4) $$r \circ s = \bar{r} \circ \bar{s} .$$

(2.5) $$mr \circ ns = mn(r \circ s), \quad m, n \in Z .$$

(2.6) $$rt \circ s = r \circ s\bar{t}, \quad tr \circ s = r \circ \bar{t}s .$$

(N.B. prove it first for $t \in \Gamma$, and then pass to the general case.)

3. *Anti-derivations.*

Fox ([3], p. 549) introduced the notion of *derivation* in $Z[\Gamma]$. This is a map $D : Z[\Gamma] \to Z[\Gamma]$ with the following two properties, where r, t are elements of $Z[\Gamma]$.

$$D(r+t) = Dr + Dt \quad \text{(linearity)} . \tag{3.1}$$

$$D(rt) = (Dr)t^0 + r(Dt) . \tag{3.2}$$

We now introduce the notion of *anti-derivation* in $Z[\Gamma]$. This is a map $D' : Z[\Gamma] \to Z[\Gamma]$ with the following two properties, where r, t are elements of $Z[\Gamma]$.

$$D'(r+t) = D'r + D't \quad \text{(linearity)} . \tag{3.3}$$

$$D'(rt) = (D'r)t^0 + (D't)\bar{r} . \tag{3.4}$$

It is easily seen that D' is an anti-derivation if and only if $\bar{D}'$ is a derivation, where $\bar{D}'(r) = \overline{D'r}$. Therefore anti-derivations have properties similar to those of derivations.

4. *Biderivations*

A *biderivation* in $Z[\Gamma]$ is a map $\Theta : Z[\Gamma] \times Z[\Gamma] \to Z[\Gamma]$ with the following two properties, where r, s, t are elements of $Z[\Gamma]$.

$$\begin{aligned} \Theta(r+t, s) &= \Theta(r,s) + \Theta(t,s) \\ \Theta(r, s+t) &= \Theta(r,s) + \Theta(r,t) \quad \text{(bilinearity)} \end{aligned} \tag{4.1}$$

$$\begin{aligned} \Theta(rt, s) &= \Theta(r,s)t^0 + r\Theta(t,s) \\ \Theta(r, st) &= \Theta(r,s)t^0 + \Theta(r,t)\bar{s} . \end{aligned} \tag{4.2}$$

Thus, Θ is a derivation with respect to the left variable, and anti-derivation with respect to the right variable. Hence the following properties hold.

$$\Theta(m,t) = \Theta(t,m) = 0, \qquad m \in Z . \tag{4.3}$$

$$\Theta\left(\sum_i m_i r_i, \sum_j n_j s_j\right) = \sum_{i,j} m_i n_j \,\Theta(r_i, s_j) \tag{4.4}$$

$$m_i, n_j \in Z, \qquad r_i, s_j \in Z[\Gamma] .$$

$$\begin{aligned} \Theta(\bar{f}, t) &= -\bar{f}\Theta(f,t), \qquad f \in \Gamma \\ \Theta(t, \bar{f}) &= -\Theta(t,f)f, \qquad t \in Z[\Gamma] . \end{aligned} \tag{4.5}$$

(4.6) $$\Theta(\bar{f}, \bar{g}) = \bar{f}\Theta(f, g)g, \qquad f, g \in \Gamma .$$

(4.7) A linear combination over Z of biderivations in $Z[\Gamma]$ is a biderivation in $Z[\Gamma]$.

5. *Biderivations in a free group ring*

Let now X be a free group on free generators $x_1, \cdots, x_n$, $(n < \infty)$. The following is a biderivation in $Z[X]$,

(5.1)
$$\Theta(r, s) = \sum_{i,j} \frac{\partial r}{\partial x_i} \theta_{ij} \overline{\frac{\partial s}{\partial x_j}}$$
$$r, s, \theta_{ij} \in Z[X], \qquad i, j = 1, \cdots, n$$

where $\frac{\partial}{\partial x_i}$ is the Fox-derivative ([3], p. 550), and

$$\overline{\frac{\partial}{\partial x_j}} = \left(\frac{\partial}{\partial x_j}\right)^- .$$

The operator, obtained from the right hand side of the equality (5.1) by deleting r and s, is a *biderivative* with matrix $\|\theta_{ij}\|$, where

(5.2) $$\theta_{ij} = \Theta(x_i, x_j), \qquad i, j = 1, \cdots, n .$$

The general biderivative in a free group ring is provided by (5.1) according to the following theorem.

THEOREM 5.3. *If* Θ *is a biderivation in the free group ring* $Z[X]$, *then* Θ *is defined by* (5.1) *and* (5.2).

Proof. Let us consider the following biderivative

$$\Theta^*(,) = \sum_{i,j} \frac{\partial}{\partial x_i} \Theta(x_i, x_j) \overline{\frac{\partial}{\partial x_j}}$$

where $i, j = 1, \cdots, n$. As usual, by the length $\ell(u)$ of an element u of X we mean the number of letters in the reduced word representing u. The following hold

$$
\begin{aligned}
\Theta^*(x_i, x_j) &= \Theta(x_i, x_j) \\
\Theta^*(x_i, \bar{x}_j) &= -\Theta^*(x_i, x_j)x_j && \text{by (4.5)} \\
&= -\Theta(x_i, x_j)x_j = \Theta(x_i, \bar{x}_j) \\
\\
\Theta^*(\bar{x}_i, x_j) &= -\bar{x}_i\Theta^*(x_i, x_j) && \text{by (4.5)} \\
&= -\bar{x}_i\Theta(x_i, x_j) = \Theta(\bar{x}_i, x_j) \\
\\
\Theta^*(\bar{x}_i, \bar{x}_j) &= \bar{x}_i\Theta^*(x_i, x_j)x_j && \text{by (4.6)} \\
&= \bar{x}_i\Theta(x_i, x_j)x_j = \Theta(\bar{x}_i, \bar{x}_j) .
\end{aligned}
$$

From the above and (4.3), it follows that $\ell(u) + \ell(v) \leqq 2$ implies $\Theta(u, v) = \Theta^*(u, v)$, where u and v are elements of X. We now proceed by induction.

(5.4) (Inductive hypothesis). Suppose that $\ell(u) + \ell(v) < m\,(> 2)$ implies $\Theta(u, v) = \Theta^*(u, v)$, for any two elements u, v of X.

Let us now suppose that u', v' are two elements of X, such that $\ell(u') + \ell(v') = m$. If either $\ell(u')$ or $\ell(v') = 0$ then, by (4.3),

$$\Theta^*(u', v') = 0 = \Theta(u', v') .$$

Thus, from now on we can suppose that both $\ell(u')$ and $\ell(v')$ are $\geqq 1$. We have to consider two cases, $\ell(u') > 1$ and $= 1$.

Let us first suppose that $\ell(u') > 1$. Then $u' = uw$, where u, w, uw are reduced words, such that

$$\ell(u') = \ell(u) + \ell(w), \qquad \ell(u) \text{ and } \ell(w) \geqq 1 .$$

The following hold

$$
\begin{aligned}
\Theta(u', v') &= \Theta(u, v') + u\Theta(w, v') && \text{by (4.2)} \\
&= \Theta^*(u, v') + u\Theta^*(w, v') && \text{by (5.4)} \\
&= \Theta^*(u', v') && \text{by (4.2)} .
\end{aligned}
$$

Let us now suppose that $\ell(u') = 1$. Then $\ell(v') > 1$. Thus $v' = vw$, where v, w, vw are reduced words, such that

$$\ell(v') = \ell(v) + \ell(w), \qquad \ell(v) \text{ and } \ell(w) \geqq 1 .$$

The following hold

$$\begin{aligned} \Theta(u', v') &= \Theta(u', v) + \Theta(u', w)\bar{v} && \text{by (4.2)} \\ &= \Theta^*(u', v) + \Theta^*(u', w)\bar{v} && \text{by (5.4)} \\ &= \Theta^*(u', v') && \text{by (4.2)} . \end{aligned}$$

Hence, for any two elements u, v of X, we have

$$\Theta(u, v) = \Theta^*(u, v) . \tag{5.5}$$

Let now r, s be two elements of $Z[X]$. Then we have the following, where m_i, n_j are elements of Z and u_i, v_j are elements of X,

$$r = \sum_i m_i u_i , \qquad s = \sum_j n_j v_j .$$

By (4.4) and (5.5), the following hold

$$\begin{aligned} \Theta(r, s) &= \sum_{i,j} m_i n_j \Theta(u_i, v_j) \\ &= \sum_{i,j} m_i n_j \Theta^*(u_i, v_j) \\ &= \Theta^*(r, s) . \end{aligned}$$

This completes the proof of our theorem.

6. *Special properties of biderivations*

Let X be a free group on free generators $x_1, \cdots, x_n$, $(n < \infty)$, and let $P = \Pi u^{\varepsilon}$ be a finite product where u is an element of X and $\varepsilon = \pm 1$. Then the following "chain rule" holds

$$\frac{\partial P}{\partial x_i} = \sum_u \frac{\partial P}{\partial u} \frac{\partial u}{\partial x_i} , \qquad i = 1, \cdots, n . \tag{6.1}$$

where u ranges over all the different to one another u's appearing in the product P. Here some explanation is needed: We consider the different to one another u's as representing free generators of a free group and we apply Fox-derivatives. Formula (6.1) follows from ([3], p. 549, (1.5) and (1.6)) and reductions.

LEMMA 6.2. *Let* Θ *be a biderivative in the free group ring* $Z[X]$, *and let*

$$P = \Pi u^{\delta}, \quad Q = \Pi v^{\varepsilon}, \quad \delta, \varepsilon = \pm 1$$

be two products, where u *and* v *are elements of* X. *Then the following holds*

$$\Theta(P, Q) = \sum_{u,v} \frac{\partial P}{\partial u}\, \Theta(u, v)\, \overline{\frac{\partial Q}{\partial v}}$$

where the sum ranges over all the different to one another u's *and* v's *appearing in* P *and* Q *respectively.*

Proof. By Theorem (5.3), and (5.2), we have

$$\Theta(P, Q) = \sum_{i,j} \frac{\partial P}{\partial x_i}\, \Theta(x_i, x_j)\, \overline{\frac{\partial Q}{\partial x_j}}, \quad i, j = 1, \cdots, n$$

$$= \sum_{i,j} \left(\sum_{u} \frac{\partial P}{\partial u} \frac{\partial u}{\partial x_i}\right) \Theta(x_i, x_j) \overline{\left(\sum_{v} \frac{\partial Q}{\partial v} \frac{\partial v}{\partial x_j}\right)} \qquad \text{by (6.1)}$$

$$= \sum_{i,j} \left(\sum_{u} \frac{\partial P}{\partial u} \frac{\partial u}{\partial x_i}\right) \Theta(x_i, x_j) \sum_{v} \overline{\frac{\partial v}{\partial x_j}}\, \overline{\frac{\partial Q}{\partial v}} \qquad \text{by (1.4)}$$

$$= \sum_{u,v} \frac{\partial P}{\partial u} \left(\sum_{i,j} \frac{\partial u}{\partial x_i}\, \Theta(x_i, x_j)\, \overline{\frac{\partial v}{\partial x_j}}\right) \overline{\frac{\partial Q}{\partial v}}$$

$$= \sum_{u,v} \frac{\partial P}{\partial u}\, \Theta(u, v)\, \overline{\frac{\partial Q}{\partial v}}\,. \qquad \text{by (5.1)}$$

This completes the proof of our lemma.

THEOREM 6.3. *Let* $\psi : X \to \Gamma$ *be a homomorphism of a free group* X *on free generators* $x_1, \cdots, x_n$, $(n < \infty)$, *in a denumerable group* Γ. *Let*

$$P = \prod_i u_i w_i^{\delta_i} \bar{u}_i, \quad Q = \prod_j v_j z_j^{\varepsilon_j} \bar{v}_j, \quad \delta_i, \varepsilon_j = \pm 1$$

be two products, $i = 1, \cdots, p$ *and* $j = 1, \cdots, q$, *where* u_i, v_j *are elements of* X, *and* w_i, z_j *are elements of* $\ker \psi$. *Then the following holds*

$$\psi \Theta(P, Q) = \sum_{i,j} \delta_i \varepsilon_j \psi(u_i \Theta(w_i, z_j) \bar{v}_j)$$

where Θ *is a biderivative in* $Z[X]$.

Proof. By Lemma 6.2 the following holds

$$\Theta(P, Q) = \sum_{u,v} \frac{\partial P}{\partial u} \Theta(u, v) \overline{\frac{\partial Q}{\partial v}}$$

where u (or v) runs over all the different to one another elements of X appearing in the finite sequence u_i, w_i (or v_j, z_j). By (1.6), we have

$$\psi \Theta(P, Q) = \sum_{u,v} \psi\left(\frac{\partial P}{\partial u}\right) \psi \Theta(u, v) \overline{\psi\left(\frac{\partial Q}{\partial v}\right)}$$

where $\psi : Z[X] \to Z[\Gamma]$ is the ring-homomorphism induced by the group homomorphism ψ.

We now observe that the set, which u (or v) ranges over, consists of all w_i's (or z_j's) which are different to one another, and all u_i's (or v_j's) which are different to one another and do not appear as w_i's (or z_j's). If u (or v) is one of the w_i's (or z_j's) say w_k (or z_ℓ), then the following holds

$$\psi\left(\frac{\partial P}{\partial u}\right) = \sum_\lambda \delta_\lambda \psi(u_\lambda), \left(\text{or } \psi\left(\frac{\partial Q}{\partial v}\right) = \sum_\mu \varepsilon_\mu \psi(v_\mu)\right)$$

where $1 \leq \lambda \leq p$ (or $1 \leq \mu \leq q$) and such that $w_\lambda = w_k$ (or $z_\mu = z_\ell$). If u is none of the w_i's (or z_j's), then the following holds

$$\psi\left(\frac{\partial P}{\partial u}\right) = 0, \quad \left(\text{or } \psi\left(\frac{\partial Q}{\partial v}\right) = 0\right).$$

From the above we obtain the following, where the sum ranges over all pairs k, ℓ such that, w_k (or z_ℓ) ranges over all different to one another w_i's (or z_j's),

$$\psi\Theta(P, Q) = \sum_{k,\ell}\left(\sum_\lambda \delta_\lambda \psi(u_\lambda)\right) \psi\Theta(w_k, z_\ell) \overline{\left(\sum_\mu \varepsilon_\mu \psi(v_\mu)\right)}$$

$$= \sum_{i,j} \delta_i \varepsilon_j \psi(u_i \Theta(w_i, z_j)\overline{v}_j)$$

where $i = 1, \cdots, p$ and $j = 1, \cdots, q$. This completes the proof of our theorem. Q.E.D.

Let now P be a product as of Theorem 6.3. We define the following *elementary transformations* of P.

(i) Deletion of a factor $u_i w_i^{\delta_i} \overline{u}_i$, where $w_i = 1$.

(ii) Insertion of a factor $uw\overline{u}$, such that $u \in X$, $w = 1$.

(iii) Deletion of two factors $u_i w_i^{\delta_i} \overline{u}_i$ and $u_h w_h^{\delta_h} \overline{u}_h$, such that $i \neq h$, $u_i = u_h$, $w_i = w_h$, and $\delta_i + \delta_h = 0$.

(iv) Insertion of two factors $uw\overline{u}$ and $u\overline{w}\overline{u}$, such that $u \in X$, $w \in \ker \psi$. (N.B. the inserted factors need not be neighboring in the final product.)

A product which is obtained from P by a finite number of the above elementary transformations is called *homologous* to P. Under the hypotheses of Theorem 6.3 the following holds.

THEOREM 6.4. *If* P'' *and* Q'' *are products homologous to* P *and* Q *respectively, then*

$$\psi\Theta(P'', Q'') = \psi\Theta(P, Q).$$

If P' *and* Q' *are products similar to* P *and* Q *respectively, then*

$$\psi\Theta(PP',Q) = \psi\Theta(P,Q) + \psi\Theta(P',Q)$$
$$\psi\Theta(P,QQ') = \psi\Theta(P,Q) + \psi\Theta(P,Q') \ .$$

Finally, we have

$$\psi\Theta(\overline{P},Q) = -\psi\Theta(P,Q), \ \ \psi\Theta(P,\overline{Q}) = -\psi\Theta(P,Q) \ .$$

The proof of the theorem follows from Theorem 6.3 and No. 4. We now observe that P and P'' (or Q and Q'') represent the same element of

$$H_1(\ker\psi, Z) = \ker\psi/[\ker\psi, \ker\psi] \ .$$

This justifies the term homologous. We finally observe that, the operator $\psi\Theta$ behaves like an intersection theory operator.

§3. Intersection Theory

7. *Topological considerations*

Let N be an orientable closed surface of genus $p \geq 2$. We denote by $\alpha_1, \beta_1, \cdots, \alpha_p, \beta_p$ a fundamental system of N based at a point o, and let $\alpha_1^*, \beta_1^*, \cdots, \alpha_p^*, \beta_p^*$ be the dual to it based at a point o^*. The first system defines a fundamental polygon of N shown in (Figure 1),

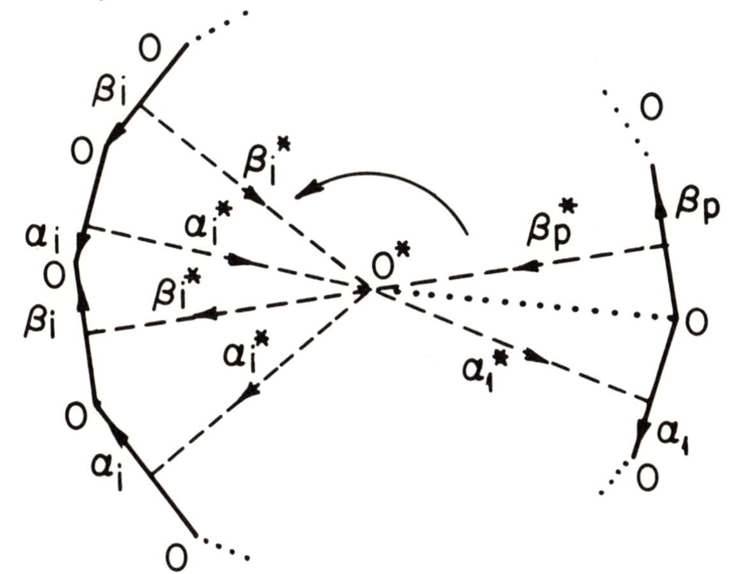

Fig. 1

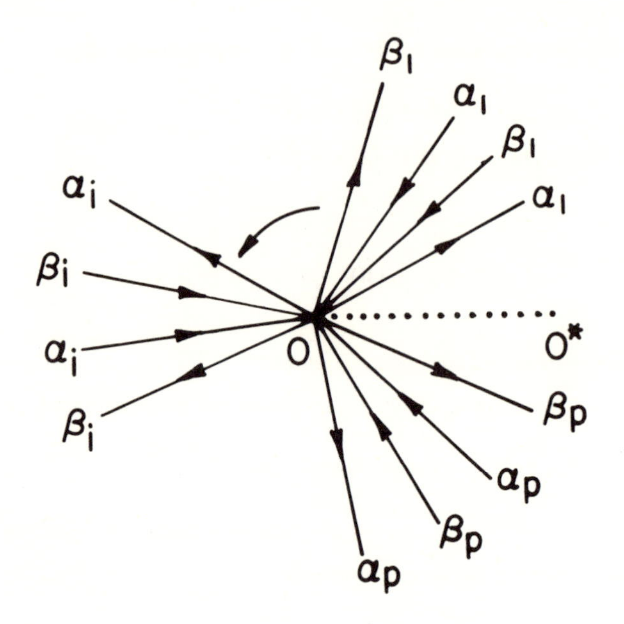
β_1
α_1
β_1
α_1
α_i
β_i
α_i
β_i
O
O*
β_p
α_p
β_p
α_p

Fig. 2

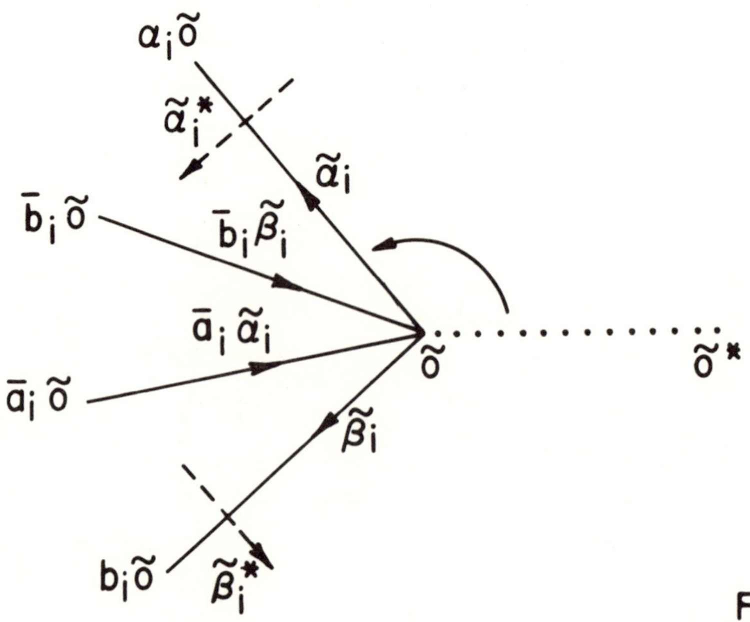
$\alpha_i\tilde{o}$
$\tilde{\alpha}_i^*$
$\tilde{\alpha}_i$
$\bar{b}_i\tilde{o}$
$\bar{b}_i\tilde{\beta}_i$
$\bar{a}_i\tilde{\alpha}_i$
$\bar{a}_i\tilde{o}$
$\tilde{o}$
$\tilde{o}^*$
$\tilde{\beta}_i$
$b_i\tilde{o}$
$\tilde{\beta}_i^*$

Fig. 3

the picture indicates also the orientation of N we are going to consider. In (Figure 2) is shown how the α, β's are arranged around o.

Let us write $F = \pi_1(N, o)$, let G be a normal subgroup of F, and $H = F/G$. We denote by $\tilde{N}$ the universal covering surface of N, and by $\hat{N}$ the regular covering surface of N corresponding to G. The surfaces $\tilde{N}$ and $\hat{N}$ have orientations induced by that of N. Thus we have the following diagram

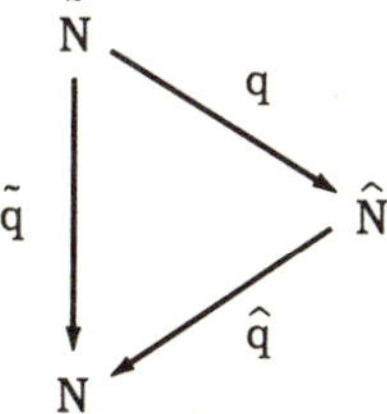

where $q, \hat{q}, \tilde{q}$ are the projections, and the corresponding covering translations are G, H, F respectively.

We select a point $\tilde{o}$ on $\tilde{N}$ lying over o, and we denote by $\tilde{\alpha}_i, \tilde{\beta}_i, \tilde{\alpha}_i^*, \tilde{\beta}_i^*$ paths on $\tilde{N}$ lying over $\alpha_i, \beta_i, \alpha_i^*, \beta_i^*$ respectively as shown in (Figure 3), where a_i, b_i are the elements of F defined by the loops α_i, β_i respectively, $i = 1, \cdots, p$. We also lift the path oo^* (see (Figure 2)) and we obtain the path $\tilde{o}\tilde{o}^*$, on $\tilde{N}$, see (Figure 3). By means of the projection q, we obtain $\hat{o}, \hat{\alpha}_i, \hat{\beta}_i, \hat{\alpha}_i^*, \hat{\beta}_i^*, \hat{o}\hat{o}^*$ on $\hat{N}$.

For the sake of convenience we denote by $\rho_1, \rho_2, \cdots, \rho_{2p-1}, \rho_{2p}$ (or $\rho_1^*, \rho_2^*, \cdots, \rho_{2p-1}^*, \rho_{2p}^*$) the loops $\alpha_1, \beta_1, \cdots, \alpha_p, \beta_p$ (or $\alpha_1^*, \beta_1^*, \cdots, \alpha_p^*, \beta_p^*$) respectively. The following hold concerning intersection numbers.

$$\text{(7.1)} \qquad \operatorname{Is}(\tilde{\rho}_i, \tilde{\rho}_j^*) = \delta_{ij}, \qquad \operatorname{Is}(\hat{\rho}_i, \hat{\rho}_j^*) = \delta_{ij}$$

$i, j = 1, \cdots, 2p$, where δ_{ij} is the Kronecker symbol.

8. *The intersection formula*

Let $\chi : F \to H$ be the natural epimorphism, and let us denote the right cosets of F mod G by $Gf_{0\kappa}$, where $f_{0\kappa}$ are the representatives, $\kappa = 1, 2, \cdots$ (the sequence is finite or infinite, and $f_{01} = 1$). We write $\chi(Gf_{0\kappa}) = h_\kappa$.

Let γ, δ^* be paths on N based at o, o^* and composed of loops ρ_i, ρ_i^* and their inverses respectively, $i = 1, \cdots, 2p$. Let $\tilde{\gamma}, \tilde{\delta}^*$ be the paths on $\tilde{N}$ based at $\tilde{o}$, $f\tilde{o}^*$ and lying over γ, δ^* respectively, where $f \in F$. We write $\hat{\gamma} = q(\tilde{\gamma})$ and $\hat{\delta}^* = q(\tilde{\delta}^*)$. These are paths on $\hat{N}$ based at $\hat{o}$, $h\hat{o}^*$ respectively, where $h = \chi(f)$. Let $\tilde{c}, \tilde{d}^*$ and $\hat{c}, \hat{d}^*$ be the 1-chains corresponding to the paths $\tilde{\gamma}, \tilde{\delta}^*$ and $\hat{\gamma}, \hat{\delta}^*$ respectively. The following hold

$$\tilde{c} = \sum_i r_i \tilde{\rho}_i \qquad \tilde{d}^* = \sum_i s_i \tilde{\rho}_i^* \tag{8.1}$$

$$\hat{c} = \sum_i \chi(r_i) \hat{\rho}_i \qquad \hat{d}^* = \sum_i \chi(s_i) \hat{\rho}_i^*$$

where $i = 1, \cdots, 2p$ and r_i, s_i or $\chi(r_i), \chi(s_i)$ are elements of the integral group ring $Z[F]$ or $Z[H]$ respectively. We observe that r_i, s_i can be written as follows

$$r_i = \sum_\lambda \left(\sum_\mu m_{i\lambda\mu} g_\mu \right) f_{0\lambda} \tag{8.2}$$

$$s_i = \sum_\lambda \left(\sum_\mu n_{i\lambda\mu} g_\mu \right) f_{0\lambda}$$

where $g_\mu \in G$, $m_{i\lambda\mu}$ and $n_{i\lambda\mu} \in Z$, and the indices λ, μ have a finite range. Thus, we have

$$\chi(r_i) = \sum_\lambda \left(\sum_\mu m_{i\lambda\mu} \right) h_\lambda, \qquad \chi(s_i) = \sum_\lambda \left(\sum_\mu n_{i\lambda\mu} \right) h_\lambda \tag{8.3}$$

$$= \sum_\lambda m_{i\lambda} h_\lambda \qquad = \sum_\lambda n_{i\lambda} h_\lambda \, .$$

By (8.1) and the above formulas the following hold

$$\hat{c} = \sum_i \left(\sum_\lambda m_{i\lambda} h_\lambda \right) \hat{\rho}_i, \qquad \hat{d}^* = \sum_i \left(\sum_\lambda n_{i\lambda} h_\lambda \right) \hat{\rho}_i^*$$

$$= \sum_{i,\lambda} m_{i\lambda} h_\lambda \hat{\rho}_i \qquad = \sum_{i,\lambda} n_{i\lambda} h_\lambda \hat{\rho}_i^* \, .$$

Concerning intersection numbers the following hold

$$\mathrm{Is}(\hat{c},\hat{d}^*) = \mathrm{Is}\left(\sum_{i,\lambda} m_{i\lambda} h_\lambda \hat{\rho}_i,\ \sum_{j,\nu} n_{j\nu} h_\nu \hat{\rho}_j^*\right)$$

$$= \sum_{i,j,\lambda,\nu} m_{i\lambda} n_{j\nu}\, \mathrm{Is}(h_\lambda \hat{\rho}_i, h_\nu \hat{\rho}_j^*)$$

where $i, j = 1, \cdots, 2p$, ν has the range of λ, and the range of λ is defined by (8.2). We now observe that

$$\mathrm{Is}(h_\lambda \hat{\rho}_i, h_\nu \hat{\rho}_j^*) = 1 \text{ or } 0$$

in case $i = j$, $\lambda = \nu$ or any other case respectively, see (7.1). Hence, we have

(8.4) $$\mathrm{Is}(\hat{c},\hat{d}^*) = \sum_{i,\lambda} m_{i\lambda} n_{i\lambda}$$

where $i = 1, \cdots, 2p$, and the range of λ is defined by (8.2).

We now consider the following sums of inner products, making use of (8.3), where $i, j = 1, \cdots, 2p$, μ has the range of λ, and the range of λ is defined by (8.2).

$$\sum_i \chi(r_i) \circ \chi(s_i) = \sum_i \left(\sum_\lambda m_{i\lambda} h_\lambda\right) \circ \left(\sum_\mu n_{i\mu} h_\mu\right)$$

$$= \sum_i \sum_{\lambda,\mu} m_{i\lambda} n_{i\mu} (h_\lambda \circ h_\mu)$$

$$= \sum_i \sum_\lambda m_{i\lambda} n_{i\lambda}$$

$$= \sum_{i,\lambda} m_{i\lambda} n_{i\lambda}\ .$$

The right hand side of the above equalities hold by (2.2), (2.5), and (2.1). Hence, by (8.4), we obtain the following *intersection formula*

$$\text{(8.5)} \qquad \mathrm{Is}(\hat{c}, \hat{d}^*) = \sum_i \chi(r_i) \circ \chi(s_i), \quad i = 1, \cdots, 2p$$

where r_i, s_i are defined by (8.1).

§4. The Main Formulas and Planar Coverings

9. *The basic biderivatives*

We now are going to introduce some biderivatives, which are indispensable in obtaining the main formulas.

Let $\Phi = (a_1, b_1, \cdots, a_p, b_p)$, $p \geqq 2$, be a free group of rank $2p$, and let $Z[\Phi]$ be the integral group ring of Φ. We write

$$\text{(9.1)} \qquad \begin{aligned} \Lambda_1(r,s) = \sum_i \Bigg[& \frac{\partial r}{\partial a_i}(1 - a_i)\overline{\frac{\partial s}{\partial a_i}} + \frac{\partial r}{\partial a_i} a_i \bar{b}_i \overline{\frac{\partial s}{\partial b_i}} \\ & + \frac{\partial r}{\partial b_i}(1 - \bar{a}_i - b_i)\overline{\frac{\partial s}{\partial a_i}} + \frac{\partial r}{\partial b_i}(1 - \bar{b}_i)\overline{\frac{\partial s}{\partial b_i}} \Bigg] \end{aligned}$$

where $i = 1, \cdots, p$ and $r, s \in Z[\Phi]$. The operator Λ_1 is a biderivative in $Z[\Phi]$, see No. 5. We also write

$$\text{(9.2)} \qquad \Lambda_2(r,s) = \sum_j \left(\sum_i d_i r \right) \overline{d_j s}$$

where $i = j+1, \cdots, p$, $j = 1, \cdots, p-1$, and the operator d_k is defined as follows

$$\text{(9.3)} \qquad d_k = \frac{\partial}{\partial a_k}(a_k - 1) + \frac{\partial}{\partial b_k}(b_k - 1), \qquad k = 1, \cdots, p \,.$$

The operator Λ_2 is a biderivative in $Z[\Phi]$, as it can be proved easily. Finally, we write

$$\text{(9.4)} \qquad \Lambda(r,s) = \Lambda_1(r,s) + \Lambda_2(r,s) \,.$$

The operator Λ is a biderivative in $Z[\Phi]$, by (4.7), and it appears in the main formula.

For any two elements r, s of the integral group ring $Z[\Phi]$ the following formula holds

$$\sum_i (d_i r)\,\overline{d_i s} = \Lambda_1(r,s) + \overline{\Lambda_1(s,r)}, \quad i = 1,\cdots,p \ . \tag{9.5}$$

This is proved by computing the two sides of the formula and comparing them. The computation makes use of (9.1), (9.3), (1.2) and (1.5), and is straightforward.

The matrix of each one of the Λ's is of the form

$$\begin{Vmatrix} (aa) & (ab) \\ (ba) & (bb) \end{Vmatrix}$$

where (aa), (ab), (ba), (bb) are "submatrices" $p \times p$. The matrix of Λ_1 has all its other entries zero, except those on the main diagonals of the "submatrices." Each one of the "submatrices" of Λ_2 has all its entries on and above the main diagonal zero. Hence, each of the "submatrices" of Λ has all its entries above the main diagonal zero, i.e., each of the "submatrices" of Λ is of triangular form.

10. *The general case*

In the present No. 10 we keep the notation and conventions introduced in Nos. 7, 8 and 9. We thus have

$$\Phi \xrightarrow{\omega} F \xrightarrow{\chi} H$$

where ω and χ are epimorphisms, and

$$F = \left(a_1, b_1, \cdots, a_p, b_p : \prod_i [a_i, b_i]\right), \quad i = 1,\cdots, p \geqq 2 \ .$$

We also have the following sequence

$$Z[\Phi] \xrightarrow{\omega} Z[F] \xrightarrow{\chi} Z[H]$$

concerning the integral group rings.

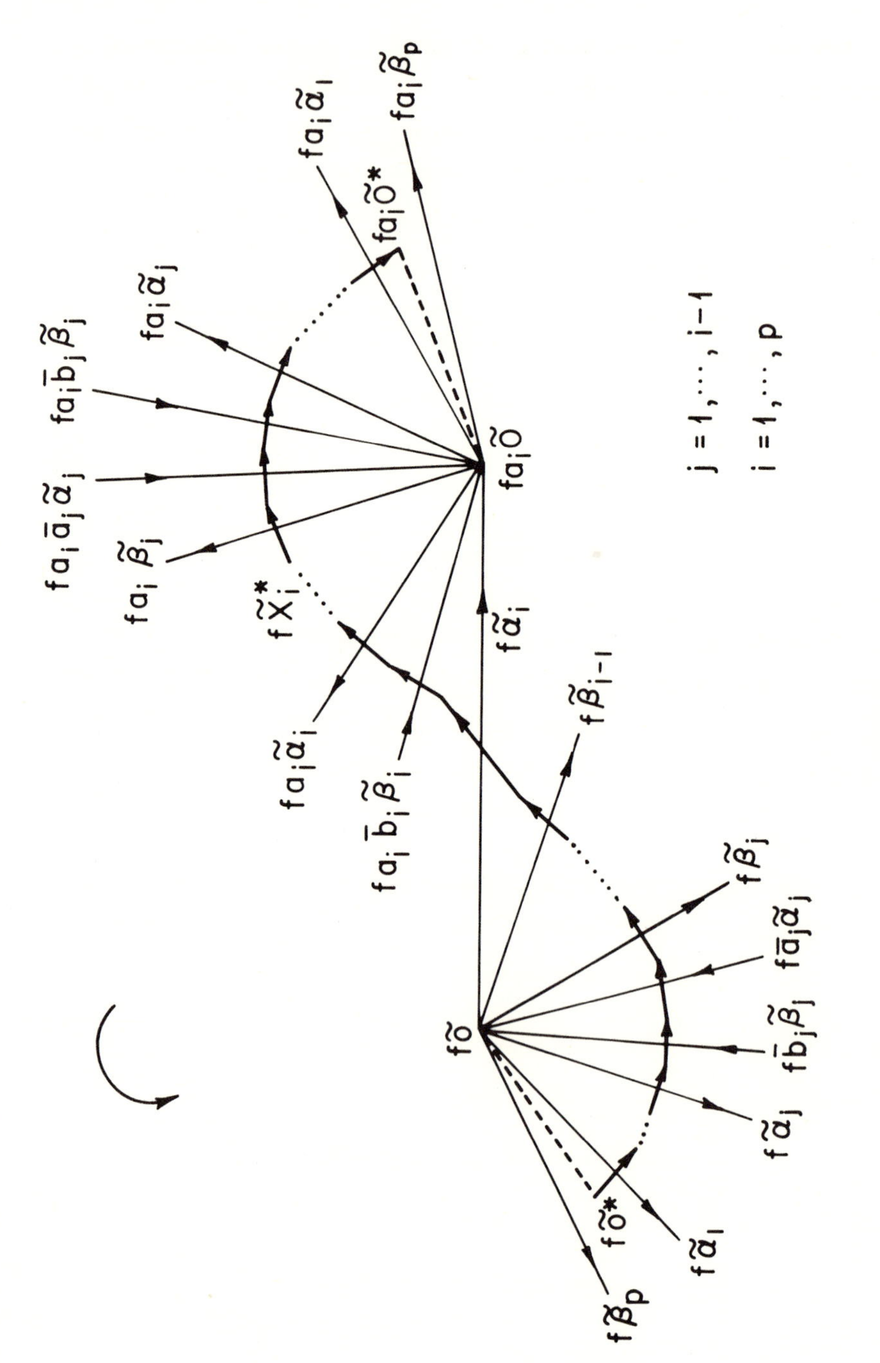

Fig. 4

Let now γ, γ' be two loops on N based at o, and composed of loops α_i, β_i and their inverses, $i = 1, \cdots, p$. Let $\tilde{\gamma}, \tilde{\gamma}'$ be the paths on $\tilde{N}$ based at $\tilde{o}$ and lying over γ, γ', and let $\tilde{c}, \tilde{c}'$ be the 1-chains corresponding to $\tilde{\gamma}, \tilde{\gamma}'$ respectively. Let finally w, w' be the words in a_i, b_i's corresponding to the loops γ, γ' respectively. By a formula due to Fox ([2], p. 521) we have

$$\tilde{c} = \sum_i [\omega(u_i)\tilde{\alpha}_i + \omega(v_i)\tilde{\beta}_i], \quad i = 1, \cdots, p$$

$$\tilde{c}' = \sum_i [\omega(u'_i)\tilde{\alpha}_i + \omega(v'_i)\tilde{\beta}_i] \tag{10.1}$$

$$u_i = \frac{\partial w}{\partial a_i}, \quad v_i = \frac{\partial w}{\partial b_i}, \quad u'_i = \frac{\partial w'}{\partial a_i}, \quad v'_i = \frac{\partial w'}{\partial b_i}.$$

To be able to apply formula (8.5), we have to deform γ on N to a loop γ^* based at o^* and composed of loops α_i^*, β_i^* and their inverses, $i = 1, \cdots, p$. This is done by deforming $\tilde{\gamma}$ on $\tilde{N}$. We observe that $\tilde{\gamma}$ is composed of paths $f\tilde{\alpha}_i^{\varepsilon}$, $f\tilde{\beta}_i^{\varepsilon}$, where $f \in F$ and $\varepsilon = \pm 1$, $i = 1, \cdots, p$. If $f\tilde{\alpha}_i^{\varepsilon}$ appears in $\tilde{\gamma}$, we then replace it by the path $f\tilde{X}_i^{*\varepsilon}$, see (Figure 4). If $f\tilde{\beta}_i^{\varepsilon}$ appears in $\tilde{\gamma}$, we then replace it by the path $f\tilde{Y}_i^{*\varepsilon}$, see (Figure 5). We thus obtain a new path $\tilde{\gamma}^*$ on $\tilde{N}$, such that $\gamma^* = \tilde{q}(\tilde{\gamma}^*)$ is a loop based at o^*, composed of loops α_i^*, β_i^* and their inverses, $i = 1, \cdots, p$, and γ^* is homotopic to γ on N. The 1-chains corresponding to $f\tilde{X}_i^*$, $f\tilde{Y}_i^*$ are $f\tilde{A}_i^*$, $f\tilde{B}_i^*$ respectively, where

$$\begin{aligned}
\tilde{A}_i^* &= (1 - a_i)\tilde{\alpha}_i^* + a_i\bar{b}_i\tilde{\beta}_i^* \\
&\quad + (1 - a_i)\sum_j [(1 - \bar{a}_j)\tilde{\alpha}_j^* + (1 - \bar{b}_j)\tilde{\beta}_j^*] \\
\tilde{B}_i^* &= (1 - \bar{a}_i - b_i)\tilde{\alpha}_i^* + (1 - \bar{b}_i)\tilde{\beta}_i^* \\
&\quad + (1 - b_i)\sum_j [(1 - \bar{a}_j)\tilde{\alpha}_j^* + (1 - \bar{b}_j)\tilde{\beta}_j^*]
\end{aligned} \tag{10.2}$$

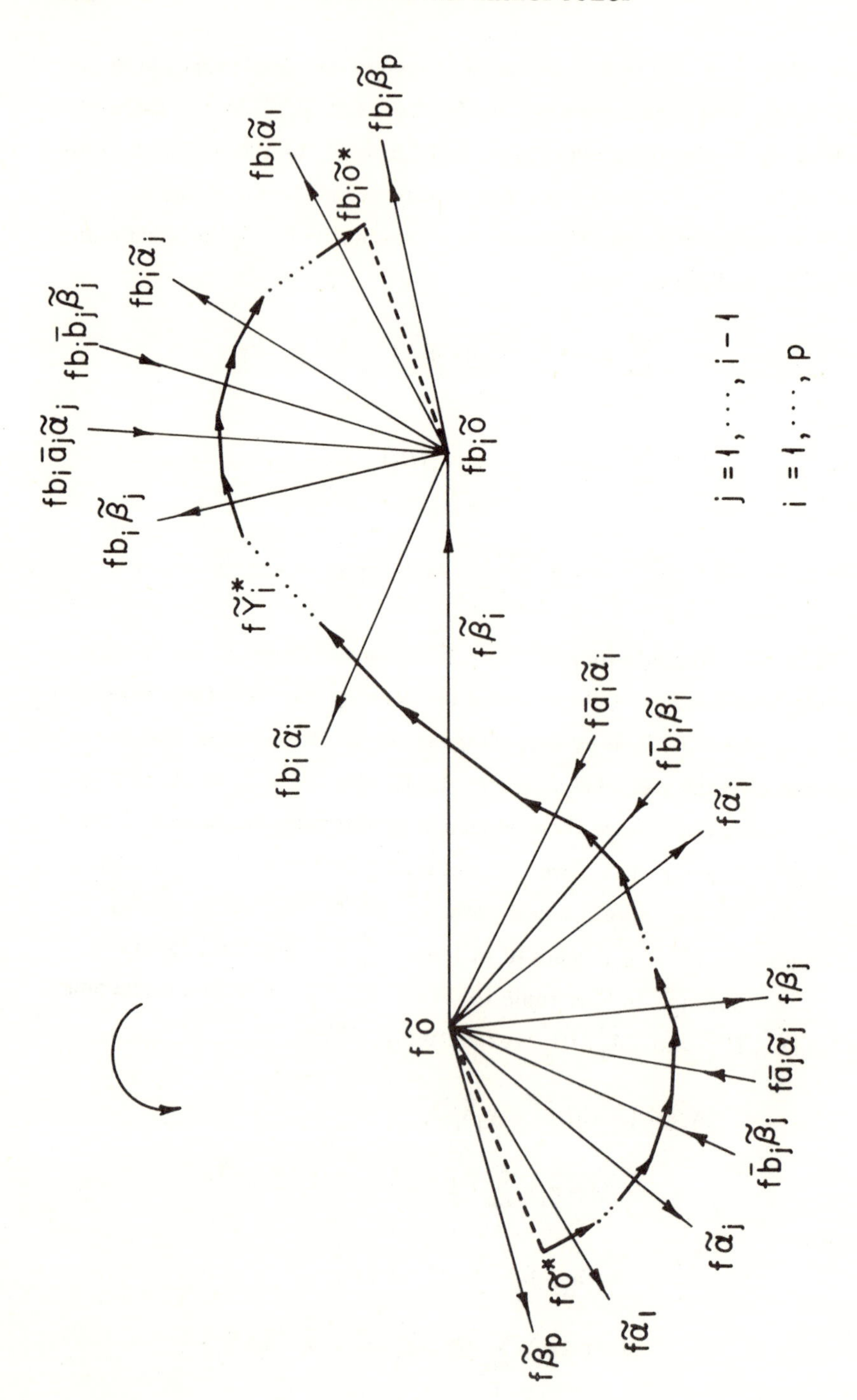

Fig. 5

where $j = 1, \cdots, i-1$. Hence the 1-chain $\tilde{c}^*$ corresponding to $\tilde{\gamma}^*$ is, by (10.1),

$$\tilde{c}^* = \sum_i [\omega(u_i)\tilde{A}_i^* + \omega(v_i)\tilde{B}_i^*], \quad i = 1, \cdots, p . \tag{10.3}$$

Before applying (8.5), we have to evaluate the right hand side of the last formula using (10.2). For the sake of convenience we write

$$\begin{aligned} \omega(w) &= z \qquad & \omega(u_i) &= x_i \qquad & \omega(v_i) &= y_i \\ & & \omega(a_i) &= a_i & \omega(b_i) &= b_i \end{aligned} \tag{10.4}$$

where $i = 1, \cdots, p$. By (10.3) and (10.2) we have

$$\begin{aligned} \tilde{c}^* = \sum_i \Big(& x_i(1-a_i)\tilde{\alpha}_i^* + x_i a_i \bar{b}_i \tilde{\beta}_i^* \\ & + y_i(1-\bar{a}_i-b_i)\tilde{\alpha}_i^* + y_i(1-\bar{b}_i)\tilde{\beta}_i^* \\ & + \sum_j [x_i(1-a_i)(1-\bar{a}_j)\tilde{\alpha}_j^* + x_i(1-a_i)(1-\bar{b}_j)\tilde{\beta}_j^* \\ & + y_i(1-b_i)(1-\bar{a}_j)\tilde{\alpha}_j^* + y_i(1-b_i)(1-\bar{b}_j)\tilde{\beta}_j^*] \Big) \end{aligned}$$

where $j = 1, \cdots, i-1$ and $i = 1, \cdots, p$. From this we obtain

$$\begin{aligned} \tilde{c}^* = \sum_i \Bigg(& \Big\{ x_i(1-a_i) + y_i(1-\bar{a}_i-b_i) \\ & + \sum_j [x_j(1-a_j) + y_j(1-b_j)](1-\bar{a}_i) \Big\} \tilde{\alpha}_i^* \\ & + \Big\{ x_i a_i \bar{b}_i + y_i(1-\bar{b}_i) \\ & + \sum_j [x_j(1-a_j) + y_j(1-b_j)](1-\bar{b}_i) \Big\} \tilde{\beta}_i^* \Bigg) \end{aligned} \tag{10.5}$$

where $j = i+1, \cdots, p$ and $i = 1, \cdots, p$. That the right hand sides of the last two equalities are equal can be seen by computing the coefficients of $\tilde{\alpha}_i^*$ and $\tilde{\beta}_i^*$. By the fundamental formula of the free differential calculus ([3], p. 551, (2.3)), we have

$$w - 1 = \sum_i \left[\frac{\partial w}{\partial a_i}(a_i - 1) + \frac{\partial w}{\partial b_i}(b_i - 1) \right] \qquad \text{by (10.1)}$$

$$= \sum_i [u_i(a_i-1) + v_i(b_i-1)]$$

where $i = 1, \cdots, p$. By applying the group ring homomorphism ω, we obtain

$$\sum_j [x_j(1-a_j) + y_j(1-b_j)] = (1-z) + [x_i(a_i-1) + y_i(b_i-1)] + \omega\left(\sum_k d_k w\right)$$

where $j = i+1, \cdots, p$ and $k = 1, \cdots, i-1$, see (9.3).

Replacing the left hand side of the above equality by the right hand side in (10.5), and performing reductions we obtain the following

$$\tilde{c}^* = \sum_i \left(\left\{ (1-z)(1-\bar{a}_i) - x_i(1-\bar{a}_i) - y_i b_i \bar{a}_i + \omega\left(\sum_j d_j w\right)(1-\bar{a}_i) \right\} \tilde{\alpha}_i^* + \left\{ (1-z)(1-\bar{b}_i) - x_i(1-a_i-\bar{b}_i) - y_i(1-b_i) + \omega\left(\sum_j d_j w\right)(1-\bar{b}_i) \right\} \tilde{\beta}_i^* \right)$$

where $j = 1, \cdots, i-1$ and $i = 1, \cdots, p$. We now write

$$U_i = (1-w)(1-\bar{a}_i) - u_i(1-\bar{a}_i) - v_i b_i \bar{a}_i + \left(\sum_j d_j w\right)(1-\bar{a}_i)$$

$$V_i = (1-w)(1-\bar{b}_i) - u_i(1-a_i-\bar{b}_i) - v_i(1-b_i) + \left(\sum_j d_j w\right)(1-\bar{b}_i)$$

where $j = 1, \cdots, i-1$ and $i = 1, \cdots, p$. From the last three equalities and (10.4) we obtain the following

(10.6) $$\tilde{c}^* = \sum_i [\omega(U_i)\tilde{\alpha}_i^* + \omega(V_i)\tilde{\beta}_i^*], \quad i = 1,\cdots,p .$$

This is the 1-chain corresponding to the path $\tilde{\gamma}^*$ on $\tilde{N}$ based at $\tilde{o}^*$, and such that $\gamma^* = \tilde{q}(\tilde{\gamma}^*)$ is a loop on N based at o^*, composed of loops α_i^*, β_i^* and their inverses $i = 1,\cdots,p$, and

(10.7) $$\gamma^* \simeq \gamma \text{ on } N$$

i.e., γ^* is homotopic to γ on N.

We now are ready to apply formula (8.5). Let us consider the paths $\tilde{\gamma}', \overline{f}\tilde{\gamma}^*$ which are based at $\tilde{o}, \overline{f}\tilde{o}^*$ where $f \in F$. We write $\hat{\gamma}' = q(\tilde{\gamma}')$ and $\hat{\gamma}^* = q(\tilde{\gamma}^*)$, whence $q(\overline{f}\tilde{\gamma}^*) = \overline{h}\hat{\gamma}^*$, where $h = \chi(f)$. We observe that $\hat{\gamma}', \overline{h}\hat{\gamma}^*$ are paths on $\hat{N}$ based at $\hat{o}, \overline{h}\hat{o}^*$ respectively. Let $\hat{c}', \overline{h}\hat{c}^*$ be the 1-chains corresponding to $\hat{\gamma}', \overline{h}\hat{\gamma}^*$ respectively. By (10.1) and (10.6)

$$\hat{c}' = \sum_i [\chi\omega(u_i')\hat{\alpha}_i + \chi\omega(v_i')\hat{\beta}_i]$$

$$\overline{h}\hat{c}^* = \sum_i [\overline{h}\chi\omega(U_i)\hat{\alpha}_i^* + \overline{h}\chi\omega(V_i)\hat{\beta}_i^*]$$

where $i = 1,\cdots,p$ and $h \in H$. By (8.5), we have

$$\begin{aligned} Is(h\hat{c}', \hat{c}^*) &= Is(\hat{c}', \overline{h}\hat{c}^*) \\ &= \sum_i [\chi\omega(u_i') \circ \overline{h}\chi\omega(U_i) + \chi\omega(v_i') \circ \overline{h}\chi\omega(V_i)] \end{aligned}$$

where $i = 1,\cdots,p$. By (2.6), (2.2) and (1.6), we obtain

$$\begin{aligned} Is(h\hat{c}', \hat{c}^*) &= \sum_i [h \circ \chi\omega(U_i)\overline{\chi\omega(u_i')} + h \circ \chi\omega(V_i)\overline{\chi\omega(v_i')}] \\ &= h \circ \sum_i [\chi\omega(U_i)\chi\omega(\overline{u}_i') + \chi\omega(V_i)\chi\omega(\overline{v}_i')] \end{aligned}$$

where $i = 1, \cdots, p$. Hence the following holds

$$\text{Is}(h\hat{c}', \hat{c}^*) = h \circ \chi\omega \left(\sum_i [U_i \bar{u}'_i + V_i \bar{v}'_i] \right) \tag{10.8}$$

where $i = 1, \cdots, p$, $h \in H$, u'_i and v'_i are given by (10.1), and U_i, V_i are given by the formulas preceding (10.6).

We now have to compute the second factor of the right hand side of (10.8). We write

$$K = \sum_i [U_i \bar{u}'_i + V_i \bar{v}'_i], \quad i = 1, \cdots, p \,. \tag{10.9}$$

From this and the formulas defining U_i and V_i, after performing some rearrangement, we obtain

$$\begin{aligned} K = (1-w) \sum_i [(1-\bar{a}_i)\bar{u}'_i + (1-\bar{b}_i)\bar{v}'_i] \\ - \sum_i [u_i(1-\bar{a}_i)\bar{u}'_i + v_i b_i \bar{a}_i \bar{u}'_i + u_i(1-a_i-\bar{b}_i)\bar{v}'_i + v_i(1-b_i)\bar{v}'_i] \\ + \sum_i \left(\sum_j d_j w \right) [(1-\bar{a}_i)\bar{u}'_i + (1-\bar{b}_i)\bar{v}'_i] \end{aligned}$$

where $j = 1, \cdots, i-1$ and $i = 1, \cdots, p$. By (1.5), (10.1) and (9.3), the following holds

$$(1-\bar{a}_i)\bar{u}'_i + (1-\bar{b}_i)\bar{v}'_i = -\overline{d_i w'}, \quad i = 1, \cdots, p \,.$$

By the fundamental formula of the free differential calculus ([3], p. 551, (2.3)) and (9.3) we have

$$w' - 1 = \sum_i d_i w', \quad i = 1, \cdots, p \,. \tag{10.10}$$

From the last two equations and (1.5) we obtain

$$\sum_i [(1-\bar{a}_i)\bar{u}'_i + (1-\bar{b}_i)\bar{v}'_i] = 1 - \bar{w}'$$

where $i = 1,\cdots,p$. By (1.5), (1.2), (10.1) and (9.1) the following holds

$$\sum_i [u_i(1-\bar{a}_i)\bar{u}'_i + v_i b_i \bar{a}_i \bar{u}'_i + u_i(1-a_i-\bar{b}_i)\bar{v}'_i + v_i(1-b_i)\bar{v}'_i] = \overline{\Lambda_1(w',w)}$$

where $i = 1,\cdots,p$. From the last five equalities, we have

$$K = (1-w)(1-\bar{w}') - \overline{\Lambda_1(w',w)} - \sum_i \left(\sum_j d_j w\right) \overline{d_i w'}$$

where $j = 1,\cdots,i-1$ and $i = 1,\cdots,p$. By (10.10), where now w' is replaced by w, we obtain

$$K = (1-w)(1-\bar{w}') - \overline{\Lambda_1(w',w)} - \sum_i \left(w-1-\sum_j d_j w\right) \overline{d_i w'}$$

where $j = i,\cdots,p$ and $i = 1,\cdots,p$. By (10.10) and (1.5), the following hold

$$K = (1-w)(1-\bar{w}') - \overline{\Lambda_1(w',w)} - \sum_i (w-1)\overline{d_i w'} + \sum_i \left(\sum_j d_j w\right) \overline{d_i w'}$$

$$K = (1-w)(1-\bar{w}') - \overline{\Lambda_1(w',w)} - (w-1)(\bar{w}'-1) + \sum_i \left(\sum_j d_j w\right) \overline{d_i w'}$$

$$= -\overline{\Lambda_1(w',w)} + \sum_i \left(\sum_j d_j w\right) \overline{d_i w'}$$

where $j = i,\cdots,p$ and $i = 1,\cdots,p$. From the above we obtain

$$K = -\overline{\Lambda_1(w',w)} + \sum_i (d_i w)\overline{d_i w'} + \sum_k \left(\sum_j d_j w\right) \overline{d_k w'}$$

where $j = k+1,\cdots,p$, $k = 1,\cdots,p-1$ and $i = 1,\cdots,p$. By (9.5), (9.2) and (9.4), we have

$$K = \Lambda(w,w') .$$

From the above equality, (10.9) and (10.8), we obtain the following *intersection formula*

$$\text{(10.11)} \qquad Is(h\hat{c}', \hat{c}^*) = h \circ \chi\omega\Lambda(w,w') .$$

From this we have the following *expansion formula*

$$\chi\omega\Lambda(w, w') = \sum Is(h\hat{c}', \hat{c}^*)h \tag{10.12}$$

where the sum runs over all elements h of H. This sum is finite, because $\hat{\gamma}$ and $\hat{\gamma}'$ are compact. The last formula is obtained by observing that the left hand side is an element of $Z[H]$, and then computing the coefficients by use of (10.11) and the formulas for inner products.

We now summarize things: Let N be an oriented closed surface of genus $p \geqq 2$, with base point o, and let

$$F = \left(a_1, b_1, \cdots, a_p, b_p : \prod_i [a_i, b_i]\right), \quad i = 1, \cdots, p$$

be the fundamental group of N, see (Figures 1 and 2). Let G be a normal subgroup of F, let $\hat{N}$ be the regular covering surface corresponding to G, and let $\hat{o}$ be a point on $\hat{N}$ lying over o. Let γ, γ' be two loops on N based at o, and let w, w' be words in a_i, b_i defining the elements of F represented by γ, γ' respectively. I.e., w, w' are elements of the free group

$$\Phi = (a_1, b_1, \cdots, a_p, b_p)$$

of rank $2p$. Let $\hat{\gamma}, \hat{\gamma}'$ be two paths on $\hat{N}$ based at $\hat{o}$ and lying over γ, γ' respectively.

THEOREM 10.13. *If the elements of* F *defined by* w, w' *belong to* G, *then* $\hat{\gamma}, \hat{\gamma}'$ *are loops, and the following formulas hold*

$$Is(h\hat{\gamma}', \hat{\gamma}) = h \circ \chi\omega\Lambda(w, w')$$

$$\chi\omega\Lambda(w, w') = \sum Is(h\hat{\gamma}', \hat{\gamma})h$$

where the sum runs over all $h \in H = F/G$.

Proof. Without any loss of generality we can suppose that γ, γ' are loops composed of α_i, β_i's in accordance with the words w, w' respectively. In the present case, because of (10.7), $\hat{\gamma}^*$ is a loop homotopic to $\hat{\gamma}$ on $\hat{N}$. Let us denote by $\hat{c}^*, \hat{c}'$ the 1-cycles corresponding to $\hat{\gamma}^*, \hat{\gamma}'$ respectively. The following hold

$$Is(h\hat{\gamma}', \hat{\gamma}) = Is(h\hat{\gamma}', \hat{\gamma}^*) = Is(h\hat{c}', \hat{c}^*) .$$

From the above, (10.11) and (10.12), follow the formulas of our theorem.

REMARK 10.14. Let us now consider the special case $G = F$. Then $H = 1$, $\hat{N} = N$, $\hat{q} =$ identity, $\hat{\gamma} = \gamma$ and $\hat{\gamma}' = \gamma'$. Thus, the formula of Theorem 10.13 becomes now

$$(\omega\Lambda(w, w'))^0 = Is(\gamma', \gamma)$$

see No. 1, or ([3], p. 549, *ll.* 18-20).

11. *Planar coverings*

We now suppose that the group G, of the previous No. 10, is the normal closure in F of the finite or infinite sequence of elements $\omega(w_1), \omega(w_2), \cdots$, where $w_1, w_2, \cdots$ are words of the free group Φ. Thus,

$$H = F/G = \left(a_1, b_1, \cdots, a_p, b_p : \prod_i [a_i, b_i], w_1, w_2, \cdots\right)$$

where $i = 1, \cdots, p$. Hence $H = \Phi/G_0$, where G_0 is the normal closure in Φ of the elements $w_0, w_1, w_2, \cdots$, and w_0 is the product of the commutators.

Let now w, w' be two elements of G_0, then the following hold

$$w = u \prod_k u_k w_{\mu_k}^{\delta_k} \bar{u}_k, \qquad w' = v \prod_j v_j w_{\lambda_j}^{\varepsilon_j} \bar{v}_j$$

where u_i, v_i are words of Φ, μ_k and λ_j are positive, δ_k and ε_j are ± 1, the range of the indices k and j is finite, and u, v are products of transforms of w_0^θ in Φ, $\theta = \pm 1$.

Let γ, γ' be two loops on N representing the elements $\omega(w)$, $\omega(w')$ of F respectively. The loops $\hat{\gamma}, \hat{\gamma}'$ are those of Theorem 10.13.

THEOREM 11.1. *If the group* G *is the normal closure in* F *of the elements* $\omega(w_1), \omega(w_2), \cdots$, *then*

$$\sum_h \mathrm{Is}(h\hat{\gamma}', \hat{\gamma})h = \sum_{k,j} \delta_k \varepsilon_j \chi\omega(u_k \Lambda(w_{\mu_k}, w_{\lambda_j})\bar{v}_j)$$

where h *runs over all elements of* $H = F/G$. *Finally,* $\hat{N}$ *is planar if and only if*

$$\chi\omega\Lambda(w_\kappa, w_\nu) = 0$$

where $\kappa, \nu = 1, 2, \cdots$.

Proof. Let us denote by $\gamma_0, \gamma_{\mu_k}, \gamma_{\lambda_j}$ loops on N based at o representing the elements $\omega(w_0), \omega(w_{\mu_k}), \omega(w_{\lambda_j})$ of F, and let $\hat{\gamma}_0, \hat{\gamma}_{\mu_k}, \hat{\gamma}_{\lambda_j}$ be the loops on $\hat{N}$ based at $\hat{o}$ and lying over $\gamma_0, \gamma_{\mu_k}, \gamma_{\lambda_j}$ respectively. Then, $\hat{\gamma}_0$ is contractible in $\hat{N}$, because $\omega(w_0) = 1$. Thus, by Theorem 10.13, we have the following

$$\chi\omega\Lambda(w_0, w_0) = \sum \mathrm{Is}(h\hat{\gamma}_0, \hat{\gamma}_0)h = 0$$

$$\chi\omega\Lambda(w_0, w_{\lambda_j}) = \sum \mathrm{Is}(h\hat{\gamma}_{\lambda_j}, \hat{\gamma}_0)h = 0$$

$$\chi\omega\Lambda(w_{\mu_k}, w_0) = \sum \mathrm{Is}(h\hat{\gamma}_0, \hat{\gamma}_{\mu_k})h = 0$$

where the sums run over all $h \in H$. By Theorem 6.3 and the above three equalities, the following holds

$$\chi\omega\Lambda(w, w') = \sum_{k,j} \delta_k \varepsilon_j \chi\omega(u_k \Lambda(w_{\mu_k}, w_{\lambda_j})\bar{v}_j) .$$

By Theorem 10.13, we have the following

$$\chi\omega\Lambda(w, w') = \sum \mathrm{Is}(h\hat{\gamma}', \hat{\gamma})h \ .$$

The last two formulas imply the first formula of our theorem.

By ([5], p. 140, ℓℓ. 26-29 and p. 165, ℓℓ. 15-17), it is easily proved that $\hat{N}$ is planar if and only if the intersection number of any two loops on $\hat{N}$ based at $\hat{o}$ is zero. If $\hat{N}$ is planar, then we apply the first formula of our theorem for the special case $w = w_\kappa$, $w' = w_\nu$ and we have

$$\chi\omega\Lambda(w_\kappa, w_\nu) = \sum \mathrm{Is}(h\hat{\gamma}', \hat{\gamma}) = 0, \quad \kappa, \nu = 1, 2, \cdots \ .$$

If the second formulas of our theorem hold, then by the first formula the following holds

$$\mathrm{Is}(\hat{\gamma}', \hat{\gamma}) = 0$$

for any two loops $\hat{\gamma}, \hat{\gamma}'$ on $\hat{N}$ based at $\hat{o}$. Hence, $\hat{N}$ is planar. This completes the proof of our theorem.

REMARK 11.2. In the case of actual computations, Theorem 6.4 may turn out to be very useful sometimes.

§5. A Conjecture

The formulas of Theorem 11.1 provide us with a solution of the problem we posed in the Introduction. However, as we have already mentioned there, the result provided by the solution is not sufficient to solve the second conjecture of ([7], p. 251). Nevertheless, our formulas may be very helpful for a solution of the conjecture we express in the sequel.

Let us suppose that, on the oriented closed surface N of genus at least two, we have two oriented simple closed curves A, B meeting at only one point with intersection number one. Let A' be an oriented simple closed curve on N homologous to B. Let X, Y and X' be the primary simple closed geodesics on N, corresponding to A, B and A' respectively ([7], p. 270, Lemma (12.2)). Let o_k, $k = 1, \cdots, 2m+1$ ($\geqq 1$), be the common points of X and X', and let X_k and X'_k be the loops with base point o_k defined by X and X' respectively. Let finally w_k

be the element of $F = \pi_1(N, o_k)$ corresponding to the commutator $[X_k, X'_k]$. Under the above hypotheses we formulate the following conjecture.

CONJECTURE. There is a k ($\geqq 1$ and $\leqq 2m+1$) such that the regular covering of N, corresponding to the normal closure of w_k in F, is *planar*.

This conjecture implies the second conjecture of ([7], p. 251), but not conversely. That conjecture is more delicate. Hence, a proof of the above conjecture would imply a proof of the Poincaré conjecture.

PRINCETON UNIVERSITY

BIBLIOGRAPHY

[1] Dyer, E., and Vasquez, A. T., Some small aspherical spaces. *J. Austral. Math. Soc.*, 16 (1973), 332-352.

[2] Fox, R. H., On the asphericity of regions in 3-space. *Bull. Amer. Math. Soc.*, 55 (1949), 521.

[3] ———, Free differential calculus I. *Ann. of Math.*, 57 (1953), 547-560.

[4] Karrass, A., Magnus, W., and Solitar, D., Elements of finite order in groups with a single defining relation. *Comm. Pure and Appl. Math.*, 13 (1960), 57-66.

[5] Kerékjártó, B. v., Vorlesungen über Topologie. Springer, Berlin, 1923.

[6] Maskit, B., A theorem on planar covering surfaces with applications to 3-manifolds. *Ann. of Math.*, 81 (1965), 341-355.

[7] Papakyriakopoulos, C. D., A reduction of the Poincaré conjecture to group theoretic conjectures. *Ann. of Math.*, 77 (1963), 250-305.

[8] ———, Attaching 2-dimensional cells to a complex. *Ann. of Math.*, 78 (1963), 205-222.

[9] Rapaport, Elvira S., Proof of a conjecture of Papakyriakopoulos. *Ann. of Math.*, 79 (1964), 506-513.

[10] Reidemeister, K., Homotopiegruppen und Schnittrelationen. *Abh. Math. Sem., Hamburgischen Univ.*, 10 (1934), 298-304.

[11] ———, Kommutative Fundamentalgruppen. *Monatsh. Math. Phys.*, 43 (1936), 20-28.

INFINITELY DIVISIBLE ELEMENTS IN 3-MANIFOLD GROUPS

Peter B. Shalen

An element g of a group is said to be *divisible* by an integer n if $g = x^n$ holds for some element x in the group. We will say that g is *infinitely divisible* if it is divisible by infinitely many different integers.

L. Neuwirth ([5], Problem S) has asked whether a knot group can have infinitely divisible elements other than the identity. We show that it cannot; more generally, we show that the fundamental group of a compact, orientable, irreducible, piecewise-linear 3-manifold M has no infinitely divisible elements $\neq 1$, provided that M is *almost sufficiently large* in the sense of Waldhausen (see Section 7 for definitions). This is the theorem of Section 7. A weaker result in this direction was obtained in [2].

The result cannot quite extend to an arbitrary compact 3-manifold M since a lens space (for example) has a finite fundamental group, and any element of finite order in a group is clearly infinitely divisible. However, one can make the

CONJECTURE (Cf. [20], p. 87). Every compact, irreducible, orientable, piecewise-linear 3-manifold with infinite fundamental group is almost sufficiently large.

Now it follows from Moise's triangulation theorem ([8], [9]; also [1]; and [13]) and a theorem of Kneser's ([5]; see [7] for a good discussion) that for any compact, orientable 3-manifold M, $\pi_1(M)$ is isomorphic to a finite free product of infinite cyclic groups and fundamental groups of compact, irreducible, orientable 3-manifolds. It is easy to see that an infinitely divisible element in a free product must be conjugate to an infinitely divisible element of some factor. Using the theorem of Section 7, the above conjecture would therefore imply that an element of $\pi_1(M)$ is infinitely divisible only if it has finite order. By considering the orientable double cover, one could extend this result to the non-orientable case (cf. Lemma 13 in Section 7 below).

It should be noted that the compactness hypothesis is essential, since there exists an open subset of Euclidean 3-space whose fundamental group is isomorphic to the additive group of rational numbers (see [3]).

The proof of the theorem depends heavily on Haken's theory of hierarchies; the only facts needed are contained in [20], and we review them in Section 7. An arbitrary sufficiently large manifold can be built up from one or two 3-cells by successive application of a boundary-gluing process. Using this, the proof of the theorem reduces to comparing the divisibility properties of elements of an incompressible (cf. Section 1) piece of the boundary of a 3-manifold, with the divisibility properties of the same elements regarded as lying in the fundamental group of the 3-manifold. This is done in Sections 2-4 for boundary pieces that are not tori, and in Section 5 for tori. The results obtained in the two cases (Prop. 2 of Section 2 and Prop. 3 of Section 5) are rather different. The problem of extending the results of Sections 2-4 directly to tori is related to the problem discussed in Section 3 after the definition of an "envelope."

I would like to thank W. Jaco and B. Evans for pointing out the proof of Proposition 1 of Section 1 based on Waldhausen's generalized loop theorem, which is much simpler than my original proof. They had obtained a similar result independently. I would also like to thank F. Waldhausen for a series of interesting discussions on the unsolved problem discussed in Section 3. Finally, I am indebted to Marcelo Kupferwasser and to the referee for correcting a good many errors in the original typescript.

§0. *Conventions*

We work in the piecewise-linear (PL) category everywhere (except in Section 4 where we briefly consider the simplicial category). Thus all *manifolds, maps, homeomorphisms, homotopies, isotopies,* etc. are understood to be PL. A *surface* is a connected 2-manifold. A *disc* is a (PL) 2-cell; an *arc* is a (PL) 1-cell; *tori, annuli,* etc. have the usual PL structures. The boundary and interior of a manifold M are written ∂M and $\mathring{M}$. On the other hand, if X is a subset of a space Y, the *frontier* and (set-theoretic) *interior* of X in Y are denoted Fr X and Int X (or $\mathrm{Fr}_Y X$ and $\mathrm{Int}_Y X$).

By a *simple curve* we mean simply a (PL) 1-sphere. On the other hand, a *singular curve* is a (PL) map of the standard 1-sphere S^1 (boundary of the standard 2-simplex) into a space (polyhedron). A singular curve which is 1–1 will sometimes be called a *parametrized simple curve*, or a *parametrization* of its image.

In any connected polyhedron P, there is a bijective correspondence between (free) homotopy classes of singular curves in P, and conjugacy classes in $\pi_1(P)$. (Here, as in other statements that are independent of the choice of a basepoint, we suppress the basepoint.) THE CONJUGACY CLASS ASSOCIATED WITH A SINGULAR CURVE σ WILL BE WRITTEN $[\sigma]$. If the singular curves σ, τ are such that $[\tau] = [\sigma]^k$, k an integer, we will say that *τ is homotopic to a k-th power of σ*. (Operations in a group such as raising to the k-th power are clearly defined on conjugacy classes.) We will say that *τ is a k-th power of σ* if $\tau(S^1) = \sigma(S^1)$, and τ is homotopic to a k-th power of σ in the space $\sigma(S^1)$. Singular curves σ and τ are said to be *anti-homotopic* if τ is homotopic to a minus-first power of σ. A singular curve is *homotopic to a power of* a simple curve γ if it is homotopic to a power of an orientation of γ. A singular curve is *contractible* if it is homotopic to a constant map; a simple curve is *contractible* if a parametrization of it is contractible.

Two embeddings $f, g: P \to Q$ are *isotopic* if there is a map $H: P \times I \to Q$, such that for all $t \in I$ the map $H_t: P \to Q$ defined by $H_t(p) = H(p,t)$ is an embedding, and such that $H_0 = f$, $H_1 = g$. They are *ambient-isotopic* if there is a homeomorphism $h: Q \to Q$, isotopic to the identity, such that

$h \circ f = g$. Two simple curves are *isotopic* if they have isotopic orientations; two arcs $\alpha, \beta \subset Q$ are *isotopic with endpoints fixed* if there is a homeomorphism $j : \alpha \to \beta$ which is isotopic in Q to the inclusion, under an isotopy which is constant on $\partial\alpha$.

An (n–1)-manifold N contained in a n-manifold M is *2-sided* if there is an embedding $c : N \times [-1,1] \to M$ such that $c(N \times [-1,1])$ is a neighborhood of N in M, and $c(n,0) = n$ for all $n \in N$. Such an embedding c is called *a collar neighborhood* of N. If N is 2-sided in M, then in particular $N \cap \partial M = \partial N$. If N is 2-sided in M and $f : M' \to M$ is a map, M' a manifold, we say that f is *transversal* to N if $f^{-1}(N) = N'$ is a 2-sided submanifold of M', and if there exist collar neighborhoods c, c' of N, N' such that $f(c'(N' \times \{t\})) \subset c(N \times \{t\})$ for all $t \in [-1,1]$. It is well-known that transversality is a "general-position" condition: for example, if M' is compact, $f : M' \to M$ is such that $f(\partial M') \subset \partial M$, N is 2-sided in M, and $f|\partial M'$ is transversal to $\partial N \subset \partial M$, then f can be approximated (in the metric sense, say) by a map which agrees with f on ∂M and is transversal to N. (These facts will be used only for dimensions ≤ 3.)

The fundamental results of Papakyriakopoulos on 3-manifolds – Dehn's lemma, the loop theorem and the sphere theorem ([11], [12]; see also [14], [16]) are crucial for the arguments in this paper. Many of the applications are made via two corollaries which we state below, with references or indications of proofs, as Principles 1 and 2.

DEFINITION. Let T be a surface in a connected 3-manifold M such that either T is 2-sided or $T \subset \partial M$. We say that T is *incompressible* in M if T is neither a disc in ∂M nor the frontier of a 3-cell in M, but $\pi_1(T) \to \pi_1(M)$ is injective. More generally, if a 2-manifold T is either 2-sided in M or contained in ∂M, T is *incompressible* in M if each of its components is incompressible.

PRINCIPLE 1. Let M be a connected 3-manifold, and let the 2-manifold T be either 2-sided in M or contained in ∂M. Assume that T is neither a disc in ∂M nor the frontier of a 3-cell in M. Then T is incompressible if, and only if, for every disc $\Delta \subset M$ such that $\Delta \cap T = \partial\Delta$, $\partial\Delta$ bounds a disc in T.

Proof. The "only if" assertion follows from the elementary fact that a contractible simple curve in a 2-manifold always bounds a disc. The "if" assertion is essentially the second sentence of 2.B.1 on p. 14 of [17]. (A slight paraphrase of this sentence is, "if T is a compact 2-sided surface in a 3-manifold M and cannot be reduced, then $\pi_1(T) \to \pi_1(M)$ is injective." The statement that T "cannot be reduced" is precisely the hypothesis of our assertion; and the compactness of T is not used in the proof of the second sentence of 2.B.1.)

DEFINITION. A *homotopy* 3-*cell* is a compact, contractible 3-manifold whose boundary is a 2-sphere.

PRINCIPLE 2. Let M be a connected, orientable 3-manifold. We have $\pi_2(M) = 0$ if and only if every 2-sphere in M bounds a homotopy 3-cell in M.

Proof. This is proved in the same way as Theorem 2 on p. 5 of [7], except that the term "cell" is replaced in both its occurrence by "homotopy 3-cell," and the reference to the Poincaré hypothesis is deleted. The fact that a simply connected 3-manifold bounded by a 2-sphere is a homotopy 3-cell follows from Poincaré duality and the Hurewicz theorem.

Finally we use the following general conventions. All unlabeled homomorphisms (e.g. $\pi_1(X) \to \pi_1(Y)$) are understood to be induced by inclusion. The Euler characteristic of a finite polyhedron P is denoted $\chi(P)$. We use c(x) to denote the conjugacy class of an element x of a group, and $<x>$ to denote the subgroup generated by x.

§1. *Divisibility of loops in boundary surfaces*

We define a notion of *divisibility* for conjugacy classes in the fundamental group of a surface, and prove a result (Proposition 1) to the effect that the divisibility associated to a curve in the boundary of a 3-manifold depends only on its homotopy class in the manifold. The proof of Proposition 1 given here is due to W. Jaco and B. Evans; it uses Waldhausen's generalization of the loop theorem [19].

LRMMA 1. *Let* G *be the fundamental group of an orientable surface. Any* $g \in G-\{1\}$ *can be written as* x^n, *where* x *is primary in* G *and* $n \geq 0$; *moreover,* n *and* x *are uniquely determined by* g.

Proof. This is obvious if G is free abelian. We can therefore assume $G = \pi_1(T, p)$, where the surface T is not a torus. Let $(\tilde{T}, \tilde{p})$ be the based covering corresponding to the centralizer of g in G. Then $\tilde{T}$ is an orientable surface, not a torus, and $\pi_1(\tilde{T})$ has non-trivial center.

In particular, $\tilde{T}$ cannot be closed. Hence $\pi_1(\tilde{T})$ is free. As $\pi_1(\tilde{T})$ has a center, it has rank 1. Thus the centralizer of g in G is infinite cyclic, and the lemma follows.

DEFINITION. In the situation of Lemma 1, x is called the *primary root*, and n the *divisibility* of g in G. As n obviously depends only on the conjugacy class $c(g)$ of g, it may also be called the *divisibility* of $c(g)$. The *divisibility* of a singular curve σ in a 2-manifold T is the divisibility of $[\sigma] \subset \pi_1(T_\sigma)$, where T_σ is the component of T containing $\sigma(S^1)$.

COROLLARY 1. *If an element of* $\pi_1(T)$, T *an orientable surface, is divisible by an integer* k, *its divisibility in* $\pi_1(T)$ *is an integer divisible by* k.

COROLLARY 2. *If* $g \in \pi_1(T)$ *has divisibility* k, *then* g^m *has divisibility* $|m|k$ *for any integer* $m \neq 0$.

In particular g^m and $g^{m'}$ have the same divisibility only if $m = \pm m'$. Hence:

COROLLARY 3. *A cyclic subgroup of* $\pi_1(T)$ *contains only two elements of a given divisibility.*

The next lemma is well-known and may be proved in the same style as Lemma 1.

LEMMA 2. *If* T *is an oriented surface, a conjugacy class in* $\pi_1(T)$, *represented by a non-contractible parametrized simple curve in* T, *is primary.*

PROPOSITION 1. *Let* T *be an incompressible 2-manifold in the boundary of an orientable 3-manifold* M. *Then any two non-contractible singular curves in* T, *that are freely homotopic in* M, *have the same divisibility in* T.

Proof. Assume the assertion false. Then there are singular curves σ_1, σ_2 in T, homotopic in M, with respective divisibilities m_1, m_2 in T, where $m_1 < m_2$. Write $T_i (i = 1,2)$ for the component of T containing $\sigma_i(S^1)$. Fix a basepoint $t \in T_2$. We may take σ_2 to be based, so that it defines an element g_2 of $\pi_1(T_2, t)$, and denote by x the primary root of g_2 in $\pi_1(T_2, t)$. The infinite cyclic subgroup $\langle x \rangle$ of $\pi_1(M)$ determines a covering space $\tilde{M}$ of M with a canonical basepoint $\tilde{t}$. Let p denote the covering projection and let $\tilde{a}_2$ be the lifting of σ_2 based at $\tilde{t}$. Identify $\pi_1(\tilde{M})$ with $\mathbf{Z}$ in such a way that $x = 1$. As $\mathbf{Z}$ is abelian, every free homotopy class of singular curves in $\tilde{M}$ defines an integer. Clearly $\tilde{\sigma}_2$ defines the integer m_2.

By the covering homotopy property, the based lifting $\tilde{\sigma}_2$ of σ_2 to $\tilde{M}$ is freely homotopic in $\tilde{M}$ to some lifting $\tilde{\sigma}_1$ of σ_1, which again represents the integer m_2 if regarded as a loop in $\tilde{M}$. The components $\tilde{T}_1, \tilde{T}_2$ of $p^{-1}(T)$ that contain $\tilde{\sigma}_1, \tilde{\sigma}_2$ are incompressible but have non-trivial fundamental groups; thus $\pi_1(\tilde{T}_1) \approx \pi_1(\tilde{T}_2) \approx \mathbf{Z}$, and by orientability $\mathring{\tilde{T}}_1$ and $\mathring{\tilde{T}}_2$ are open annuli.

I claim that there exist non-contractible parametrized simple curves $\tilde{\xi}_1, \tilde{\xi}_2$ in $\tilde{T}_1, \tilde{T}_2$ that are homotopic in $\tilde{M}$. This is trivial if $\tilde{T}_1 = \tilde{T}_2$. If $\tilde{T}_1 \neq \tilde{T}_2$, the open subsets $\mathring{\tilde{T}}_1, \mathring{\tilde{T}}_2$ of $\partial\tilde{M}$ are disjoint and contain the mutually homotopic closed curves $\tilde{\sigma}_1, \tilde{\sigma}_2$ which are non-contractible in $\tilde{M}$. The claim therefore follows from the theorem of [19].

Now since $\tilde{T}_2$ is an annulus and $\pi_1(\tilde{T}_2) \to \pi_1(\tilde{M})$ is surjective by construction, $\tilde{\xi}_2$ represents an integer $u = \pm 1$. Hence so does $\tilde{\xi}_1$. Let $\tilde{\sigma}_1(1) = \tilde{q}$, and let $\tilde{\xi}_1$ be based at $\tilde{q}$; then the loop $\tilde{\xi}_1^{m_2 u} * \tilde{\sigma}_1^{-1}$ represents 0, and is therefore null-homotopic in $\tilde{T}_1$. It follows that $[\tilde{\sigma}_1]$ is divisible by m_2 in $\pi_1(\tilde{T}_1)$, and therefore that $[\sigma_1]$ is divisible by m_2 in $\pi_1(T_1)$; thus $m_2 | m_1$ by Corollary 1 to Lemma 1, and the assumption $m_1 < m_2$ is contradicted.

§2. *Divisibility of boundary curves in the interior*

DEFINITION. A 3-*manifold pair* is a pair (M, T), where M is a connected 3-manifold and $T \subset \partial M$ is a surface. (M, T) is *acceptable* if $\pi_2(M) = 0$, M is orientable, and T is compact and incompressible (Section 0).

N.B. We may have $\partial T \neq \emptyset$.

DEFINITION. Let (M, T) be a 3-manifold pair, and let $c(g) \neq \{1\}$ be a conjugacy class in $\pi_1(T)$ having divisibility k. If the image of $c(g)$ in $\pi_1(M)$ is divisible by some integer $\ell > 2k$, $c(g)$ will be called *special* (with respect to (M, T)).

The object of Sections 3 and 4 is to prove:

PROPOSITION 2. *Let* (M, T) *be an acceptable pair. Then:*

(a) *Any special conjugacy class in* $\pi_1(T)$ *can be represented by a power (Section 0) of a parametrized simple curve in* T.

(b) *If* T *is not a torus, then any finite set of special conjugacy classes in* $\pi_1(T)$ *can be represented by a set of powers of disjoint parametrized simple curves.*

Now it is well-known and easy to prove that *for any compact orientable surface* T, *there is an integer* N(T) *such that* T *cannot contain more than* N(T) *disjoint, non-homotopic, non-contractible simple curves.*

Assuming the truth of Proposition 2, a set of special conjugacy classes $c_1, \cdots, c_n$ in $\pi_1(T)$ can be represented by powers of disjoint parametrized simple curves $\gamma_1, \cdots, \gamma_n$ in T. If we assume in addition that the c_i all have the same divisibility k, Corollary 3 to Lemma 1 of Section 1 implies that at most two of the γ_i can lie in any given cyclic subgroup of $\pi_1(T)$; hence by the fact just recalled, $n \leq 2\,N(T)$. In particular we obtain the

COROLLARY TO PROPOSITION 2. *If* (M, T) *is an acceptable pair and* T *is not a torus,* $\pi_1(T)$ *contains only finitely many special conjugacy classes having a prescribed divisibility in* $\pi_1(T)$.

It is this group-theoretical conclusion that is used in the proof of our main result.

§3. *Cutting and pasting*

This section is preliminary to the proof of Proposition 2. It is assumed in this section and the next that (M, T) is an acceptable pair. The following technical notion is central to the argument:

DEFINITION. Let σ be a non-contractible singular curve in T. An *envelope* for σ is a compact 3-manifold $K \subset M$, such that

(i) every component of ∂K is a torus and

(ii) the conjugacy class $[\sigma] \subset \pi_1(B)$, where B is the component of ∂K containing σ, is special with respect to the 3-manifold pair (K, B).

Set $A = K \cap \partial M$. The envelope K is called *normal* if the following extra conditions hold:

(iii) $A = T$ if T is a torus, and A is an annulus otherwise; and

(iv) the 2-sided 2-manifold $(\partial K) - \mathring{A}$ is incompressible in M.

If σ has an envelope, it is said to be *enveloped*.

The obvious examples of enveloped curves are obtained as follows: Let (M_0, T_0) be a 3-manifold pair, and let $B \subset T_0$ be an annulus. Let γ be a non-contractible simple curve in $\partial(D^2 \times S^1)$ such that $[\gamma] \subset \pi_1(D^2 \times S^1)$ has divisibility > 2. In the disjoint union of M_0 and $D^2 \times S^1$, identify B with a regular neighborhood of γ in $\partial(D^2 \times S^1)$; this gives a 3-manifold M, and we can set $T = (T_0 \cup \partial(D^2 \times S^1)) - \mathring{B}$. It is easy to find conditions guaranteeing that (M, T) is acceptable. In this case $D^2 \times S^1$ is clearly a (normal) envelope for any non-contractible σ in $\partial(D^2 \times S^1) - B$.

The most vexed (and vexing) question left open in this paper is whether every enveloped singular curve has an envelope homeomorphic to $D^2 \times S^1$. This could perhaps be settled by F. Waldhausen's unpublished "torus-annulus theorem."

LEMMA 3. *An enveloped singular curve has a normal envelope.*

Proof. Let σ have the envelope K. By enlarging K if necessary, we may assume that $K \cap \partial M$ is a 2-manifold. We may further assume that $K \cap \partial M$ is connected and is contained in T; for if this is not the case, we can modify K by a (non-ambient) isotopy which is constant on a regular neighborhood N of $\sigma_1(S^1)$ in $K \cap \partial M$, and moves $(\partial K) - N$ into $\mathring{M}$.

To prove the lemma we must choose K so that (a) the surface $A = K \cap \partial M$ is a torus or annulus, (b) $A = T$ if T is a torus, and (c) each component of $(\partial K) - \mathring{A}$ is incompressible.

We claim, first, that if either (a) or (c) fails to hold, then σ has an envelope K', such that $K' \cap \partial M$ is again a surface in T, but such that $M - K'$ has fewer components than $M - K$; and that if K satisfied (b), then so will K'.

First suppose that (a) does not hold. Let B_0 denote the component of ∂K that contains A; then since B_0 is a torus, $\pi_1(A) \to \pi_1(B_0)$ cannot be injective. Hence some component C of ∂A bounds a disc $\Delta \subset B_0 - \mathring{A}$; as T is incompressible, C must also bound a disc $D \subset T$. But since A is connected and contains the non-contractible curve σ, we must have $D \subset T - \mathring{A}$. Now $\Delta \cup D$ is a 2-sphere, and by Principle 2 of Section 0 it bounds a homotopy 3-cell E; since K is connected and contains the non-contractible curve σ, it can intersect E only in Δ. We may set $K' = K \cup E$, proving the claim in this case.

Now suppose that (a) holds but that (c) does not. Then some component S of $(\partial K) - \mathring{A}$ fails to be incompressible. Now it follows from (a) that A is incompressible; for T is incompressible, $\pi_1(T)$ is torsion free, and A contains a non-contractible curve. Hence if A is an annulus, the annulus $A' \subset (\partial K) - \mathring{A}$ which has the same boundary as A is also incompressible, since its generating curve defines the same conjugacy class in $\pi_1(M)$ as the generating curve of A. Therefore S must be a torus. As S is not incompressible, we may use Principle 1 of Section 0 to replace a non-contractible annulus in S by two discs, thus producing a 2-sphere $S' \subset M$. Then S' is homologically trivial, since $\pi_2(M) = 0$, and therefore S is also homologically trivial. It follows that S bounds a compact PL 3-manifold $R \subset \overline{M-K}$. We can set $K' = K \cup R$, and the claim is proved in this case as well.

It follows from the claim just proved that σ has an envelope K_1 satisfying (a) and (c). Then K_1 is normal unless (b) fails to hold, i.e. unless T is a torus but $A_1 = K_1 \cap \partial M$ is an annulus. In this case, let K_2 be a regular neighborhood of $K_1 \cup T$. K_2 clearly is an envelope for σ and satisfies (b).

Hence by the claim proved above, σ has an envelope K_3 satisfying (a) and (c), and (b) as well.

COROLLARY 1. *If* σ *is enveloped then* $[\sigma]$ *is special with respect to* (M, T).

A converse to Corollary 1 will be proved as Lemma 7 of Section 4.

COROLLARY 2. *An enveloped curve is homotopic in* T *to a power (Section 0) of a parametrized simple curve.*

Proof. Every singular curve in a torus or annulus is homotopic to a power of a parametrized simple curve

LEMMA 4. *Let* K_1, K_2 *be normal envelopes for singular curves* σ_1, σ_2 *in* $\mathring{T}$. *Set* $A_i = K_i \cap \partial M$ $(i=1,2)$ *and suppose that* ∂A_1 *and* ∂A_2 *intersect each other transversally. If* $(\partial A_1) \cap (\partial A_2) \neq \emptyset$, *then there exists a disc* $\Delta \subset \mathring{T}$ *whose boundary is of the form* $b_1 \cup b_2$, *where* $b_i \subset \partial A_i$ *is an arc* $(i=1,2)$, *and* $b_1 \cap b_2 = \partial b_1 = \partial b_2$.

Proof. By taking K_1 and K_2 in general position, without altering A_1 and A_2, we may assume that $\mathrm{Fr}\, K_1$ and $\mathrm{Fr}\, K_2$ (which by normality are 2-sided 2-manifolds) intersect transversally.

By hypothesis A_1 and A_2 have non-empty boundaries. Since the envelopes K_1 and K_2 are normal, it follows that A_1 and A_2 are annuli and that $\mathring{T}$ is *not* a torus. Let $\gamma_i \subset A_i$ be a simple curve carrying a generator of $H_1(A_i)$. Any 2-sided surface in K_i has an integer intersection number with γ_i, defined up to sign. We claim that

(*) No 2-sided surface in K_i $(i=1,2)$ can have intersection number ± 1 or ± 2 with γ_i.

To see this, fix orientations of γ_i and of the 2-sided surface $J \subset K_i$, so that the intersection number $\gamma_i \cdot J$ is a well-defined integer. Let $k > 0$ denote the divisibility (Section 1) of the singular curve σ_i in ∂K_i. Since σ_i is in A_i, it is a $\pm k$-th power (Section 0) of γ_i. Hence $\sigma_i \cdot J = \pm k(\gamma_i \cdot J)$. On the other hand, since σ_i is *special* in K_i by the definition of an envelope, $[\sigma_i]$ is divisible by some integer $\ell > 2k$ in $\pi_1(K_i)$;

therefore $\sigma_i \cdot J$ is divisible by ℓ. So if $\gamma_i \cdot J$ were ± 1 or ± 2, then $\ell > 2k > 0$ would divide either k or 2k, which is impossible. Thus (*) is proved.

Write $A'_i (i=1,2)$ for the 2-sided annulus $B_i - \mathring{A}_i \subset M$, where B_i is the component of ∂K_i containing A_i. We will say that a 2-sided arc a in an annulus A *traverses* A if the endpoints of a lie in different components of ∂A. The lemma is proved by distinguishing two cases, according to whether (I) each arc, which is a component of $A'_1 \cap A'_2$, traverses both A'_1 and A'_2, or (II) some component of $A'_1 \cap A'_2$ is an arc which does not simultaneously traverse A'_1 and A'_2.

Proof in Case I. By hypothesis, $\partial A_1 = \partial A'_1$ has non-empty and transversal intersection with $\partial A_2 = \partial A'_2$; in particular, ∂A_1 intersects A_2, and an arbitrary component of $(\partial A_1) \cap A_2$ is an arc β, 2-sided in A_2. The endpoints of β lie in $\partial A_1 \cap \partial A_2 = \partial A'_1 \cap \partial A_2$. The components a, a^* of $A'_1 \cap A'_2$ containing these endpoints must be arcs by transversality (A'_1 and A'_2 are components of Fr K_1 and Fr K_2). Since we are in case I, a and a^* must traverse A'_1. It follows that $a \cup a^*$ is the frontier in A'_1 of a disc ["rectangle"] $R \subset A'_1$, such that $\beta \subset R$. The boundary of R consists of a, β, a^* and another arc $\beta^* \subset \partial A'_1 = \partial A_1$.

We claim that $\partial R \subset K_2$. In fact, by construction we know that $a \cup a^* \subset A'_2$ and $\beta \subset A_2$. Hence, by transversality, K_2 contains some neighborhood of $a \cup \beta \cup a^*$ in R. Thus if β^* were not contained in $\mathring{A}_2 = K_2 \cap \mathring{\partial M}$, β^* would intersect $\partial A_2 = \partial A'_2$; and for any $x \in \beta^* \cap \partial A'_2$, the component a_0 of $A'_1 \cap A'_2$ containing x would be an arc, again by transversality. But a_0 would be contained in R, since it could not intersect a or a^*. Furthermore, since we are in case I, a_0 would traverse A'_1. Hence a_0 would have an endpoint in $\mathring{\beta}$. This is impossible; for since β is 2-sided in A_2, we have $\mathring{\beta} \cap \partial A'_2 = \emptyset$. Thus the claim is proved.

By transversality, the component J of $A'_1 \cap K_2$ containing ∂R is a 2-sided surface in K_2, and $J \subset R$. Since $A'_1 \supset J$ is 2-sided in M we have $J \cap A_2 \subset J \cap \partial M \subset \partial R$. This shows that the intersection number $\gamma_2 \cdot J$ in K_2 is numerically equal to the intersection number $\gamma_2 \cdot \partial R$ in B_2 (both are defined up to sign).

We can now show that β and β^* do not *both* traverse A_2. To do this, consider the simple curve $\partial R \subset B_2$. Here B_2 is the union of the annuli A_2 and A'_2, whose intersection is their common boundary; and ∂R consists of the disjoint arcs α and α^*, which are 2-sided in A'_2, and the disjoint arcs β and β^*, which are 2-sided in A_2. Furthermore, since we are in case I, α and α^* both traverse A'_2. If, in addition, both β and β^* were to traverse A_2, ∂R would have intersection number ± 2 with γ_2, since γ_2 represents a generator of $H_1(A_2)$. By the last paragraph we would have $\gamma_2 \cdot J = \pm 2$ in K_2, contradicting the above observation (*).

Let b_1 be one of the arcs β, β^*, chosen so as *not* to traverse A_2. Then b_1 is an arc contained in $A_2 \cap \partial A_1$ and 2-sided in A_2, and its endpoints lie in a single component C of ∂A_2. Hence there is an arc $b_2 \subset C$ such that $b_1 \cup b_2$ bounds a disc $\Delta \subset A_2$, and the lemma is proved in Case I.

Proof in case II. In this case we may assume, by symmetry, that some component of $A'_1 \cap A'_2$ is an arc α which does not traverse A'_1. Hence there is an arc $b_1 \subset \partial A'_1$ such that $\alpha \cup b_1$ bounds a disc ["hemi-disc"] $H_1 \subset A'_1$. We may suppose α to be chosen so that H_1 is minimal, i.e. contains no other arcs which are components of $A'_1 \cap A'_2$. Thus (by transversality) $\mathring{b}_1$ will contain no points of $\partial A'_1 \cap \partial A'_2 = \partial A_1 \cap \partial A_2$.

It will be shown presently that α cannot traverse A'_2. This will imply the lemma via the following argument. Since α does not traverse A'_2, there is an arc $b_2 \subset \partial A'_2$ such that $\alpha \cup b_2$ bounds a disc $H_2 \subset A'_2$. Now $b_1 \cup b_2 \subset T$ is a simple curve, since $\mathring{b}_1$ contains no points of $\partial A'_1 \cap \partial A'_2$. On the other hand, the existence of the discs H_1 and H_2 guarantees that b_1 and b_2 are each isotopic in M, with endpoints fixed, to α. Hence $b_1 \cup b_2$ is contractible in M; since T is incompressible, $b_1 \cup b_2$ must bound a disc $\Delta \subset T$, as required.

It remains to show that α cannot traverse A'_2. For this purpose, observe that since $\mathring{b}_1 \subset \partial A_1$ contains no points of $\partial A_1 \cap \partial A_2$, we have either (i) $b_1 \subset A_2$, or (ii) $b_1 \subset T - \mathring{A}_2$. We will assume that α traverses A'_2, and derive separate contradictions in the subcases (i) and (ii).

First suppose that (i) holds. Then α and b_1 are 2-sided arcs in A'_2 and A_2 respectively, and they have the same endpoints in $\partial A_2 = \partial A'_2$ (since $\alpha \cup b_1 = \partial H_1$). Hence if we assume that α traverses A'_2, it follows that b_1 traverses A_2, and the simple curve $\partial H_1 = \alpha \cup b_1$ has intersection number ± 1 with γ_2 in the torus $A_2 \cup A'_2 = B_2 \subset K_2$. Now the component J of $A'_1 \cap K_2$ that contains ∂H_1 is a 2-sided surface in K_2, by transversality, and $J \subset H_1$. Since $A'_1 \supset J$ is 2-sided in M, we have $J \cap A_2 \subset J \cap \partial M \subset \partial H_1$. It follows that J also has intersection number ± 1 with γ_2. But this contradicts (*) once again.

Finally, suppose that (ii) holds. Let L be a regular neighborhood of b_1 in $T - \mathring{A}_2$, such that $L \cap \partial A_2$ is a regular neighborhood of $b_1 \cap \partial A_2 = \partial b_1$ in ∂A_2. Then L is a disc and intersects A_2 in two arcs, which lie in different components of ∂A_2. Since T is contained in the boundary of the orientable manifold M, and is therefore orientable, it follows that $L \cup A_2$ is a disc with one handle.

On the other hand, let L^* be a regular neighborhood of α in A'_2. Since $\partial \alpha = \partial b_1$, we may choose L^* so that $L^* \cap \partial A_2 = L \cap \partial A_2$. The frontier of L (resp. L^*) in $T - \mathring{A}_2$ (resp. A'_2) consists of two arcs ρ_1, ρ_2 (resp. ρ_1^*, ρ_2^*); we may index these arcs so that $\partial \rho_i = \partial \rho_i^* (i = 1,2)$. Now the existence of the disc H_1 guarantees that α is isotopic to b_1 in M, with endpoints fixed. It follows that ρ_i is isotopic to ρ_i^* in M with endpoints fixed, for $i = 1, 2$.

Assume that α traverses A'_2. Then $\overline{A'_2 - L^*}$ is a disc, whose boundary consists of ρ_1^*, ρ_2^*, and two other arcs $\tau_1, \tau_2 \subset \partial A'_2$. It is clear that the boundary of the disc-with-handle $L \cup A_2$ is precisely $\rho_1 \cup \tau_1 \cup \rho_2 \cup \tau_2$. Since ρ_i is isotopic to $\rho_i^*, \partial(L \cup A_2)$ is isotopic to the boundary of the disc $\overline{A'_2 - L^*}$, and is therefore contractible in M.

But T is incompressible, and so $\partial(L \cup A_2)$ must bound a disc in T. Thus T contains a disc and a disc with one handle having the same boundary. This is impossible, since we observed at the beginning of this argument that T cannot be a torus. This contradiction completes the proof.

COROLLARY. *Assume that* T *is not a torus. Let* $\sigma_1, \cdots, \sigma_n$ *be enveloped singular curves. Then the* σ_i *are homotopic to powers of disjoint simple curves.*

Proof. It is enough to show that the σ_i are homotopic to enveloped curves σ'_i, which have normal envelopes K_i $(1 \leq i \leq n)$ such that the sets $\partial(K_i \cap \partial M)$ are disjoint. For then, since T is not a torus, the normality of the K_i will imply that the sets $K_i \cap \partial M$ are annuli; and since σ'_i lies in $K_i \cap \partial M$, σ'_i will be homotopic to a power of either component of $\partial(K_i \cap \partial M)$.

Inductively we may assume that $\sigma_1, \cdots, \sigma_{n-1}$ already have normal envelopes $K_1, \cdots, K_{n-1}$ such that $\partial(K_1 \cap \partial M), \cdots, \partial(K_{n-1} \cap \partial M)$ are disjoint. Now by Lemma 3, σ_n has a normal envelope K_n. By taking K_n in general position we may suppose that $\partial(K_n \cap \partial M)$ intersects $\partial(K_1 \cap \partial M) \cup \cdots \cup \partial(K_{n-1} \cap \partial M)$ transversally. If the number ν of points in the latter intersection is > 0, we will show how to homotop σ_n to a curve σ_n^*, which has an envelope K_n^* such that $\partial(K_n^* \cap \partial M)$ intersects $\partial(K_1 \cap \partial M) \cup \cdots \cup \partial(K_{n-1} \cap \partial M)$ transversely in fewer than ν points. By induction on ν, this will prove the corollary.

Since $\nu > 0$, Lemma 4 gives a disc $\Delta \subset \mathring{T}$ with $\partial\Delta = b_j \cup b_n$, where $b_j \subset \partial(K_j \cap \partial M)$ for some $j < n$, $b_n \subset \partial(K_n \cap \partial M)$, and $b_j \cup b_n = \partial b_j = \partial b_n$. Among all such discs let Δ be taken to be minimal with respect to inclusion. Then Δ is disjoint from $\partial(K_1 \cap \partial M) \cup \cdots \cup \partial(K_n \cap \partial M)$.

Hence if U is a small neighborhood of Δ in T, there is a homeomorphism $J : T \to T$, isotopic to the identity rel $(T - U)$, such that

$J(\partial(K_n \cap \partial M))$ has exactly $\nu-2$ intersections with $\partial(K_1 \cap \partial M) \cup \cdots \cup \partial(K_{n-1} \cap \partial M)$, all transversal. Extend J to a PL homeomorphism $\bar{J}: M \to M$, and set $\sigma_n^* = \bar{J} \circ \sigma_n$, $K_n^* = \bar{J}(K_n)$.

§4. *A tower*

We use a "tower" argument – following ideas of Papakyriakopoulos ([12]), as refined by Shapiro, Whitehead and Stallings ([14], [16]) – to produce a converse to Corollary 1 to Lemma 3 of Section 3. Combined with the corollary to Lemma 4, Section 3, this will prove Proposition 2, which was stated in Section 2.

It will be useful in what follows to distinguish between *simplicial complexes* and *polyhedra*: by a (finite-dimensional) polyhedron we understand a subset of a Euclidean space which is the underlying set $|L|$ of some (locally finite, geometric) simplicial complex L. Similarly, we distinguish between *simplicial* maps and *piecewise-linear* (PL) maps: a map $f: P \to P'$ of polyhedra is PL if there are simplicial complexes L, L', with $|L| = P$, $|L'| = P'$, such that f "is" a simplicial map from L to L'.

DEFINITION. Let L and L' be finite simplicial complexes, and let $\phi: L \to L'$ be a simplicial map. The *complexity of* ϕ is the number of unordered pairs $\{v, w\}$ of vertices of L such that $\phi(v) = \phi(w)$.

Whereas:

DEFINITION. Let P, Q be polyhedra, and suppose that P is compact. The *complexity* of a PL map $f: P \to Q$ is the smallest integer ν for which there exist simplicial complexes L, L' with $|L| = P$, $|L'| = f(P) \subset Q$, such that f "is" a simplicial map of complexity ν from L to L'. We will write $\nu(f)$ for the complexity of f.

The significance of this notion of complexity (a measure of the failure of a map to be 1–1) lies in the following lemma, essentially due to Stallings. Recall that a lifting of a PL map to a PL covering spaces is PL. A covering space is *trivial* if the covering projection is a homeomorphism.

LEMMA 5. *Let* $f : P \to Q$ *be a* PL *map of polyhedra, where* P *is compact, and suppose that* $\pi_1(f(P)) \to \pi_1(Q)$ *is surjective. If* f *has a lifting* $\tilde{f}$ *to a given non-trivial* PL *covering space* $\tilde{Q}$ *of* Q, *then* $\nu(\tilde{f}) < \nu(f)$.

Proof. Let L and L' be simplicial complexes such that $|L| = P$, $|L'| = f(P)$, and $f : L \to L'$ is a simplicial map of complexity $\nu(f)$. Let $p : \tilde{Q} \to Q$ be the covering projection, and let $\widetilde{f(P)}$ denote the component of $p^{-1}(f(P))$ that contains $\tilde{f}(P)$. Then $\widetilde{f(P)}$ is a covering space of $f(P)$. Hence $\widetilde{f(P)}$ can be identified (piecewise-linearly) with $|\tilde{L}'|$, where $\tilde{L}'$ is a simplicial complex, in such a way that $p|\widetilde{f(P)}$ is a simplicial map from $\tilde{L}'$ to L'; and the lifting $\tilde{f} : L \to \tilde{L}'$ is automatically simplicial.

To prove the lemma, it suffices to show that this simplicial map $\tilde{f} : L \to \tilde{L}'$ has complexity less than $\nu(f)$, which is the complexity of the simplicial map $f : L \to L'$. Since $f = \tilde{f} \circ p$, any two vertices of L which have the same image under $\tilde{f}$ also have the same image under f; so $\tilde{f} : L \to \tilde{L}'$ has complexity $\leq \nu(f)$. If equality held, then $p|\tilde{f}(P)$ would be 1–1 and would therefore map $\pi_1(\tilde{f}(P))$ isomorphically onto $\pi_1(f(P))$. Since by hypothesis $\pi_1(f(P)) \to \pi_1(Q)$ is surjective, it would follow that $\pi_1(\tilde{Q}) \to \pi_1(Q)$ were surjective, contradicting our hypothesis that $\tilde{Q}$ is connected and non-trivial.

It is assumed for the remainder of this section that (M, T) is an acceptable pair. Note that if $\tilde{M}$ is a finite covering space of M and $\tilde{T}$ is a component of the induced covering space of T, then $(\tilde{M}, \tilde{T})$ is again an acceptable pair.

LEMMA 6. *Let* σ *be a non-contractible singular curve in* T. *Let* $\tilde{M}$ *be a 2-sheeted covering space of* M, *and let* $\tilde{T}$ *be a component of the induced covering space of* T. *Assume that* σ *has a lifting* $\tilde{\sigma}$ *to* $\tilde{T}$, *and that* $[\tilde{\sigma}] \subset \pi_1(\tilde{T})$ *has the same divisibility as* $[\sigma] \subset \pi_1(T)$. *Finally, assume that* $\tilde{\sigma}$ *is enveloped. Then* σ *is homotopic in* T *to an enveloped singular curve.*

Proof. Let $\tau : \tilde{M} \to \tilde{M}$ denote the (non-identical) covering transformation. Then $\tau^2 = 1$, but τ has no fixed points. Let $p : \tilde{M} \to M$ denote the covering projection.

By Lemma 3 of Section 3, $\tilde{\sigma}$ has a normal envelope $\tilde{K}$. Set $\tilde{A} = \tilde{K} \cap \partial\tilde{M} \subset \tilde{T}$. By taking $\tilde{K}$ in general position we may assume that $\partial\tilde{A}$ and $\partial(\tau\tilde{A})$ intersect transversally. (Of course this condition, and the following seven paragraphs, are vacuous in the case that $\tilde{A}$ is a torus.)

We claim that if $\partial\tilde{A} \cap \partial(\tau\tilde{A}) \neq \emptyset$, then there is a disc $\Delta \subset \tilde{T}$ such that (i) $\partial\Delta$ is a union of two arcs $b_1 \subset \partial\tilde{A}$ and $b_2 \subset \partial(\tau\tilde{A})$, (ii) $\mathring{\Delta}$ is disjoint from $\partial\tilde{A}$ and from $\partial(\tau\tilde{A})$, and (iii) $\Delta \cap \tau\Delta = \emptyset$.

To prove this, first apply Lemma 4 of Section 3, placing tildes on M and T, and taking $\sigma_1 = \tilde{\sigma}$, $\sigma_2 = \tau\tilde{\sigma}$, $K_1 = \tilde{K}$, $K_2 = \tau\tilde{K}$. This shows that there is a disc Δ satisfying (i). If Δ is taken to be minimal among all discs satisfying (i), then by transversality [cf. proof of Corollary to Lemma 4, Section 3] it will satisfy (ii) as well.

The proof of the claim will be completed by showing that (iii) follows from (i) and (ii).

If (iii) does not hold, then either $\tau\mathring{\Delta}$ or $\partial(\tau\Delta)$ intersects Δ. In the first case, since $\mathring{\Delta}$ is a component of the set $\partial\tilde{M} - (\partial\tilde{A} \cup \partial(\tau\tilde{A}))$, which is invariant under τ, we must have $\tau\mathring{\Delta} = \mathring{\Delta}$. Hence $\tau\Delta = \Delta$, and the Brouwer fixed-point theorem implies that τ has a fixed point. This is a contradiction.

Now suppose that $\partial(\tau\Delta)$ intersects Δ. Then either τb_1 or τb_2 intersects Δ. Suppose for example that $\tau b_1 \cap \Delta \neq \emptyset$. Since b_1 is the closure of a component of $\partial\tilde{A} - \partial(\tau\tilde{A})$, τb_1 is the closure of a component of $\partial(\tau\tilde{A}) - \partial\tilde{A}$. The only sets which intersect Δ and which may be components of $\partial(\tau\tilde{A}) - \partial\tilde{A}$ are $\mathring{b}_2$ and $\gamma - b_2$, where γ is the component of $\partial(\tau\tilde{A})$ containing b_2. If $\tau b_1 = b_2$, then the simple curve $b_1 \cup b_2$ is invariant under τ. But Δ is the only disc in $\partial\tilde{M}$ whose boundary is $b_1 \cup b_2$; for otherwise the component of $\partial\tilde{M}$ containing $\tilde{\sigma}$ would be a 2-sphere, and $\tilde{\sigma}$ would be contractible. Hence we again conclude $\tau\Delta = \Delta$, and again we have a contradiction.

Finally, suppose that $\tau b_1 = \gamma - \mathring{b}_2$, where γ is a component of $\partial(\tau\tilde{A})$. Note that since $\tau \circ \tilde{\sigma}$ is a non-contractible singular curve in the annulus $\tau\tilde{A}$, it is homotopic in $\tau\tilde{A}$ to a k-th power $(k > 0)$ of some parametrization γ_0 of γ. Since $[\gamma_0] \subset \pi_1(\tilde{T})$ is primary by Lemma 2 of Section 1, k is the divisibility of $[\tilde{\sigma}]$ in $\pi_1(\tilde{T})$. On the other hand, since $b_1 \cup b_2$ bounds a disc $\Delta \subset T$, γ is isotopic to the simple curve $b_1 \cup (\gamma - \mathring{b}_2)$; and the latter is invariant under τ, since $\tau b_1 = \gamma - \mathring{b}_2$. Hence $[p \circ \gamma_0] \subset \pi_1(T)$ is divisible by 2, and $[p \circ \tilde{\sigma}] = [\sigma] \subset \pi_1(T)$ is divisible by 2k. But by hypothesis, $[\sigma]$ has the same divisibility in $\pi_1(T)$ as $[\tilde{\sigma}]$ in $\pi_1(\tilde{T})$, namely k. But this contradicts Corollary 1 to Lemma 1 of Section 1.

From the claim just proved we can deduce that if $\partial\tilde{A} \cap \partial(\tau\tilde{A}) \neq \emptyset$, then there is a singular curve $\tilde{\sigma}'$ homotopic to $\tilde{\sigma}$ on $\tilde{T}$, and a normal envelope $\tilde{K}'$ for $\tilde{\sigma}'$, such that if we set $\tilde{A}' = \tilde{K}' \cap \partial M, \partial\tilde{A}'$ and $\tau(\partial\tilde{A}')$ intersect transversally and in fewer points than $\partial\tilde{A}$ and $\tau(\partial\tilde{A})$. In fact, the claim implies that there is a PL homeomorphism $\mathfrak{h}: \tilde{T} \to \tilde{T}$, isotopic to the identity rel $\partial\tilde{T}$, such that $\mathfrak{h}(\partial\tilde{A}) \cap \mathfrak{h}(\partial(\tau\tilde{A}))$ contains four points fewer than $\partial\tilde{A} \cap \partial(\tau\tilde{A})$. We can extend $\mathfrak{h}$ to a PL homeomorphism $\bar{\mathfrak{h}}: \tilde{M} \to \tilde{M}$ which is PL isotopic to the identity. Then $\tilde{\sigma}' = \mathfrak{h} \circ \sigma$ and $\tilde{K}' = \bar{\mathfrak{h}}(\tilde{K})$ are the required curve and envelope.

We may therefore assume that

$$\partial\tilde{A} \cap \partial(\tau\tilde{A}) = \emptyset \ . \tag{1}$$

By taking $\tilde{K}$ in general position, we may further assume that

$$\mathrm{Fr}\,\tilde{K} \text{ and } \mathrm{Fr}(\tau\tilde{K}) \text{ intersect transversally .} \tag{2}$$

We now claim that if some component γ of $(\mathrm{Fr}\,\tilde{K}) \cap \mathrm{Fr}(\tau\tilde{K})$ bounds a disc in $\mathrm{Fr}\,\tilde{K}$ *or* in $\mathrm{Fr}(\tau\tilde{K})$, then there is a homotopy 3-cell $E \subset \tilde{M}$ such that (i′) ∂E is a union of two discs $D_1 \subset \mathrm{Fr}\,\tilde{K}$ and $D_2 \subset \mathrm{Fr}(\tau\tilde{K})$, (ii′) $\mathring{D}_1$ is disjoint from $\mathrm{Fr}(\tau\tilde{K})$, and (iii′) $(\partial D_1) \cap \tau(\partial D_1) = \emptyset$.

To prove this, recall that since the envelope $\tilde{K}$ is normal, $\mathrm{Fr}\,\tilde{K}$ is an incompressible 2-manifold. Now if $\gamma \subset \mathrm{Fr}\,\tilde{K} \cap \mathrm{Fr}(\tau\tilde{K})$ bounds a disc $D_1 \subset \mathrm{Fr}\,\tilde{K}$, say, we may take D_1 to be minimal, so that $\overset{\circ}{D}_1$ contains no components of $\mathrm{Fr}\,\tilde{K} \cap \mathrm{Fr}(\tau\tilde{K})$. By incompressibility, γ also bounds a disc $D_2 \subset \mathrm{Fr}(\tau\tilde{K})$; and the minimality of D_1 implies that $D_1 \cup D_2$ is a 2-sphere. By Principle 2 of Section 0, $D_1 \cup D_2$ bounds a homotopy 3-cell $E \subset \tilde{M}$, which therefore satisfies (i′). The minimality of D_1 implies (ii′).

The proof of the claim will be completed by showing that (iii′) follows from (i′) and (ii′). Note that ∂D_1 and $\tau(\partial D_1)$ are components of $\mathrm{Fr}\,K \cap \mathrm{Fr}(\tau K)$. Hence if (iii′) does not hold, we must have $\tau(\partial D_1) = \partial D_1$. Now D_1 is the unique disc contained in $\mathrm{Fr}\,K$ and bounded by ∂D_1; and D_2 is the unique disc contained in $\mathrm{Fr}\,\tau K$ and bounded by $\partial D_2 = \partial D_1 = \tau(\partial D_1)$. It follows that $\tau D_1 = D_2$, and hence that $D_1 \cup D_2 = \partial E$ is invariant under τ. But E is the only homotopy 3-cell in $\tilde{M}$ bounded by ∂E, and so $\tau E = E$. By a familiar application of the Lefschetz fixed point theorem, τ must then have a fixed point, which is impossible. Thus (iii′) is established.

From the last claim we can deduce that if some component of $\mathrm{Fr}\,\tilde{K} \cap \mathrm{Fr}\,\tau\tilde{K}$ bounds a disc in $\mathrm{Fr}\,\tilde{K}$ or in $\mathrm{Fr}\,\tau\tilde{K}$, then there is an envelope $\tilde{K}$ for $\tilde{\sigma}$ such that $\tilde{K}' \cap \partial M = \tilde{A}$, but such that $\mathrm{Fr}\,\tilde{K}'$ and $\mathrm{Fr}\,\tau\tilde{K}'$ intersect transversally in fewer components than $\mathrm{Fr}\,\tilde{K}$ and $\mathrm{Fr}\,\tau\tilde{K}$. In fact, if E is the homotopy 3-cell given by the claim, let P be a small regular neighborhood of E such that $(\mathrm{Fr}\,\tilde{K}) \cap P$ and $\mathrm{Fr}(\tau\tilde{K}) \cap P$ are discs $D_1^* \supset D_1$ and $D_2^* \supset D_2$. Let $D_1^{*\prime} \subset \partial P$ be a disc which is disjoint from D_2^* and which has the same boundary as D_1^*. Then $D_1^* \cup D_1^{*\prime}$ bounds a homotopy 3-cell $E^* \subset P$. Either $\overset{\circ}{E}{}^* \cap \tilde{K} = \emptyset$ or $E^* \subset \tilde{K}$; define $\tilde{K}'$ to be, respectively, $\tilde{K} \cup E^*$ or $\overline{\tilde{K} - E^*}$. Clearly $\tilde{K}'$ is a 3-manifold and $\mathrm{Fr}\,\tilde{K}' = (\mathrm{Fr}\,\tilde{K} - D_1^*) \cup D_1^{*\prime}$. In view of (ii′) and (iii′), this shows that if P is a small enough neighborhood of E then $\mathrm{Fr}\,\tilde{K}' \cap \mathrm{Fr}(\tau\tilde{K}')$ has fewer components than $\mathrm{Fr}\,\tilde{K} \cap \mathrm{Fr}(\tau\tilde{K})$ – the components ∂D_1 and $\partial(\tau D_1)$

having been removed. On the other hand it is clear that $\partial\tilde{K}' = (\partial\tilde{K} - D_1^*) \cup D_1^{*\prime}$, so that $\partial\tilde{K}'$ is homeomorphic to $\partial\tilde{K}$, and therefore consists of tori. To show that $\tilde{K}'$ is an envelope for $\tilde{\sigma}$, it remains to show that $[\tilde{\sigma}]$ is special with respect to $(\tilde{K}', \tilde{A})$. This is obvious if $\tilde{K} \subset \tilde{K}'$. The other possibility is that $\tilde{K} = \tilde{K}' \cap E^*$, where $\tilde{K}' \cap E^* = D_1^{*\prime}$. But then, since E^* is a homotopy 3-cell, $\pi_1(\tilde{K}') \to \pi_1(\tilde{K})$ is an isomorphism by van Kampen's theorem, and it follows that $\tilde{\sigma}$ is special in $\tilde{K}'$.

We may therefore assume that

(3) No component of $\mathrm{Fr}\,\tilde{K} \cap \mathrm{Fr}(\tau\tilde{K})$ bounds a disc in $\mathrm{Fr}\,\tilde{K}$ or in $\mathrm{Fr}\,\tau\tilde{K}$.

From (1), (2), and (3), it follows that every component of $(\mathrm{Fr}\,\tilde{K}) \cap \mathrm{Fr}(\tau\tilde{K})$ is a simple curve, non-contractible both in $\mathrm{Fr}\,\tilde{K}$ and in $\mathrm{Fr}(\tau\tilde{K})$. Since each component of $\mathrm{Fr}\,\tilde{K}$ or $\mathrm{Fr}(\tau\tilde{K})$ is an annulus or a torus, it now follows that the closures of the components of $\mathrm{Fr}\,\tilde{K} - \mathrm{Fr}(\tau\tilde{K})$ and $\mathrm{Fr}(\tau\tilde{K}) - \mathrm{Fr}\,\tilde{K}$ are all annuli and tori. But by (2), $\tilde{K} \cup \tau\tilde{K}$ is a 3-manifold; and we have shown that its boundary is a union of annuli and tori, meeting only pairwise and only in components of their own boundaries. Hence each component of $\partial(\tilde{K} \cup \tau\tilde{K})$ has Euler characteristic zero.

Set $K = p(\tilde{K} \cup \tau\tilde{K})$. Then K is covered by $\tilde{K} \cup \tau\tilde{K}$; hence it is a 3-manifold whose boundary components all have Euler characteristic zero. Since $K \subset M$ must be orientable, the components of ∂K must be tori. Finally, by hypothesis, σ and $\tilde{\sigma}$ have the same divisibility k; and since $\tilde{K}'$ is an envelope, $[\tilde{\sigma}] \subset \pi_1(\tilde{K})$ is divisible by some integer $\ell > 2k$. Hence $[\sigma] \subset \pi_1(K)$ is divisible by ℓ, and is therefore special with respect to (K, A). It follows that K is an envelope for σ, and the lemma is proved.

In the proof of the next lemma, which is the crucial result of this section, we use a space constructed as follows. Let A be an annulus, let b and b' be the components of ∂A, and let S and S' be 1-spheres. In the disjoint union $S' \cup A \cup S$ make the identifications $x \sim j(x)\,(x \in b)$,

$x \sim j'(x')(x' \in b')$, where $j: b \to S$ and $j': b' \to S'$ are covering maps of degrees k and ℓ respectively. The resulting space, which we will denote by P_k^ℓ, can be identified homeomorphically with a polyhedron in such a way that the 1-spheres S and S' are subpolyhedra. Let S and S' be identified with the standard S^1 via orientations that are compatible in the obvious sense. It is then clear that if σ and σ' are singular curves in a polyhedron Q, such that $[\sigma]^k = [\sigma']^\ell \subset \pi_1(Q)$, then there is a map $f: P_k^\ell \to Q$ such that $f|S = \sigma$, $f|S' = \sigma'$; and conversely, that if such an f exists then $[\sigma]^k = [\sigma']^\ell$.

LEMMA 7. *Every special conjugacy class in* $\pi_1(T)$ *is represented by an enveloped curve.*

Proof. Let $c(x)$ be special in $\pi_1(T)$ and let k denote its divisibility in $\pi_1(T)$. Then $c(x)$ is divisible in $\pi_1(M)$ by an integer $\ell > 2k$. Set $c(x) = c(y)^k$, where $c(y)$ is primary in $\pi_1(T)$, and $c(x) = c(u)^\ell$, $c(u) \subset \pi_1(M)$. Then by the above discussion there is a PL map $f: P_k^\ell \to M$ such that $f|S$ represents $c(y)$ in $\pi_1(T)$ and $f|S'$ represents $c(u)$ in $\pi_1(M)$. We may suppose f to be chosen so that $f^{-1}(\partial M) = S$. We will prove by induction on the complexity $\nu(f)$ that $c(x)$ is represented by an enveloped curve.

We can always find a neighborhood N of $f(P_k^\ell)$ in M, and a neighborhood U of $f(S)$ in T, such that $U \subset N$, (N, U) is an acceptable pair, and $\pi_1(f(P_k^\ell)) \to \pi_1(N)$ is surjective. To see this, let U_0 be a regular neighborhood of $f(S)$ in T. If $\pi_1(U_0) \to \pi_1(T)$ is not injective, there is a disc $D \subset T$ such that $D \cap U_0 = \partial D$; then $U_0 \cup D$ is a 2-manifold with fewer boundary components than U_0. Hence by repeating this process a finite number of times, we obtain a surface $U \subset T$ such that $U_0 \subset U$ and $\pi_1(U) \to \pi_1(T)$ is injective. It follows that $\pi_1(U) \to \pi_1(M)$ is injective. It is clear from the construction that $\pi_1(f(P_k^\ell)) \to \pi_1(f(P_k^\ell) \cup U)$ is surjective. Now if N_0 is a regular neighborhood of

$f(P_k^\ell) \cup U$ in M, the pair (N_0, U) has all the properties required of (N, U) except that $\pi_2(N_0)$ may be non-zero.

But if $\pi_2(N_0) \neq 0$, then by Principle 2 of Section 0 there is a 2-sphere $\Sigma \subset N_0$ which does not bound a homotopy 3-cell in N_0. Since $\pi_2(M) = 0$, Principle 2 implies that Σ bounds a simply-connected 3-manifold $B \subset M$. Now $(N_0 \cup B, U)$ still has all the properties required of (N, U), except that $\pi_2(N_0 \cup B)$ may still be non-zero; but $N_0 \cup B$ has fewer boundary components than N_0, since $\mathring{B}$ must contain a component of ∂N_0. Hence it is again sufficient to repeat the process a finite number of times.

Now it is immediate from the definition of (PL) complexity that f still has complexity $\nu(f)$ if it is regarded as a map of P_k^ℓ into N. Hence in doing the induction step we may replace (M, T) by (N, U); i.e. we may assume that $\pi_1(f(P_k^\ell)) \to \pi_1(M)$ is surjective.

If $H_1(M; \mathbf{Q})$ has rank ≤ 1, then ∂M has total genus ≤ 1. If ∂M contains a 2-sphere, it follows from Principle 2 of Section 0 that M is a homotopy 3-cell; this is impossible since ∂M contains the non-contractible singular curve $f|S$. The boundary of M is therefore exactly a torus.

Hence if $H_1(M; \mathbf{Q})$ has rank ≤ 1, M is itself an envelope for σ, and the lemma is therefore true in this case.

Now suppose that $H_1(M; \mathbf{Q})$ has rank > 1. Note that $H_1(P_k^\ell; \mathbf{Q})$ has rank 1, since $\pi_1(P_k^\ell)$ has a presentation $\langle a, b : a^k = b^\ell \rangle$. Hence $f_* : H_1(P_k^\ell; \mathbf{Q}) \to H_1(M; \mathbf{Q})$ cannot be surjective; thus $H_1(M; \mathbf{Z})/\mathrm{im}(f_* : H_1(P_k^\ell; \mathbf{Z}) \to H_1(M; \mathbf{Z}))$ is infinite, and therefore admits a homomorphism onto a group of order two. It follows that $\pi_1(M; \mathbf{Z})$ has a subgroup H of index 2 which contains the image of $f_\# : \pi_1(P_k^\ell) \to \pi_1(M)$ (basepoints being irrelevant since H is necessarily normal). This means that f lifts to a map $\tilde{f} : P_k^\ell \to \tilde{M}$, where $\tilde{M}$ is some 2-sheeted covering of M.

Let $\tilde{T}$ denote the component of the induced covering space of T which contains $\tilde{f}(S)$. Let $c(\tilde{y})$ denote the conjugacy class in $\pi_1(T)$ determined by $\tilde{f}|S$, and $c(u)$ the class in $\pi_1(\tilde{M})$ determined by $\tilde{f}|S'$. Set

$c(\tilde{x}) = c(\tilde{y})^k$. Then $c(\tilde{y})$ is primary, since $c(y)$ is, and $c(\tilde{x})$ therefore has divisibility k in $\pi_1(\tilde{T})$. But the existence of the map $\tilde{f}$ on P_k^ℓ shows that $c(\tilde{x}) \subset c(\tilde{u})^\ell$. In particular $c(\tilde{x})$ is special, so that the hypotheses of the lemma are satisfied by $(\tilde{M}, \tilde{T})$ and $c(\tilde{x})$; in this context $\tilde{f}$ obviously has the property required of f above. But since we have assumed that $\pi_1(f(P_k^\ell)) \to \pi_1(M)$ is surjective, Lemma 5 implies that $\nu(\tilde{f}) < \nu(f)$. By the induction hypothesis, therefore, $\tilde{x}$ is represented by an enveloped curve $\tilde{\sigma}$ in $\tilde{T}$. Since $c(\tilde{x})$ and $c(x)$ both have divisibility k, Lemma 6 now shows that the projection of $\tilde{\sigma}$ in T, which represents $c(x)$, is homotopic to an enveloped curve. This completes the induction.

We can at last give the

Proof of Proposition 2. Statement (a) follows from Lemma 7 above and Corollary 2 to Lemma 3 of Section 3. Statement (b) follows from Lemma 7 and the corollary to Lemma 4 of Section 3.

§5. *Boundary tori*

We must deal separately with acceptable pairs (M, T) for which T is a torus; Proposition 2 gives no useful information in this case.

Note that since a torus T has an abelian fundamental group, it is natural to speak of *elements* of $\pi_1(T)$ where until now we have spoken of conjugacy classes.

DEFINITION. Let (M, T) be an acceptable pair. A non-contractible singular curve σ in T is called *distinguished* (relative to (M, T)) if there is a singular curve in an incompressible component of ∂M which is homotopic to σ in M, but is neither homotopic nor anti-homotopic (Section 0) to σ in ∂M.

DEFINITION. An oriented 3-manifold M is called *exceptional* if M is compact, and if each component of ∂M is a torus T such that $\operatorname{im}(\pi_1(T) \to \pi_1(M))$ has index ≤ 2 in $\pi_1(M)$.

REMARK. It may be shown, using the Stallings fibration theorem, than an exceptional oriented 3-manifold which is irreducible (cf. Section 6) is a regular neighborhood of a 1-sided Klein bottle or a 2-sided torus. This fact will not be needed.

PROPOSITION 3. *Let* (M, T) *be an acceptable pair such that* T *is a torus but* M *is not exceptional. Then any two distinguished singular curves in* T *which have the same divisibility in* T *are either homotopic or anti-homotopic (Section 0) in* T.

Proof. Let σ_1 and σ_2 be distinguished and let each have divisibility k. Let ξ_i be a singular curve in T $(i=1,2)$ such that $[\xi_i] \in \pi_1(T)$ is primary and $[\xi_i]^k = [\sigma_i]$; we may assume that ξ_i is a parametrized simple curve, for every primary element of $\pi_1(S^1 \times S^1)$ is represented by such a curve.

It is enough to show that the simple curves $\xi_1(S^1)$ and $\xi_2(S^1)$ are isotopic to disjoint simple curves; for then ξ_1 and ξ_2 are either homotopic or anti-homotopic, and hence so are σ_1 and σ_2. We suppose $\xi_1(S^1)$ and $\xi_2(S^1)$ to intersect each other transversally, and to have been chosen within their isotopy classes so as to minimize the number of points in their intersection. Under these conditions we will show that $\xi_1(S^1) \cap \xi_2(S^1) = \emptyset$.

Note that

(*) there is no disc $\Delta \subset T$ whose boundary has the form $a_1 \cup a_2$, where $a_i \subset \xi_i(S^1)$ is an arc and $a_1 \cap a_2 = \partial a_1 = \partial a_2$.

For if such a Δ existed we could take it to be minimal with respect to inclusion; and $\xi_1(S^1)$ would then be isotopic, under an ambient isotopy constant outside a small neighborhood of Δ, to a curve which would intersect $\xi_2(S^1)$ transversally in a smaller number of points.

We will study ξ_1 and ξ_2 by lifting them to an appropriate covering space. Fix a basepoint $x \in T$, and let $\tilde{M}$ be the covering space of M determined by the subgroup $\operatorname{im}(\pi_1(T,x) \to \pi_1(M,x))$ of $\pi_1(M,x)$. Let $\tilde{x}$ be the canonical basepoint of $\tilde{M}$, and let $p: \tilde{M} \to M$ be the projection. Then since $\pi_1(T) \to \pi_1(M)$ is injective, $\pi_1(\tilde{T}) \to \pi_1(\tilde{M})$ is an isomorphism, where $\tilde{T}$ is the component of $p^{-1}(T)$ containing $\tilde{x}$. On the other hand, $\pi_2(\tilde{M}) \approx \pi_2(M) = 0$, since (M, T) is acceptable. Since the 3-manifold $\tilde{M}$ with non-empty boundary is necessarily without homology in dimensions > 2, the Hurewicz theorem now implies that $\pi_i(\tilde{M}) = 0$ for all $i > 1$. We can conclude that $\tilde{T} \hookrightarrow \tilde{M}$ is a homotopy equivalence; this follows, for example, from Whitehead's theorem ([15], p. 405) that a map between connected polyhedra is a homotopy equivalence if it induces isomorphisms of homotopy groups in all dimensions.

It is clear from the construction that $\tilde{T}$ is a degree-one covering of T. We claim that no component $\tilde{B} \neq \tilde{T}$ of $\partial\tilde{M}$ can be a torus. First of all, since $\tilde{T} \hookrightarrow \tilde{M}$ is a homotopy equivalence, the generator of $H_2(\tilde{T};\mathbf{Z}_2)$ maps onto a generator of $H_2(\tilde{M};\mathbf{Z}_2)$; hence if $\tilde{B}$ is a torus, a generator of $H_2(\tilde{B};\mathbf{Z}_2)$ must either map to zero in $H_2(\tilde{M};\mathbf{Z}_2)$, or else have the same image as the generator of $H_2(\tilde{T};\mathbf{Z}_2)$. Thus either $\tilde{B}$ or $\tilde{T} \cup \tilde{B}$ bounds a compact 3-manifold, which by connectedness must be all of $\tilde{M}$. But $\tilde{B}$ cannot bound $\tilde{M}$, since $\tilde{T} \subset \partial\tilde{M}$. Hence $\tilde{M}$ is compact and $\partial\tilde{M} = \tilde{T} \cup \tilde{B}$. It follows that $p^{-1}(T)$ is either $\tilde{T}$ or $\tilde{T} \cup \tilde{B}$. On the other hand, since $\tilde{T} \hookrightarrow \tilde{M}$ is a homotopy equivalence, the exact homology sequence of $(\tilde{M}, \tilde{T})$ shows that $H_i(\tilde{M}, \tilde{T}; \mathbf{Z}) = 0$ for all i. Now $\tilde{M}$ is orientable, since M is, and Poincaré-Lefschetz duality ([15], p. 298) shows that $H^i(\tilde{M}, \tilde{B}; \mathbf{Z}) = 0$ for all i. By the universal coefficient theorem, $H_i(\tilde{M}, \tilde{B}; \mathbf{Z}) = 0$ for all i. Again by the exact homology sequence, $H_1(\tilde{B}) \to H_1(\tilde{M})$ is an isomorphism; since $\pi_1(\tilde{B})$ and $\pi_1(\tilde{M})$ are abelian, this means that $\pi_1(\tilde{B}) \to \pi_1(\tilde{M})$ is an isomorphism. In the case that $p^{-1}(T) = \tilde{T} \cup \tilde{B}$, it follows that $p|\tilde{B}$ induces an isomorphism of $\pi_1(\tilde{B})$ onto $\pi_1(T)$; thus $\tilde{B}$, as a covering space of T, has degree one. This implies that $p^{-1}(T) = \tilde{T} \cup \tilde{B}$ is a degree-two covering space of T. On the other hand, in the case that $p^{-1}(T) = T$, $p^{-1}(T)$ is of course a degree-one covering of T. Hence the degree of $\tilde{M}$ as a covering space of M, which is equal to the degree of $p^{-1}(T)$ as a covering space of T, is at most two in any case.

Now let T_1 be *any* component of ∂M. Then T_1 is covered either by $\tilde{T}$ or by $\tilde{B}$ (possibly by both), with degree one. In particular T_1 is a torus. Moreover, since $\pi_1(\tilde{T})$ and $\pi_1(\tilde{B})$ are mapped isomorphically onto $\pi_1(\tilde{M})$ via inclusion, the subgroup $\operatorname{im}(\pi_1(T_1) \to \pi_1(M))$ of $\pi_1(M)$ (defined a priori up to conjugacy) corresponds to the covering space $\tilde{M}$ and hence has index at most two. This means that M is exceptional, a contradiction to the hypothesis. Thus the claim is proved.

Note, however, that any incompressible component B of $\partial\tilde{M}$ has abelian fundamental group since $\tilde{M}$ does. Since we have shown that $\tilde{B}$ is not a torus if $\tilde{B} \neq \tilde{T}$, it must be an open disc or an open annulus.

Now, since the singular curve σ_i is distinguished for $i = 1, 2$, there is a singular curve ρ_i in ∂M which is homotopic to σ_i in M, but not in ∂M. By the covering homotopy property for covering spaces, the unique lifting $\tilde{\sigma}_i$ of σ_i to $\tilde{T}$ is homotopic to some lifting $\tilde{\rho}_i$ of ρ_i to $\partial\tilde{M}$; but $\tilde{\sigma}_i$ and $\tilde{\rho}_i$ cannot be homotopic in $\partial\tilde{M}$. If $\tilde{\rho}_i$ were to lie in $\tilde{T}$, then since $\tilde{T} \hookrightarrow \tilde{M}$ is a homotopy equivalence, $\tilde{\rho}_i$ would be homotopic to $\tilde{\sigma}_i$ in $\tilde{T}$; hence $\tilde{\rho}_i$ must lie in a component $\tilde{B}_i \neq \tilde{T}$ of $\partial\tilde{M}$. By the above remarks $\tilde{B}_i$ is an open annulus or disc; since it contains the non-contractible singular curve $\tilde{\rho}_i$, it is an open annulus.

Since σ_i is homotopic in T to a k-th power of $\xi_i(S^1)$, $\tilde{\sigma}_i$ – and hence $\tilde{\rho}_i$ – are homotopic in $\tilde{M}$ to a k-th power of the unique lifting $\tilde{\xi}_i$ of ξ_i to $\tilde{T}$. Thus if N_i is a regular neighborhood of $\tilde{\xi}_i(S^1)$ in $\tilde{T}$, there are non-contractible singular curves in the disjoint open subsets $\mathring{N}_i$ and $\mathring{\tilde{B}}_i$ of $\partial\tilde{M}$ which are homotopic in $\tilde{M}$. The generalized loop theorem ([19]) then asserts that there are simple curves $x_i \subset N_i$, $r_i \subset \tilde{B}_i$, which bound an annulus $A_i \subset \tilde{M}$. Since x_i is necessarily ambient isotopic to $\tilde{\xi}_i(S^1)$ in N_i, and since ambient-isotopic curves in $\tilde{T}$ are ambient-isotopic in $\tilde{M}$, we may assume that $x_i = \tilde{\xi}_i(S^1)$. Furthermore, since $\tilde{B}_1$ and $\tilde{B}_2$ are either disjoint annuli or the same annulus, r_1 and r_2 are ambient-isotopic to disjoint curves, and may therefore be assumed disjoint. Finally, we may assume that $A_i \cap \partial\tilde{M} = \partial A_i$; and since $\tilde{\xi}_1(S^1)$ and $\tilde{\xi}_2(S^1)$ intersect transversally, we may take A_1 and A_2 to intersect each other transversally by putting them in general position.

We are at last ready to prove that $\xi_1(S^1) \cap \xi_2(S^1) = \emptyset$. Since $\tilde{T}$ is a degree-one covering of T, it suffices to show that $\tilde{\xi}_1(S^1) \cap \tilde{\xi}_2(S^1) = \emptyset$. Assume to the contrary that $\tilde{\xi}_1(S^1) \cap \tilde{\xi}_2(S^1)$ contains a point y. The component of $A_1 \cap A_2$ containing y is an arc c (by transversality, since $y \in \partial\tilde{M}$) and the other endpoint z of c must lie in $(\partial A_1) \cap (\partial A_2)$. But Z cannot lie in r_1 or r_2 since $r_1 \cap r_2 = \emptyset$ and since $\tilde{B}_1, \tilde{B}_2$ are disjoint from $\tilde{T}$. Hence $z \in \tilde{\xi}_1(S^1) \cap \tilde{\xi}_2(S^1)$. In particular, for $i = 1, 2$, c is a 2-sided arc in the annulus A_i, and the two points of $\partial c = c \cap \partial A_i$ lie in the same component $\tilde{\xi}_i(S^1)$ of ∂A_i; hence c is the frontier of a disc $D_i \subset A_i$, and $(\partial D_i) - \mathring{c}_i$ is an arc $a_i \subset \xi_i(S^1)$.

Each choice of a point $y \in \tilde{\xi}_1(S^1) \cap \tilde{\xi}_2(S^1)$ determines discs $D_1 \subset A_1$, $D_2 \subset A_2$ in this way. Let y be chosen so as to make the disc D_1 minimal with respect to inclusion. Then $\mathring{a}_1$ contains no point $y' \in \tilde{\xi}_1(S^1) \cap \xi_2(S^1)$, for y' would determine a disc $D_1' \subset D_1$. In particular, $\mathring{a}_1 \cap a_2 = \emptyset$; since a_1 and a_2 have the same endpoints, $a_1 \cap a_2 \subset \tilde{T}$ is a simple curve. It is contractible in $\tilde{M}$, for a_1 can be (non-ambiently) isotoped through D_1 to c, and then through D_2 to a_2. Since $\tilde{T}$ is incompressible, $a_1 \cup a_2$ must actually contract in $\tilde{T}$, and must therefore bound a disc $\Delta \subset \tilde{T}$. This contradicts the statement (*) proved above, and thus completes the proof.

COROLLARY 1. *Let* (M, T) *be as in Proposition 3. Then any two singular curves in* T *which are homotopic in* M *are either homotopic or anti-homotopic in* T.

Proof. If the singular curves σ and σ' in T are homotopic in M but are *not* homotopic or anti-homotopic in T, then by definition they are both distinguished. On the other hand, it follows from Proposition 1 (Section 1) that σ and σ' have the same divisibility. Then Proposition 3 asserts that σ and σ' are homotopic or anti-homotopic, after all.

COROLLARY 2. *Let* (M, T) *be any acceptable pair such that* T *is a torus. Then any conjugacy class in* $\pi_1(M)$ *is represented by at most two elements of* $\pi_1(T)$.

Proof. If M is not exceptional this is contained in Corollary 1. If M is exceptional we can identify $\pi_1(T)$ with its image in $\pi_1(M)$, which is of index ≤ 2. Now for any $x \in \pi_1(T)$, the number of conjugates of x in $\pi_1(M)$ is equal to the index of the centralizer of x in $\pi_1(M)$, which contains $\pi_1(T)$ since the latter is abelian. Thus any conjugacy class which intersects $\pi_1(T)$ contains at most two elements.

We will also need

LEMMA 8. *If in the acceptable pair* (M, T), M *is an exceptional 3-manifold and* T *is a component of* ∂M, *then* $\pi_1(T)$ *contains no special elements (Section 2).*

Proof. Identify $\pi_1(T)$ with its image in $\pi_1(M)$. Since $\pi_1(T)$ has index ≤ 2 in $\pi_1(M)$, it is normal; in particular, the square of any element of $\pi_1(M)$ is in $\pi_1(T)$. Now if $x \in \pi_1(T)$ is special and has divisibility k in $\pi_1(T)$, it has the form $x = y^{\ell}$, where $1 \neq y \in \pi_1(M)$ and $\ell > 2k$. Then $(y^2)^{\ell} = x^2$ has divisibility 2k in $\pi_1(T)$ by Corollary 2 to Lemma 1 of Section 1, but is divisible by $\ell > 2k > 0$ in $\pi_1(T)$, since $y^2 \in \pi_1(T)$ by the above. This contradicts Corollary 1 to Lemma 1 of Section 1.

§6. *Free products with amalgamation*

This section contains the only group theory required for the proof of the theorem of Section 7.

Let F, G, and H be groups, and let $i: H \to F$ and $j: H \to G$ be monomorphisms, regarded as identifying H with subgroups of F and G. Recall that the *free product* of F and G *with amalgamated subgroup* H is the quotient of the free product $F*G$ by the relations $i(h) = j(h)$ for all $h \epsilon H$. Recall the fundamental property of $F \underset{H}{*} G$, as proved for example on pp. 198-199 of [6]: if Φ, Γ are complete sets of left coset representatives for F, G, then every element of $F \underset{H}{*} G$ has a unique expression in the *canonical form* $h\alpha_1 \cdots \alpha_n$, where $h(= i(h) = j(h)) \epsilon H$, $\alpha_i \epsilon \Phi \cup \Gamma$ but $\alpha_i \not\epsilon H (1 \leq i \leq n)$, and $\alpha_{i+1} \epsilon \Phi$ if and only if $\alpha_i \epsilon \Gamma (1 \leq i < n)$. We will call the integer $n \geq 0$ the *length* of the given element. The element will be called a *cyclically reduced word* if $n \leq 1$, or if one of the elements α_1 and α_n is in Φ and the other is in Γ.

LEMMA 9. *In a free product with amalgamation* $F \underset{H}{*} G$,

(i) *every element is conjugate to a cyclically reduced word;*

(ii) *two cyclically reduced words which are conjugate in* $F \underset{H}{*} G$ *have the same length, provided that one of them has length* > 1;

(iii) *if* w *is a cyclically reduced word of length* $n \geq 2$, *then* w^m $(m \geq 0)$ *is a cyclically reduced word of length* mn.

Proof. Part (i) is the initial statement of Theorem 4.6 from p. 212 of [6]. Part (ii) follows immediately from Part (iii) of the theorem just quoted. Part (iii) appears on the bottom of p. 208 and the top of p. 209 of [6].

COROLLARY. *If* $w \epsilon F \underset{H}{*} G$ *is such that* w^m *is infinitely divisible for some* $m > 0$, w *is conjugate to an element of* F *or* G.

Proof. If the conclusion is false, then by part (i) of the lemma, w is conjugate to a cyclically reduced word w' of length $\ell > 1$. By part (iii) of the lemma, w'^m, which is infinitely divisible, is a cyclically reduced word of length $m\ell > 1$.

For infinitely many integers $n > 0$ there exist elements x_n of $F \underset{H}{*} G$ such that $x_n^n = w'^m$. By part (i) of the lemma, x_n is conjugate to a cyclically reduced word x'_n of some length λ_n. If $\lambda_n > 1$, then $(x'_n)^n$ is cyclically reduced of length $n\lambda_n$ by part (iii) of the lemma; hence by part (ii), $n\lambda_n = m\ell$. Since this is possible for only finitely many values of n, some x_n must be conjugate to an element of F or G; hence w'^m must also be conjugate to an element of F or G. But since w'^m is cyclically reduced of length > 1, this contradicts part (ii) of the lemma.

LEMMA 10. *Let* F *and* G *be subgroups of groups* F' *and* G'. *Let* H *be a group that is identified isomorphically with subgroups of* F *and* G, *so that* $F \underset{H}{*} G$ *and* $F' \underset{H}{*} G'$ *are defined. Then the natural homomorphism* $\mu : F \underset{H}{*} G \to F' \underset{H}{*} G'$ *is injective, and for any* $w \in F \underset{H}{*} G$, $\mu(w)$ *has the same length as* w. *Furthermore, if* w *is cyclically reduced then so is* $\mu(w)$.

Proof. Let w be written in the above canonical form as an element of $F \underset{H}{*} G$. Then using the identifications described in the hypothesis, we can regard this as the canonical form of $\mu(w)$ considered as an element of $F' \underset{H}{*} G'$. The lemma follows, since the length of an element, and the properties of being cyclically reduced and of being the identity, can be read off from the canonical form of the element.

The final result of this section interprets the preceding group theory in a topological context. Its proof is conveniently worded in terms of a construction that will be used in a stronger way in Section 7.

Let $\mathcal{T}$ be a 2-sided surface in a 3-manifold $\mathfrak{M}$. Then it is easy to construct a 3-manifold M, possibly disconnected, and disjoint surfaces T, T' in ∂M, such that $\mathfrak{M}$ is obtained from M by identifying T with T' via some (PL) homeomorphism, and such that $\mathcal{T} = T = T'$ under the identification. Moreover, the pair $(M, T \cup T')$ is determined up to homeomorphism by $\mathfrak{M}$ and $\mathcal{T}$. We will say that M is obtained from $\mathfrak{M}$ by *splitting along* $\mathcal{T}$.

LEMMA 11. *Let $\mathcal{T}$ be an incompressible 2-sided surface in a 3-manifold $\mathfrak{M}$. Then any conjugacy class $c(x) \subset \pi_1(\mathfrak{M})$, such that x^m is infinitely divisible for some $m > 0$, is represented by a curve in $\mathfrak{M} - \mathcal{T}$.*

Proof. If $\mathcal{T}$ separates $\mathfrak{M}$, then since $\mathcal{T}$ is incompressible, van Kampen's theorem provides an identification of $\pi_1(\mathfrak{M})$ with a free product with amalgamation $F = \pi_1(A) \underset{\pi_1(\mathcal{T})}{*} \pi_1(B)$, where A and B are the components of the manifold obtained by the splitting $\mathfrak{M}$ at $\mathcal{T}$. Hence by the corollary to Lemma 9, $c(x)$ is represented by a (singular) curve in A or B, and hence by one in $\mathring{A}$ or $\mathring{B}$.

Now suppose that $\mathcal{T}$ does not separate $\mathfrak{M}$. Since $\mathfrak{M}$ is orientable we can define a homomorphism from $H_1(\mathfrak{M}; \mathbf{Z})$ to $\mathbf{Z}$ as intersection number with the surface $\mathcal{T}$ (or with its fundamental class in $H_2(\mathfrak{M}, \partial\mathfrak{M}; \mathbf{Z})$). This induces a homomorphism from $\pi_1(\mathfrak{M})$ to $\mathbf{Z}$, whose kernel L determines an infinite cyclic covering space $\tilde{\mathfrak{M}}$ of $\mathfrak{M}$. Write $p: \tilde{\mathfrak{M}} \to \mathfrak{M}$ for the projection, and $\tau: \tilde{\mathfrak{M}} \to \tilde{\mathfrak{M}}$ for a generator of the covering group. If M is the closure of a component of $\tilde{\mathfrak{M}} - p^{-1}(\mathcal{T})$, then M is homeomorphic to the manifold obtained by splitting $\mathfrak{M}$ at $\mathcal{T}$; its frontier in $\tilde{\mathfrak{M}}$ consists of two surfaces $\tilde{\mathcal{T}}$ and $\tau\tilde{\mathcal{T}}$, each of which is mapped homeomorphically onto $\mathcal{T}$ by p. We have $\tilde{\mathfrak{M}} = \bigcup_{n \in \mathbf{Z}} \tau^n M$, $\tau^{n-1} M \cap \tau^n M = \tau^n \tilde{\mathcal{T}}$, and $\tau^n M \cap \tau^{n'} M = \emptyset$ for $|n' - n| > 1$. Note also that $\tilde{\mathcal{T}}$ is incompressible in $\tilde{\mathfrak{M}}$, since $\mathcal{T}$ is incompressible in $\mathfrak{M}$.

The image of the conjugacy class $c(x)$ under the intersection number homomorphism is an integer ν such that $m\nu$ is infinitely divisible in $\mathbf{Z}$; this implies $\nu = 0$, i.e. $c(x) \subset L$. Moreover, for any conjugacy class $c(y) \subset \pi_1(\mathfrak{M})$, such that $c(y)^p = c(x)^m$, the same argument shows that $c(y) \subset L$. It follows that $c(x)^m$ is actually infinitely divisible *in* L. Hence a singular curve σ representing $c(x)$ has a lifting $\tilde{\sigma}$ in $\tilde{\mathfrak{M}}$, and the conjugacy class $c(\tilde{x})$ determined by $\tilde{\sigma}$ in $\pi_1(\tilde{\mathfrak{M}})$ has infinitely divisible m-th power.

By compactness we can find integers $n_1 \leq n_2$ such that $\tilde{\sigma}(S^1) \subset \tau^{n_1}(M) \cup \cdots \cup \tau^{n_2}(M)$. Suppose this to have been done in such a way that $n_2 - n_1 \geq 0$ has the smallest possible value. Then we claim that $n_1 = n_2$.

Assume, to the contrary, that $n_1 < n_2$. Then since $\tilde{\mathcal{T}}$ is incompressible, van Kampen's theorem allows us to identify $\pi_1(\tau^{n_1}(M) \cup \cdots \cup \tau^{n_2}(M))$ with an amalgamated free product $F \underset{H}{*} G$, where $F = \pi_1(\tau^{n_1}(M) \cup \cdots \cup \tau^{n_2-1}(M))$, $G = \pi_1(\tau^{n_2}(M))$, and $H = \pi_1(\tau^{n_2}(\tilde{\mathcal{T}}))$. Then the conjugacy class determined by $\tilde{\sigma}$ in $\pi_1(\tau^{n_1}(M) \cup \cdots \cup \tau^{n_2}(M))$ is represented by a cyclically reduced word w in $F \underset{H}{*} G$, by part (i) of Lemma 9. Let ℓ denote the length of w. Now set $F' = \pi_1(A)$, $G' = \pi_1(B)$, where A and B are the closures of the components of $\tilde{\mathfrak{M}} - \tau^{n_2}(\tilde{\mathcal{T}})$ containing $\tau^{n_2-1}(\mathring{M})$ and $\tau^{n_2}(\mathring{M})$ respectively. We can identify $\pi_1(\tilde{\mathfrak{M}})$ with $F' \underset{H}{*} G'$. Furthermore, the natural map $F \to F'$ is injective, for F' can be identified with $F \underset{\pi_1(\tau^{n_1}(\tilde{\mathcal{T}}))}{*} \pi_1(A - (\tau^{n_1}(M) \cup \cdots \cup \tau^{n_2-1}(M)))$; similarly the natural map $G \to G'$ is injective. Identifying F and G with their images under these injections we see that F, G, F', G', and H satisfy the hypotheses of Lemma 11. Hence $c(\mu(\omega))$, which is the conjugacy class $c(\tilde{x})$ determined by $\tilde{\sigma}$ in $\pi_1(\tilde{\mathfrak{M}}) = F' \underset{H}{*} G'$, is a cyclically reduced word of length ℓ in $F' \underset{H}{*} G'$. But we observed above that $c(\tilde{x})^m$ is infinitely divisible in $\pi_1(\tilde{\mathfrak{M}})$. Thus by the corollary to Lemma 9, $\tilde{x}$ is conjugate in $F' \underset{H}{*} G'$ to an element of F' or G', i.e. to a cyclically reduced word of length ≤ 1. Part (ii) of Lemma 9 therefore shows that $\ell \leq 1$.

Recalling that $w \in \pi_1(\tau^{n_1}(M) \cup \cdots \cup \tau^{n_2}(M)) = F \underset{H}{*} G$ is a cyclically reduced word of length ℓ, we now know that w, which represents the conjugacy class in $F \underset{H}{*} G$ determined by $\tilde{\sigma}$, is an element of F or G; i.e. $\tilde{\sigma}$ is homotopic in $\tau^{n_1}(M) \cup \cdots \cup \tau^{n_2}(M)$ to a (singular) curve in $\tau^{n_1}(M) \cup \cdots \cup \tau^{n_2-1}(M)$ or in $\tau^{n_2}(M)$. This contradicts the assumed minimality of $n_2 - n_1 > 0$; thus we must have $n_1 = n_2$.

In other words, $\tilde{\sigma}$ lies in a region $\tilde{\tau}^{n_1}(M)$, and by a homotopy it may be assumed to lie in $\tilde{\tau}^{n_1}(\mathring{M}) \subset \tilde{\mathfrak{M}} - p^{-1}(\tilde{\mathfrak{T}})$. Then $p \circ \tilde{\sigma}$ is a curve in $\mathfrak{M} - \mathfrak{T}$ representing $<x>$.

§7. *Hierarchies; the main theorem*

DEFINITION. A 3-manifold M is *irreducible* if every 2-sphere in M bounds a 3-cell.

DEFINITION (cf. [18], [20]). A compact, orientable, irreducible 3-manifold is *sufficiently large* if it contains an incompressible 2-sided surface. A compact irreducible 3-manifold M is *almost sufficiently large* if some orientable, irreducible finite covering of M is sufficiently large.

In [18], the sufficiently large manifolds are characterized among the compact, orientable irreducible 3-manifolds by their fundamental groups. In particular it is shown that M is sufficiently large if $H_1(M; \mathbf{Z})$ is infinite. This is true for example if M is the complement of an open regular neighborhood of a knot in S^3.

We now state our main result.

THEOREM. *If the compact, irreducible, orientable 3-manifold* M *is almost sufficiently large then* $\pi_1(M)$ *has no infinitely divisible elements.*

The proof of this theorem occupies the rest of the present section.

The following standard argument shows that for $\mathfrak{M}$ almost sufficiently large, $\pi_1(\mathfrak{M})$ is torsion-free. Since $\mathfrak{M}$ is irreducible and orientable, Principle 2 of Section 0 implies that $\pi_2(\mathfrak{M}) = 0$. On the other hand, since some finite cover $\tilde{\mathfrak{M}}$ of $\mathfrak{M}$ is sufficiently large, $\pi_1(\tilde{\mathfrak{M}})$ is either a nontrivial free product with amalgamation or admits a homomorphism onto the integers: this is shown in [18]. In either case, $\pi_1(\mathfrak{M})$ is infinite. By applying the Hurewicz theorem to the universal covering space of $\mathfrak{M}$, one concludes that $\mathfrak{M}$ is aspherical ($\pi_n(\mathfrak{M}) = 0$ for $n > 1$). By a theorem of

P. A. Smith's (Theorem 16.1 on p. 287 of [4] applied to the universal covering of $\mathfrak{M}$), finiteness of dimension then implies that $\pi_1(\mathfrak{M})$ is torsion-free.

The proof of the above theorem now reduces to the case where $\mathfrak{M}$ is orientable and sufficiently large via the following fact:

LEMMA 12. *If a torsion-free group* G *has an infinitely divisible element* $\neq 1$, *so does each of its subgroups of finite index.*

Proof. If α is infinitely divisible in G, so is α^m for any $m > 0$. If $\alpha \neq 1$, then $\alpha^m \neq 1$ since G is torsion-free.

The proof in the case that $\mathfrak{M}$ is sufficiently large depends on Haken's theory of hierarchies; we review the relevant results from [20].

DEFINITION. A *hierarchy* for a 3-manifold $\mathfrak{M}$ is a sequence of 3-manifolds $\mathfrak{M} = M_0, \cdots, M_n$, not necessarily connected, such that

(i) each component of M_n is a 3-cell, and

(ii) for $0 \leq i < n$, M_{i+1} is obtained by splitting M_i along a 2-sided incompressible surface T_i (Section 6).

The integer $n \geq 0$ is called the *length* of the hierarchy.

REMARK. Any component of a manifold obtained by splitting an irreducible manifold $\mathfrak{M}$ at an incompressible surface $\mathcal{T}$ is irreducible.

We extract the following result from [16]. It seems to be essentially due to Haken.

LEMMA 13. *Every sufficiently large, compact, irreducible, orientable, connected 3-manifold* $\mathfrak{M}$ *has a hierarchy.*

Proof. If $\partial\mathfrak{M} \neq \emptyset$, this is contained in Theorem 1.2, p. 60 of [20]. If $\mathfrak{M}$ is closed, it has an incompressible surface $\mathcal{T}$; we can split $\mathfrak{M}$ at $\mathcal{T}$ to obtain a 3-manifold M. Each component of M is irreducible, by the remark following the definition of a hierarchy, and has non-empty boundary. Hence each component of M has a hierarchy, and it follows that $\mathfrak{M}$ has one.

To prove the theorem when $\mathfrak{M}$ is sufficiently large, we argue by induction on the length of a hierarchy of $\mathfrak{M}$. By definition, if $\mathfrak{M}$ has a hierarchy of length n, then $\mathfrak{M}$ can be split at some incompressible 2-sided surface $\mathfrak{T}$ to produce a manifold $\mathfrak{M}'$ which has a hierarchy of length $< n$. Arguing inductively, we assume that

(*) for any component M of $\mathfrak{M}'$, $\pi_1(M)$ is without infinitely divisible elements $\neq 1$.

Assuming in addition that

(†) $\pi_1(\mathfrak{M})$ has an infinitely divisible element a,

we will produce a contradiction.

Let T_0 and T_1 denote the surfaces in $\partial\mathfrak{M}'$ that are identified to produce $\mathfrak{T}$; for $i = 0, 1$, let M_i denote the component of $\mathfrak{M}'$ containing T_i (so that $M_0 \neq M_1$ if and only if $\mathfrak{T}$ separates $\mathfrak{M}$).

LEMMA 14. *For* $i = 0, 1$, (M_i, T_i) *is an acceptable pair.*

Proof. Since $\mathfrak{T}$ is incompressible in $\mathfrak{M}$, T_i is clearly incompressible in M_i. On the other hand, M_i is orientable, and is irreducible by the remark following the definition of a hierarchy. Hence by Principle 2 of Section 0, $\pi_2(M_i) = 0$.

Let $\phi : \mathfrak{M}' \to \mathfrak{M}$ denote the identification map.

DEFINITION. A *lifting* of a singular curve σ in $\mathfrak{M}$ is a singular curve $\tilde{\sigma}$ in $\mathfrak{M}'$ such that $\phi \circ \tilde{\sigma} = \sigma$.

The following elementary fact will be used twice.

LEMMA 15. *If* β *and* β' *are non-contractible (singular) curves in* $\mathfrak{M} - \mathfrak{T}$ *which are homotopic in* $\mathfrak{M}$, *then there are singular curves* $\beta = \beta_0, \beta_1, \cdots, \beta_s = \beta'$ $(s > 1)$ *such that*

(i) $\beta_1, \cdots, \beta_{s-1}$ *are in* $\mathfrak{T}$,

and

(ii) β_i *and* β_{i+1} *admit homotopic liftings in* $\mathfrak{M}'$ *for* $0 \leq i < s$.

Proof. Let $f: S^1 \times I \to \mathfrak{M}$ be a PL homotopy between β and β'; thus $f(x,0) = \beta(x)$ and $f(x,1) = \beta'(x)$, for all $x \in S^1$. We may take f to be transversal to $\mathfrak{T}$. Suppose that in addition we can choose f so that no component of $f^{-1}(\mathfrak{T})$ bounds a disc in $S^1 \times \mathring{I}$. Then it will be possible to index the components of $f^{-1}(\mathfrak{T}) \cup (S^1 \times \partial I)$ as $S^1 \times \{0\} = S_0, S_1, \cdots, S_{s-1}, S_s = S^1 - \{1\}$, in such a way that $S_i \cup S_{i+1}$ bounds an annulus A_i, with $\mathring{A}_i \cap f^{-1}(\mathfrak{T}) = \emptyset$, for $0 \leq i < n$. The lemma will then follow, for we can set $\beta_i = f|S_i$, where S_i is identified with S^1 via an appropriate homeomorphism.

It is therefore enough to show that if some component γ of $f^{-1}(\mathfrak{T})$ bounds a disc $\Delta \subset S^1 \times \mathring{I}$, then there is a PL homotopy $f': S^1 \times I \to \mathfrak{M}$, transversal to $\mathfrak{T}$ and agreeing with f on $S^1 \times \partial I$, but such that $f'^{-1}(\mathfrak{T})$ has fewer components than $f^{-1}(\mathfrak{T})$. To do this, let Δ' be a regular neighborhood of Δ in $S^1 \times \mathring{I}$, such that $\Delta' - \Delta$ is disjoint from $f^{-1}(\mathfrak{T})$ and such that $f(\Delta' - \mathring{\Delta})$ is contained in a regular neighborhood N of $\mathfrak{T}$. Then $f|\partial\Delta'$ is homotopic to a constant in $\mathfrak{M}$, and therefore also in $N - \mathfrak{T}$ since $\mathfrak{T}$ is 2-sided and incompressible. Hence we can extend $f|((S^1 \times I) - \mathring{\Delta}')$ to a PL map $f'|S^1 \times I \to \mathfrak{M}$ such that $f'(\Delta') \subset N - \mathfrak{T}$. Clearly f' has the required properties.

COROLLARY. *The conjugacy class* $c(\alpha)$ *(see † above) is represented by a singular curve* σ_0 *in* $\mathfrak{T}$.

Proof. Since α is infinitely divisible, there exists an element x_n of $\pi_1(\mathfrak{M})$, for each of infinitely many integers $n > 0$, such that $x_n{}^n = \alpha$.

By Lemma 11 of Section 6, $c(\alpha)$ is represented by a singular curve σ in $\mathfrak{M} - \mathfrak{T}$, and each $c(x_n)$ is represented by a curve ξ_n in $\mathfrak{M} - \mathfrak{T}$. Now by the above assumption (*) σ cannot be homotopic in $\mathfrak{M} - \mathfrak{T}$ to the n-th power (Section 0) of ξ_n for infinitely many n; fix n so that σ is not homotopic to the n-th power η_n of ξ_n in $\mathfrak{M} - \mathfrak{T}$. Since σ and η_n are homotopic in $\mathfrak{M}$, Lemma 15 applies with $\beta = \alpha$, $\beta' = \eta_n$. The integer s appearing in the conclusion of Lemma 15 must be > 1, since otherwise σ and η_n would be homotopic in $\mathfrak{M} - \mathfrak{T}$. Hence we can define σ_0 to be the singular curve β_1. Since σ and σ_0 admit homotopic liftings in M, they are certainly homotopic in $\mathfrak{M}$.

Let k denote the divisibility (Section 1) of $[\sigma_0] \subset \pi_1(\mathfrak{T})$.

LEMMA 16. *For each of infinitely many integers* $n > 0$, *there exists a singular curve* σ_n *in* $\mathfrak{T}$ *such that*

(i) $[\sigma_n] \subset \pi_1(\mathfrak{T})$ *has divisibility* k;

(ii) *for some lifting* $\tilde{\sigma}_n$ *of* σ_n *to some* $T_j (j=0$ *or* $1)$, $[\tilde{\sigma}_n] \subset \pi_1(T_j)$ *is special (Section 2) with respect to the pair* (M_j, T_j), *and is divisible by* n *in* $\pi_1(M_j)$;

(iii) σ_n *is either homotopic to* σ_0 *in* $\mathfrak{T}$, *or else has a lifting* $\tilde{\sigma}'_n$ *to some* $T_{j'} (j' = 0$ *or* $1)$ *which is distinguished (Section 5) with respect to the pair* $(M_{j'}, T_{j'})$;

and

(iv) *if* $\mathfrak{T}$ *is a torus and separates* $\mathfrak{M}$, *then* σ_n *has a lifting which is homotopic in* $\mathfrak{M}' = M_0 \cup M_1$ *[disjoint] to some lifting of* σ_0.

Proof. Since α is infinitely divisible in $\pi_1(\mathfrak{M})$, there are infinitely many integers $n > 2k$ such that $\alpha = x_n{}^n$ for some $x_n \in \pi_1(\mathfrak{M})$. It follows, moreover, from Lemma 11 of Section 6, that each $c(x_n)$ is represented by a curve $\xi_n \subset \mathfrak{M} - \mathfrak{T}$. On the other hand, any lifting $\tilde{\sigma}_0$ of σ_0 to $\mathfrak{M}'$ is certainly homotopic to a curve $\tilde{\sigma}'_0$ in $\mathring{\mathfrak{M}}'$; and if η_n is an n-th power (Section 0) of ξ_n in $\mathfrak{M} - \mathfrak{T}$, $\sigma'_0 = \phi \circ \tilde{\sigma}'_0$ and η_n are homotopic in $\mathfrak{M}$. So by Lemma 15, there are singular curves $\sigma'_0 = \beta_0, \beta_1, \cdots, \beta_s = \eta_n$, such that $\beta_1, \cdots, \beta_{s-1}$ are in $\mathcal{T}$, and β_i and β_{i+1} have homotopic liftings to $\mathfrak{M}'$ for $0 \leq i < s$. If we now define $\beta'_0 = \sigma_0$, $\beta'_i = \beta_i (1 < i \leq n)$, it is still true that β'_i and β'_{i+1} have homotopic liftings to $\mathfrak{M}' (0 \leq i < m)$; and $\beta'_0, \cdots, \beta'_{s-1}$ are in $\mathcal{T}$.

We claim that the conclusions of the lemma are true if we set $\sigma_n = \beta'_{s-1}$.

First of all, since for $1 \leq i < s-1$, β'_i and β'_{i+1} have liftings in $T_0 \cup T_1 \subset \partial \mathfrak{M}'$ which are homotopic in $\mathfrak{M}'$, Proposition 1 of Section 1 shows that β'_i and β'_{i+1} have the same divisibility in $\mathcal{T}$; hence $\sigma_n = \beta_{s-1}$ has the same divisibility as $\beta_0 = \sigma_0$, namely k. This is conclusion (i). On the other hand, since some lifting $\tilde{\sigma}_n$ of $\sigma_n = \beta_{s-1}$ to T_j, $j = 0$ or 1, is homotopic in M_j to $\beta_s = \eta_n$, which is an n-th power of ξ_n in M_j, $[\tilde{\sigma}_n] \subset \pi_1(M_j)$ is divisible by n. But since, by conclusion (i), $\tilde{\sigma}_n \subset \pi_1(T_j)$ has divisibility k, our restriction of n to values $> 2k$ guarantees that σ_n is special. Thus (ii) is proved.

We may assume that β'_i and β'_{i+1} are never homotopic or anti-homotopic (Section 0) in $\mathcal{T}$ for $1 \leq i < m-1$; for if they are we can replace the sequence $\beta'_0, \cdots, \beta'_s$ by a sequence with fewer terms but having the same properties. Now if $s > 1$, this assumption implies in particular that $\sigma_n = \beta'_{s-1}$ and β'_{s-2} are not homotopic or anti-homotopic in $\mathcal{T}$, although they have homotopic liftings $\tilde{\sigma}'_n$ and $\tilde{\beta}'_{s-2}$ to $\mathfrak{M}'$. Thus $\tilde{\sigma}'_n$ is not homotopic or anti-homotopic to $\tilde{\beta}'_{s-2}$ in $\partial M_{j'}$, where j' is defined by $\tilde{\sigma}'_n(S^1) \subset T_{j'}$. This says that $\tilde{\sigma}'_n$ is distinguished with respect to the pair $(M_{j'}, T_{j'})$. On the other hand, if $s = 1$, then obviously $\sigma_n = \beta'_0 = \sigma_0$. This proves (iii).

Finally, suppose that $\mathcal{T}$ is a torus and separates $\mathfrak{M}$. Then we can identify M_0 and M_1 with submanifolds of $\mathfrak{M}'$, and within each M_i we can identify T_i with $\mathcal{T}$. Note also that T_1 is disjoint from M_0 (in $\mathfrak{M}'$), and T_0 from M_1.

By (ii), $[\tilde{\sigma}_n]$ is special with respect to some (M_j, T_j) for $j = 0$ or 1; by symmetry we can take $j = 0$. Then the manifold M_0 is not exceptional, according to Lemma 8 of Section 5. (Note that M_1, on the other hand, may very well be exceptional.) Our assumption that β'_i and β'_{i+1} are never homotopic or anti-homotopic in $\mathcal{T}$ therefore implies, by Corollary 1 to Proposition 3 of Section 5, that they are never homotopic in M_0. But our original condition on the β_i means, in this case, that β'_i and β'_{i+1} are homotopic in M_0 or in M_1 for $1 \leq i < s-1$. Thus $\sigma_0 = \beta'_0$ and $\sigma_n = \beta'_{s-1}$ are homotopic in M_1, and conclusion (iv) is proved.

LEMMA 17. *The singular curves* σ_n *given by Lemma 16 represent only finitely many homotopy classes in* $\mathcal{T}$.

Proof. First consider the case where $\mathcal{T}$ is *not* a torus. By conclusion (ii) of Lemma 16, each σ_n has a lifting in T_j, $j = 0$ or 1, which represents a special conjugacy class in $\pi_1(T_j)$. But in this case, by the corollary to Proposition 2, Section 2, each $\pi_1(T_j)$ contains only finitely many special conjugacy classes (relative to (M_j, T_j)). The lemma follows in this case.

Next suppose that $\mathcal{T}$ is a torus but does *not* separate $\mathfrak{M}$. Then the split manifold $\mathfrak{M}'$ is connected and has T_0 and T_1 among its boundary components. By conclusion (ii) of Lemma 16, there are special curves $\tilde{\sigma}_n$ with respect to one of the pairs $(\mathfrak{M}', T_0)$ and $(\mathfrak{M}', T_1)$. It therefore follows from Lemma 8 of Section 5 that $\mathfrak{M}'$ is not exceptional. Hence by Proposition 3 of Section 5, each of T_0 and T_1 contains at most two homotopy classes of distinguished curves of divisibility k. But by conclusion (iii) of Lemma 16, each σ_n either is homotopic to σ_0 in $\mathcal{T}$, or else has a lifting $\tilde{\sigma}'_n$ to T_0 or T_1 which is distinguished, and which, by conclusion (i) of the same lemma, has divisibility k. It follows that in this case the σ_n represent at most five different homotopy classes in $\mathcal{T}$.

Finally, suppose that $\mathcal{T}$ is a torus and separates $\mathfrak{M}$. Then we can identify M_0 and M_1 with submanifolds of $\mathfrak{M}$, and T_i with $\mathcal{T}$, within M_i. By conclusion (iv) of Lemma 16, each σ_n is homotopic to σ_0 in M_0 or in M_1. But by Corollary 2 to Proposition 3 of Section 5, there is at most one homotopy class of curves in T_0 which are homotopic to σ_0

in M_0, apart from the class of σ_0 itself; and similarly in M_1. Hence in this case the σ_n represent at most three distinct homotopy classes.

Proof of the theorem concluded. Lemma 16 gives singular curves σ_n in $\mathcal{T}$ for an infinity of integers $n > 0$. By Lemma 17, these represent only finitely many homotopy classes in $\mathcal{T}$; thus by restricting n to a smaller infinite set of integers we may assume that the σ_n are all homotopic in $\mathcal{T}$. Furthermore, by (ii) of Lemma 16, each σ_n has a lifting $\tilde{\sigma}_n$ to some T_j ($j = 0$ or 1) such that $[\tilde{\sigma}_n]$ is divisible by n in $\pi_1(M_j)$. By restricting n to a still smaller infinite set of integers, and perhaps re-indexing, we may assume that these j are all equal to 0. Then the $\tilde{\sigma}_n$ all represent the same non-trivial conjugacy class in $\pi_1(M_0)$, which is divisible by each of integers n in our infinite set. This contradicts our induction hypothesis (*), and the theorem is thereby proved.

COLUMBIA UNIVERSITY

BIBLIOGRAPHY

[1] Bing, R. H., *An alternative proof that 3-manifolds can be triangulated.* Ann. of Math. (2) 69 (1959), 37-65.

[2] Evans, B., and Jaco, W., *Varieties of groups and 3-manifolds.* Topology 12 (1973), 83-97.

[3] Evans, B., and Moser, L., *Solvable fundamental groups of compact 3-manifolds.* Trans. Amer. Math. Soc. 168 (1972), 189-210.

[4] Hu, S.-T., *Homotopy Theory.* Academic Press, New York, 1959.

[5] Kneser, H., *Geschlossen Flächen in dreidimensionalen Mannigfaltigkeiten,* Jahresbericht der Deutschen Mathematiker Vereinigung, 38 (1929), 248-260.

[6] Magnus, W., Karrass, A., and Solitar, D., *Combinatorial Group Theory.* Interscience, New York, 1966.

[7] Milnor, J. W., *A unique decomposition theorem for 3-manifolds.* Amer. J. Math. 84 (1962), 1-7.

[8] Moise, E. E., *Affine structures in 3-manifolds, V. The triangulation theorem and Hauptvermutung.* Ann. of Math. (2) 56 (1952), 96-114.

[9] Moise, E. E., *Affine structures in 3-manifolds, VIII. Invariance of the knot types; local tame imbedding*. Ann. of Math. (2) 59 (1954), 159-170.

[10] Neuwirth, L. P., *Knot Groups*. Princeton University Press, 1965.

[11] Papakyriakopoulos, C. D., *On solid tori*. Proc. London Math. Soc. (3) 7 (1957), 281-299.

[12] ——————, *On Dehn's lemma and the asphericity of knots*. Ann. of Math. (2) 62 (1957), 1-26.

[13] Shalen, P. B., *A "piecewise-linear" method for triangulating 3-manifolds*, to appear.

[14] Shapiro, A., and Whitehead, J. H. C., *A proof and extension of Dehn's lemma*. Bull. Amer. Math. Soc. 64 (1958), 174-178.

[15] Spanier, E. H., *Algebraic Topology*. McGraw-Hill, New York, 1966.

[16] Stallings, J. R., *On the loop theorem*. Ann. of Math. (2) 72 (1960), 12-19.

[17] ——————, *Group theory and three-dimensional manifolds*. Yale Mathematical Monograph #4, Yale University Press, 1971.

[18] Waldhausen, F., *Gruppen mit Zentrum und 3-dimensionale Mannigfaltigkeiten*. Topology 6 (1967), 505-517.

[19] ——————, *Eine Verallgemeinerung des Schleifensatzes*. Topology 6 (1967), 501-504.

[20] ——————, *On irreducible 3-manifolds which are sufficiently large*. Ann. of Math. (2) 87 (1968), 56-88.

Library of Congress Cataloging in Publication Data
Main entry under title:

Knots, groups, and 3-manifolds.

(Annals of mathematics studies ; no. 84)
1. Knot theory--Addresses, essays, lectures.
2. Groups, Theory of--Addresses, essays, lectures.
3. Manifolds (Mathematics)--Addresses, essays, lectures. I. Fox, Ralph Hartzler, 1913- II. Neuwirth, Lee Paul. III. Series.
QA612.2.K66 514'.224 75-5619
ISBN 0-691-08170-0